The Kinetics of Environmental Aquatic Photochemistry

The Kinetics of Environmental Aquatic Photochemistry

Theory and Practice

Asa Leifer

ACS Professional Reference Book

Library of Congress Cataloging-in-Publication Data

Leifer, Asa, 1930–
The kinetics of environmental aquatic photochemistry.
(ACS professional reference book)

Bibliography: p.
Includes index.

1. Environmental chemistry. 2. Photochemistry.
3. Water chemistry. 4. Humic acid. I. Title.
II. Series.

TD193.L45 1988 551.4 88–16718
ISBN 0–8412–1464–6

PRINTED IN THE UNITED STATES OF AMERICA

About the Author

Asa Leifer joined the U.S. Environmental Protection Agency (U.S. EPA) in 1977 when the Office of Toxic Substances was just beginning to regulate industrial chemicals. He developed an innovative program for determining the fate of chemicals in the environment. His work has led to the current environmental fate test guidelines used in regulating new and existing industrial chemicals.

Dr. Leifer is active in the fields of atmospheric and aquatic chemistry, particularly in environmental photochemistry. He has developed methods for determining rates of direct and indirect (sensitized) photoreactions in aquatic media. He has made major contributions to developing data bases and determining structure–reactivity relationships for many environmental transport and transformation properties that are used by the Office of Toxic Substances. He is an authority on the environmental fate of polychlorinated biphenyls.

Since earning his Ph.D. in physical chemistry from the University of Maryland in 1958, Dr. Leifer has contributed to several fields of spectroscopy, particularly IR, UV–visible, and NMR spectroscopy. He has worked extensively in photochemistry, electrophotography, silver halide photography, and dye sensitization. He has several years of experience in pollution abatement in the paper industry. He has published 14 research papers in major scientific journals and more than 100 technical reports for the chemical industry and the U.S. EPA.

Dr. Leifer has taught basic chemistry courses at the college level and graduate-level courses in industry. He received two bronze medal awards from the U.S. EPA for regulating industrial chemicals. He is a member of the American Chemical Society and the American Physical Society.

Contents

Part III.A. Indirect Photoreaction Test Methods

Preface

Most of the earth's surface is covered by water in the form of oceans, seas, estuaries, rivers, lakes, streams, ponds, and puddles. As a result, many chemicals are likely to enter aquatic environments from a variety of sources, such as production, processing, use, and disposal of industrial chemicals. Chemical wastes are discharged directly into natural water bodies. Chemicals leach into natural water bodies from landfills. Pesticides are applied directly to water bodies, soils, and vegetation, and they subsequently leach into water bodies from soils and vegetation.

Many chemicals present in aqueous media can undergo photochemical transformation in sunlight via direct or indirect photoreaction. Direct photoreaction refers to those reactions in which a chemical absorbs sunlight and undergoes a chemical reaction. Indirect (or sensitized) photoreaction refers to those reactions in which another material absorbs sunlight and initiates a chemical reaction that transforms the chemical.

Chemicals that are present in aqueous media undergo direct photoreaction with rates that depend on the solar irradiance, the chemical's molar absorptivity, and its reaction quantum yield. Short direct photoreaction half-lives in sunlight have been observed for chemicals in natural waters having a wide variety of molecular structures. Thus, direct photoreaction in sunlight can represent an important mode of chemical transformation in the environment. In addition, some chemicals that do not undergo direct photoreaction have been found to undergo rapid indirect photoreaction in natural waters. The increased rate of photoreaction of these chemicals is a function of the molecular structures of chemicals and naturally occurring dissolved humic acids in natural waters.

To assess the role of direct and indirect photochemical reactions in the aquatic environment, procedures have been developed that use data measured in the laboratory or in sunlight, to predict how rapidly photoreaction occurs over a wide range of environmental conditions. Unfortunately, the theory and practice of determining direct and indirect photoreaction rate constants and half-lives are found scattered throughout many journal articles, photochemical texts, and U.S. Environmental Protection Agency reports, so this information is not readily accessible. Frequently, the theory and experimental procedures are only briefly discussed.

This book reviews in detail the relevant theory and practice of the kinetics of environmental aquatic photochemistry so that the environmental scientist can determine relevant direct and indirect photoreaction rate constants and half-lives in sunlight, as a function of latitude and season of the year, anywhere in the northern hemisphere. Because the subject is highly mathematical, detailed derivations are given to enable the environmental scientist to grasp the concepts easily. In this text, emphasis has been placed on the theory of the kinetics of direct and indirect photoreaction in aqueous solution and on the fundamental principles of photoreaction that guide the selection of experimental approaches. Particular emphasis is given to the photoreaction of chemicals at the very low concentrations, conditions that normally prevail in aquatic media in the environment. Procedures for obtaining the requisite kinetic data are discussed in detail so that these experiments can be carried out easily and precisely. Procedures for the extrapolation of these data to a wide variety of environmental conditions are discussed. Comprehensive tables of solar irradiance are given, along with examples that illustrate the use of all the experimental data from these procedures. For easier reading, a glossary of symbols has been provided.

This book is devoted to the theory and practice for determining photoreaction rate constants and half-lives in aqueous media in sunlight. No attempt has been made to discuss in detail electronic spectra and the structure of excited states, the production of excited states, and the time-dependent and time-independent properties of these excited states. These areas require detailed knowledge of electronic spectroscopy and quantum mechanics that are beyond the scope of this text. The reader is referred to numerous books already published on these subjects.

This book should be useful to a wide variety of scientists and engineers. Environmental scientists and engineers, photochemists, and graduate students working at all levels of research will benefit from the theoretical discussions, which include detailed mathematical derivations, and the complete description of photochemical procedures. This text should be extremely useful to environmental scientists in industry and government who regulate industrial chemicals and pesticides.

Acknowledgments

I thank the American Chemical Society and Pergamon Press for permission to reproduce certain figures for which credits are given at the appropriate places in the text. In addition, I express my appreciation to the editors of Academic Press and to the family of the late Philip A. Leighton for allowing me to use pertinent information on Leighton's theory of radiation transfer in the atmosphere in Chapter 2 of this text.

Many people have contributed directly or indirectly to the preparation of this book, and I gratefully acknowledge their assistance. I am especially indebted to Theodore Mill, Richard Zepp, John Winterle, and William Mabey for many discussions on environmental photochemistry and to Theodore Mill and John Winterle for offering valuable suggestions in the preparation of the final manuscript. Of course, I alone am responsible for any errors and omissions in the final draft. I would like to express my appreciation to Electa Thompson for skillfully typing the entire manuscript and to Joan Sutter for carefully proofreading the text and making valuable suggestions in formatting the manuscript.

Finally, I must recognize what is all too often forgotten—that no one lives or works alone, but is molded by those who share his life. Therefore, I dedicate this book to my wife Gina, whose constant encouragement guided the way and made it a reality.

Disclaimer

This book has been written by Asa Leifer in his private capacity. No official support or endorsement by the U.S. Environmental Protection Agency or any other agency of the federal government is intended or should be inferred.

ASA LEIFER, Ph.D.
Senior Scientist
U.S. Environmental Protection Agency
Office of Toxic Substances
Washington, DC 20460

October 3, 1987

Introduction

Photochemistry is the study of the effect of radiant energy on atoms and molecules and the rates and mechanisms by which reactions proceed when initiated by light. In conventional photochemistry, the radiation initiating photochemical reactions can be monochromatic or polychromatic and usually lies in the visible and ultraviolet regions (i.e., in the wavelength region from 200 to 700 nm). In environmental photochemistry, however, the source of radiation is sunlight, which is polychromatic, and has a spectral distribution in the range of approximately 290 to 800 nm. Solar radiation entering aquatic media contains a large amount of energy. Chemicals present in these water bodies, both natural and anthropogenic, absorb sunlight; as a result a complex series of photochemical reactions is initiated.

Water bodies on the surface of the earth are diverse. They include oceans, seas, estuaries, rivers, lakes, streams, and ponds. These surface waters contain a variety of natural material, both organic and inorganic, dissolved and suspended, living and dead. In addition, these water bodies can contain significant quantities of anthropogenic chemicals. These chemicals enter aquatic media from a wide variety of sources, including: production, processing, use, and disposal of industrial chemicals.

Natural water bodies, which contain a wide range of chromophoric substances, both natural and anthropogenic, are large photochemical reactor systems. As solar radiation enters these aquatic media, these chromophores absorb light, a number of reactive species are generated, and a diverse number of chemical reactions occur. These chemical changes involve primary and secondary reactions that are as varied as those of ordinary ground-state chemistry. These photoreactions involve homolytic and heterolytic cleavages, redox reactions, energy-transfer reactions, reactions with water, and isomerizations. Because oxygen is present in natural waters, it participates as an acceptor of energy or electrons and plays a role in a number of secondary reactions.

In the last two decades, a large number of research papers described a wealth of diverse photochemical processes that occur in natural aquatic media. Furthermore, significant advances have been made in the last 10 years in the theory of photoreaction of chemicals in natural water bodies in the environment. Zepp and Cline (*1*) have derived equations that describe the direct photoreaction of chemicals in aquatic media in the environment. These equations translate readily obtained laboratory data into rates and half-lives for direct photoreaction of chemicals in aquatic media in sunlight as a function of a number of environmental variables. Numerous papers have verified the method (*1–7*).

Significant advances have been made in chemical actinometry at very low concentrations in aqueous solution. Procedures have been developed for measuring the reaction quantum yield of a chemical in aqueous solution by carrying out experiments outdoors in sunlight or in the laboratory using monochromatic light (*5, 6, 8–18*). The theory of indirect (or sensitized) photoreaction in aqueous media, though complex and still in the early stages of development, has been formulated so that general rate equations can be defined. These expressions have been used to analyze the kinetic data for a number of photosensitized reactions (*4, 18–24*). Indirect photoreaction screening tests have been developed by carrying out experiments outdoors in sunlight (*25, 26*).

Overview

The purpose of this book is to summarize the theory and practice of the kinetics of environmental photochemistry in aquatic media that have been developed in the last decade.

The book has been divided into three broad parts to accommodate the discussion of the detailed theory and practice for determining photochemical reaction rates in aquatic media in the environment. Theoretical aspects are addressed in Part I. Theory, Chapters 1–5. Experimental procedures are described in Part II. Direct Photoreaction Test Methods (Chapters 6–11) and in Part III. Indirect Photoreaction Test Methods (Chapters 12–15). The theoretical aspects have been developed systematically and with mathematical rigor, and all mathematical derivations have been developed step by step so that environmental scientists can follow and grasp the concepts easily.

After reviewing some fundamental principles of photochemistry in Chapter 1, detailed discussions are presented in the following two chapters on the absorption and transmission of solar radiation through the atmosphere, through the air–water interface, and through a natural water body. These processes represent the elements essential to understanding the fundamentals of measuring environmentally relevant photochemical reaction rates. Thus, Chapter 2 gives a detailed account of the nature of sunlight and the absorption and transmission of solar radiation through the atmosphere. Chapter 3 then describes the absorption and transmission of solar radiation from the atmosphere through the air–water interface and through a water body in the environment as developed by Zepp and Cline (*1*).

The theory of absorption and transmission of solar radiation in a water body is then used in Chapter 4 to derive equations that translate readily available laboratory photoreaction data into sunlight-direct photoreaction rate constants and half-lives as a function of a number of environmental variables (e.g., time of day, season, latitude, depth in water bodies, and the ozone layer thickness). A number of photochemical studies are presented to verify the method of Zepp and Cline.

A generalized theory of the kinetics of direct and indirect photochemical reactions in aquatic media in the environment is presented in Chapter 5 and is based on the first and second laws of photochemistry. This generalized theory lays the groundwork for the development of the experimental photochemical procedures described in detail in Chapters 6 through 15.

Screening tests for determining rates of direct photochemical reaction by carrying out experiments outdoors in sunlight are described in Chapters 6 through 8. Chapters 9 through 11 describe upper tier tests for carrying out experiments in the laboratory using monochromatic light. Particular emphasis is given to the photoreaction of chemicals at very low concentrations in aquatic media, conditions that normally prevail in the environment. Detailed discussions are presented on the theory and practice of chemical actinometry in aqueous solution. Tables of solar irradiance are listed in an appendix, and examples are given that illustrate the use of these procedures and experimental data to calculate environmentally relevant direct photoreaction rate constants and half-lives as a function of latitude and season of the year in the northern hemisphere.

Chapter 12 reviews the published literature on the determination of rates of indirect photochemical reaction in humic water bodies in the environment. Procedures are described for carrying out indirect photoreaction experiments outdoors in aquatic humic media in sunlight or in the laboratory using monochromatic light.

Chapters 13 through 15 describe a screening test for carrying out parallel photoreaction experiments outdoors in sunlight with a chemical in humic aqueous solution and with an aqueous actinometer solution. Examples describe the use of all the experimental data to calculate an environmentally relevant indirect photoreaction rate constant and a net photoreaction rate constant (the sum of the direct and indirect photoreaction rate constants).

Parts II and III have been written essentially as independent sections that summarize all the pertinent theory and describe detailed experimental photochemical procedures. Thus, if an environmental scientist wants to carry out indirect photochemical screening tests, he or she need only read Chapters 13–15 of Part III and find all the necessary theory and practice for determining environmentally relevant indirect photoreaction rate constants without having to refer to numerous other sections of the test. Although this format introduces some redundancies in the text, it makes the book more versatile and, overall, more useful.

Photochemical Test Method Schemes

The schemes depicted in Figures I.1 and I.2 represent an overview of all the direct and indirect photoreaction test methods described in Chapters 6 through 15. Figure I.1 summarizes the hierarchy of direct photoreaction test methods, showing the relationship between the various screening and upper-tier tests. Figure I.2 summarizes the corresponding hierarchy for the indirect photoreaction test methods.

Direct Photochemical Reaction Test Scheme

Let us consider the direct photochemical reaction test method scheme which is depicted in Figure I.1. Three parameters are important in determining rates of direct photoreaction in aquatic media in the environment:

1. the absorption of light by a test chemical, or the molar absorptivity (ϵ_λ),
2. the overlap of ϵ_λ with the solar irradiance (L_λ or $E^{av}_{o\lambda}$) in aquatic media in the environment, and
3. the photoreaction quantum yield ϕ_d, the efficiency with which the absorbed light transforms the chemical into products.

A first tier screening test has been developed to estimate the maximum direct photoreaction rate constant $(k_{dE})_{max}$ and the corresponding minimum half-life $[t_{(1/2)E}]_{min}$ in aquatic media in the environment by assuming that the reaction quantum yield is 1, the maximum value. The maximum direct photoreaction rate constant can then be calculated by combining the molar absorptivity of a test chemical (measured by ultraviolet–visible absorption spectroscopy) with the appropriate

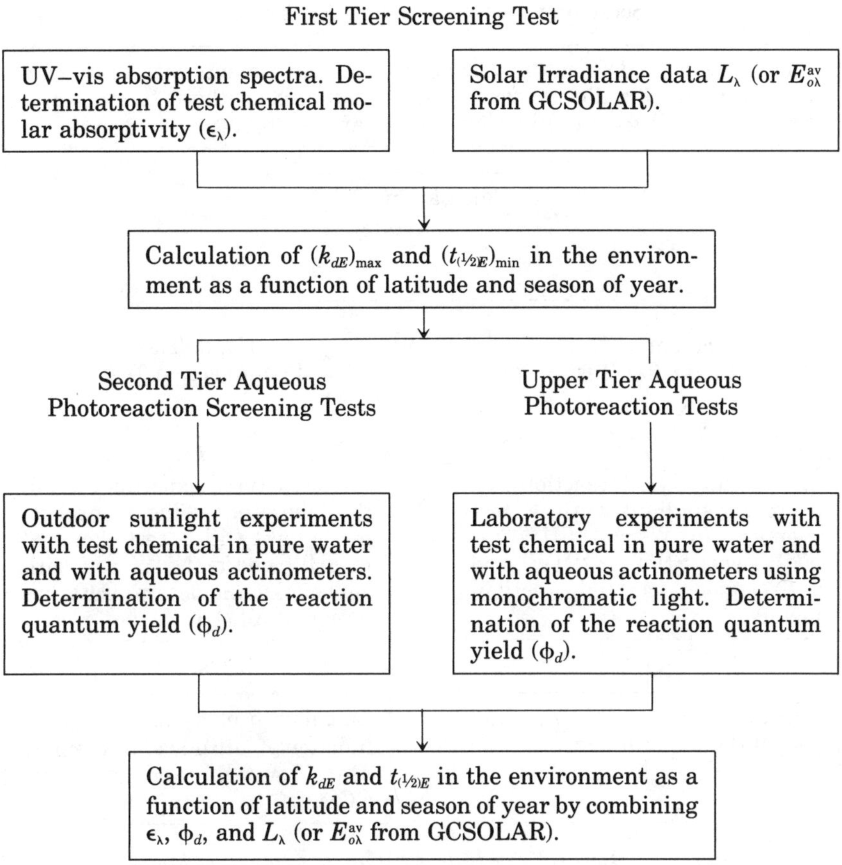

Figure I-1. Hierarchy of Direct Photoreaction Test Methods

solar irradiance data (L_λ) at shallow depths, given in Appendix A. The same calculation can be carried out in a GCSOLAR computer program using ϵ_λ and the solar irradiance $E_{o\lambda}^{av}$. Because direct photoreaction rates are first-order, the minimum direct photoreaction half-life $[t_{(1/2)E}]_{min}$ can be calculated directly from $(k_{dE})_{max}$.

If the value of $(k_{dE})_{max}$ indicates that direct photoreaction is an important mode of transformation in the environment, then second-tier photoreaction screening tests or upper tier aqueous photoreaction tests should be carried out. The second tier photoreaction screening tests provide cheap and easy methods of determining the value of ϕ_d and involve the simultaneous irradiation of a test chemical solution and an actinometer solution in pure water outdoors in sunlight. If a more precise value of ϕ_d is required, then upper-tier photoreaction tests

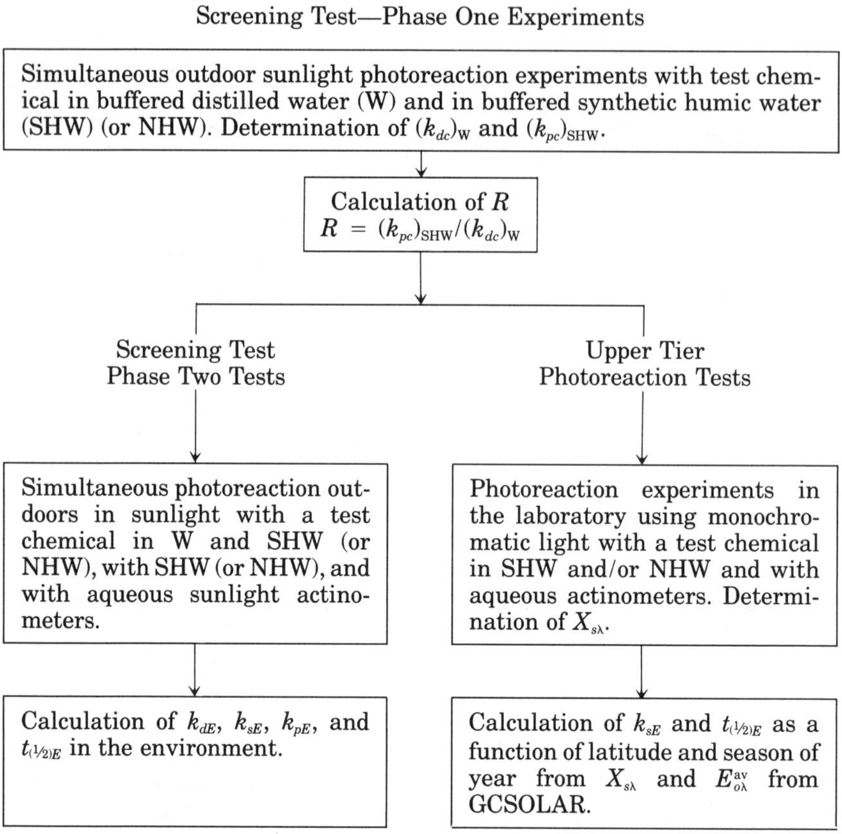

Figure I-2. Hierarchy of Indirect Photoreaction Test Methods

should be carried out by irradiating solutions of test chemical and actinometer in pure water in the laboratory using monochromatic light.

The reaction quantum yield and the molar absorptivity data can then be combined with solar irradiance data (L_λ) to calculate k_{dE} and $t_{(1/2)E}$ as a function of latitude and season of year at shallow depths in water bodies in the environment. Similar calculations can be carried out using the GCSOLAR computer program.

Indirect Photochemical Reaction Test Scheme

Let us now consider the indirect photoreaction test scheme depicted in Figure I.2. A preliminary idea of the importance of indirect photoreaction of a test chemical can be obtained by carrying out phase one experiments of a screening test. This test involves the simultaneous photoreaction of a test chemical in buffered distilled water (W) and in buffered synthetic humic acid water (SHW) outdoors in sunlight. This test method has been designed for use with synthetic humic water; however, it is easily adapted for use with a natural humic water (NHW). From the experimental data, the first-order direct photoreaction rate constant $(k_{dc})_W$ and the first-order net photoreaction rate constant $(k_{pc})_{SHW}$ (which represents the sum of the direct and indirect photoreaction rate constants) can be calculated. The ratio of these rate constants (R), which is $(k_{pc})_{SHW}/(k_{dE})_W$, represents a measure of the ability of a test chemical to undergo indirect photoreaction. If $R \geq 2$, then indirect photoreaction is an important mode of transformation in the environment.

If $R \geq 2$, then a more precise measure of the rate of indirect photoreaction can be determined from either phase two experiments in a screening test or from upper tier photoreaction tests. Phase two screening experiments involve the simultaneous photoreaction of test chemical in SHW (or NHW) and in W and the photoreaction of SHW (or NHW). To correct for variations in solar irradiance during photoreaction, an aqueous sunlight actinometer solution is simultaneously insolated. From the experimental data, the first-order direct photoreaction rate constant (k_{dE}), the first-order indirect photoreaction rate constant (k_{sE}), and the net photoreaction rate constant (k_{pE}) in the environment can be calculated.

If an upper-tier photoreaction test is used, then laboratory experiments with a test chemical in SHW or in NHW, or in both, are carried out with monochromatic light and aqueous actinometers. From these data, the response function $X_{s\lambda}$ can be calculated as a function of the wavelength. The response function is then combined with solar irradiance data $E^{av}_{o\lambda}$ from the GCSOLAR computer program to calculate the indirect photoreaction rate constant (k_{sE}) and the half-life $t_{(1/2)E}$ as a functional latitude and season in the environment. It should be noted that upper tier photoreaction tests are only applicable when R >> 2 and the test chemical undergoes essentially all indirect photoreaction.

References

1. Zepp, R. G.; Cline, D. M. *Environ. Sci. Technol.* **1977**, *11*, 359.
2. Broderius, S. J.; Smith, L. L., Jr. *Direct Photolysis of Hexacyanoferrate Complexes;* U.S. Environmental Protection Agency. U.S. Government Printing Office: Washington, DC, 1980; EPA–600/3–80–003.
3. Mill, T.; Mabey, W. R.; Lau, B. Y.; Baraze, A. *Chemosphere* **1981**, *10*, 1281.
4. Skurlatov, Y. I.; Zepp, R. G.; Baughman, G. L. *J. Agric. Food Chem.* **1983**, *31*, 1065.
5. Smith, J. H.; Mabey, W. R.; Bohonos, N.; Holt, B. R.; Lee, S. S.; Chou, T.-W.; Bomberger, D. C.; Mill, T. *Environmental Pathways of Selected Chem-*

icals in Fresh Water Systems. Part I: Background and Experimental Procedures; U.S. Environmental Protection Agency. U.S. Government Printing Office: Washington, DC, 1977; EPA 600/7–77–113.
6. Smith, J. H.; Mabey, W. R.; Bohonos, N.; Holt, B. R.; Lee, S. S.; Chou, T.-W.; Bomberger, D. C.; Mill, T. *Environmental Pathways of Selected Chemicals in Fresh Water Systems. Part II: Laboratory Studies;* U.S. Environmental Protection Agency. U.S. Government Printing Office: Washington, DC, 1978; EPA 600/7–78–074.
7. Wolfe, N. L.; Zepp, R. G.; Baughman, G. L.; Fincher, R. C.; Gordon, J. A. *Chemical and Photochemical Transformation of Selected Pesticides in Aquatic Systems;* U.S. Environmental Protection Agency. U.S. Government Printing Office: Washington, DC, 1976; EPA 600/3–76–067.
8. Dulin, D.; Mill, T. *Environ. Sci. Technol.* **1982,** *16,* 815.
9. Mill, T.; Davenport, J. E.; Dulin, D. E.; Mabey, W. R.; Bawol, R. *Evaluation and Optimization of Photolysis Screening Protocols;* U.S. Environmental Protection Agency. U.S. Government Printing Office: Washington, DC, 1981; EPA 560/5–81–003.
10. Mill, T.; Mabey, W. R.; Bomberger, D. C.; Chou, T–W.; Hendry, D. G.; Smith, J. H. *Laboratory Protocols for Evaluation of the Fate of Organic Chemicals in Air and Water;* U.S. Environmental Protection Agency. U.S. Government Printing Office: Washington, DC, 1982; EPA 600/3–82–002.
11. Mill, T.; Mabey, W. R.; Winterle, J. S.; Davenport, J. E.; Barich, V. P.; Dulin, D. E.; Tse, D.; Lee, G. "Design and Validation of Screening and Detailed Methods for Environmental Processes"; Final Report; U.S. Environmental Protection Agency, Office of Toxic Substances: Washington, DC, 1982; EPA Contract No. 68–01–6325.
12. Mill, T.; Davenport, J.; Winterle, J. S.; Mabey, W. R.; Drossman, H.; Tse, D.; Liu, A. "Toxic Substances Process Data Generation and Protocol Development"; Work assignment 12; Test standard development; Final Report, Washington, DC, 1983; EPA Contract No. 68–03–2981.
13. Mill, T.; Winterle, J. S.; Fischer, A.; Tse, D.; Mabey, W. R.; Drossman, H.; Liu, A.; Davenport, J. E. "Toxic Substances Process Data Generation and Protocol Development"; Work assignment 18, Test standard development, Final Report; Washington, DC, 1985; EPA Contract No. 68–03–2981.
14. *Fed. Regist.* **1985,** *50,* 39472; 40 CFR 796.1050.
15. *Fed. Regist.* **1985,** *50,* 39285; 40 CFR 796.3700.
16. *Fed. Regist.* **1985,** *50,* 39296; 40 CFR 796.3780.
17. Zepp, R. G. *Environ. Sci. Technol.* **1978,** *12,* 327.
18. Zepp, R. G. In *The Handbook of Environmental Photochemistry*; Hutzinger, O., Ed.; Springer–Verlag: New York, 1981; Vol. 2, Part B, p 19.
19. Haag, W. R.; Hoigne, J.; Gassman, E.; Braun, A. M. *Chemosphere* **1984,** *13,* 631.
20. Haag, W. R.; Hoigne, J.; Gassman, E.; Braun, A. M. Ibid, p 641.
21. Miller, G. C.; Zisook, R.; Zepp, R. *J. Agric. Food Chem.* **1980,** *28,* 1953.
22. Zepp, R. G. In *The Role of Solar Ultraviolet Radiation in Marine Ecosystems*; Calkins, J., Ed.; Plenum: New York, 1982; p 283.
23. Zepp, R. G.; Baughman, G. L.; Schlotzhauer, P. F. *Chemosphere* **1981,** *10,* 109.
24. Zepp, R. G.; Baughman, G. L.; Schlotzhauer, P. F. Ibid, p 119.
25. Winterle, J. S.; Mill, T. "Toxic Substances Process Data Generation and Protocol Development"; Work assignment 18, EPA Special Report; *Indirect Photoreaction Protocol*; Washington, DC, 1985; EPA Contract No. 68–03–2981.
26. *Fed. Regist.* **1986,** *51,* 472; 40 CFR 796.3765.

Part I. Theory

Chapter 1

Some Fundamental Principles of Photochemistry

Before discussing the theory and experimental procedures for the determination of environmentally relevant rate constants and half-lives for chemicals present in natural water bodies in sunlight, it is necessary to introduce some fundamental principles of photochemistry. The Lambert law is introduced to describe the transmission and absorption of light as it traverses through a medium such as air or water, and the Beer–Lambert law is introduced to describe the transmission of light as it traverses through the medium containing a chemical, specifically the absorption of light by the chemical. The Grotthus–Draper and Stark–Einstein laws are described because they are the two basic laws of photochemistry. The Grotthus–Draper law, combined with the Beer–Lambert law, defines the absorption of light by the chemical, while the Stark–Einstein law defines the efficiency with which the absorbed light transforms the chemical into products. These laws are used to define the kinetics of photoreaction of a chemical.

1.A. Lambert's Law

The manner in which the intensity of light is reduced in passing through an absorbing medium is given by Lambert's law. Mathematically, this law states that the rate of decrease of intensity of light with thickness of the medium, $-dI_{t\lambda}/dl$, at a fixed wavelength λ, is directly proportional to the intensity of light. Therefore,

$$-(dI_{t\lambda}/dl) = \sigma_\lambda I_{t\lambda} \quad (1.1)$$

where $I_{t\lambda}$ is the intensity of light transmitted in photons cm^{-2} s^{-1} at thickness l (in cm) and σ_λ is the proportionality constant. Rearranging equation 1.1 and integrating between the boundary conditions ($l = 0$,

1464–6/88/0003$06.00/1

$I_{o\lambda}$) and (l, $I_{t\lambda}$) yield

$$\ln(I_{t\lambda}/I_{o\lambda}) = -\sigma_\lambda l \quad (1.2)$$

$$I_{t\lambda} = I_{o\lambda}\exp(-\sigma_\lambda \mathrm{l}) \quad (1.3)$$

Most spectrophotometers measure $I_{t\lambda}/I_{o\lambda}$ in the decadic logarithmic system, so equation 1.2 can be converted to

$$\log_{10}(I_{t\lambda}/I_{o\lambda}) = -\alpha_\lambda l \quad (1.4)$$

where $\alpha_\lambda = \sigma_\lambda/2.303$ and is called the attenuation coefficient of the medium, at a fixed wavelength λ, in the units cm^{-1}. The symbol l is now the cell path length. The absorbance A_λ at wavelength λ, as measured in a spectrophotometer, is defined by

$$A_\lambda = \log_{10}(I_{o\lambda}/I_{t\lambda}) \quad (1.5)$$

so that equation 1.4 becomes

$$A_\lambda = \alpha_\lambda l \quad (1.6)$$

Because the fraction of light transmitted through the medium is defined as

$$T_\lambda = I_{t\lambda}/I_{o\lambda} \quad (1.7)$$

equation 1.5 becomes

$$A_\lambda = \log_{10}(1/T_\lambda) \quad (1.8)$$

1.B. Beer–Lambert Law

For absorbing chemicals in an absorbing medium, the decrease in intensity of light with thickness as it traverses through the medium, at a fixed wavelength λ, is given by the relationship

$$-(dI_{t\lambda}/dl) = \sigma_\lambda I_{t\lambda} + k_\lambda[C]I_{t\lambda} \quad (1.9a)$$

$$-(dI_{t\lambda}/dl) = (\sigma_\lambda + k_\lambda[C])I_{t\lambda} \quad (1.9b)$$

where σ_λ and k_λ are proportionality constants, $I_{t\lambda}$ is the intensity of light transmitted, and $[C]$ is the molar concentration of a chemical in the medium. Rearranging equation 1.9b and integrating between the boundary conditions ($l = 0, I_{o\lambda}$) and ($l, I_{t\lambda}$) yield

$$\ln(I_{o\lambda}/I_{t\lambda}) = (\sigma_\lambda + k_\lambda[C])l \quad (1.10)$$

Converting equation 1.10 to the decadic logarithmic system yields

$$\log_{10}(I_{o\lambda}/I_{t\lambda}) = (\alpha_\lambda + \epsilon_\lambda[C])l \quad (1.11)$$

$$I_{t\lambda} = I_{o\lambda}\{10^{-(\alpha_\lambda + \epsilon_\lambda[C])l}\} \quad (1.12)$$

where $\alpha_\lambda = \sigma_\lambda/2.303$ and represents the attenuation coefficient of the medium, at a fixed wavelength λ, in the units cm^{-1}; $\epsilon = k_\lambda/2.303$ and represents the molar absorptivity of the chemical, at a fixed wavelength λ, in the units $M^{-1}\ cm^{-1}$; $[C]$ is the molar concentration of the chemical; and l is the cell path length in cm.

Using the definitions for A_λ and T_λ defined by equations 1.5 and 1.7 in equation 1.11 yields

$$A_\lambda = \log_{10}(1/T_\lambda) = (\alpha_\lambda + \epsilon_\lambda[C])l \quad (1.13)$$

Equation 1.13 is the mathematical expression of the Beer–Lambert law. For systems in which the medium does not absorb light, $\alpha_\lambda = 0$, and equation 1.13 becomes

$$A_\lambda = \epsilon_\lambda[C]l \quad (1.14)$$

which is the conventional form of the Beer–Lambert law.

1.C. Photochemistry

1.C.1. Introduction

Ordinary or thermal chemical reactions are brought about by atomic or molecular collisions. Atoms or molecules move at high speeds and consequently have considerable kinetic energy; the higher the temperature, the greater the kinetic energy. When these atoms or molecules collide, chemical reactions can occur. Thermal activation is not the only means by which the energy of atoms or molecules can be raised sufficiently to cause reaction. Atoms or molecules can also be energized to an excited state by the absorption of light. If these excited atoms or molecules have sufficient energy, they may react.

Photochemical reactions are distinctly different from thermal reactions. For example, photochemical reactions are selective because by using selective monochromatic light, atoms or molecules can be excited without exciting the surrounding medium. In photochemical reactions, the structures of the ground and excited states of atoms or molecules are generally different, so that these states can exhibit significantly different chemical behavior. Excited-state species generated by the absorption of light are high-energy species such as triplet excited species, radicals, biradicals, or strained-ring molecules that often are not readily formed by thermal excitation of ground-state species. Finally, thermal reactions can only proceed if the net change in free energy is negative. However, in photochemical reactions, the radiant energy supplied to a system may increase the free energy of the reactants sufficiently to make the net free energy change negative. An outstanding example of such a process is photosynthesis. In sunlight, in the presence of chlorophyll, CO_2 and H_2O are combined in plants to produce complex carbohydratelike materials and O_2. This process is an example of photosensitization or indirect photoreaction.

Photochemistry is the study of the effect of radiant energy on atoms and molecules and on rates and mechanisms by which reactions proceed when initiated by light. In the environment, sunlight, which is partly composed of ultraviolet (UV) and visible light, is the primary source of radiation that can transform chemicals. When a molecule absorbs light, it is transferred from the ground energy state (E_1) to an excited state (E_2). The increase in energy is given by the equation

$$\Delta E = E_2 - E_1 \tag{1.15}$$

The energy gained by a molecule (in joules) when it absorbs a quantum of light is given by the Bohr equation

$$E = h\nu \tag{1.16}$$

where h is Planck's constant and is equal to 6.626×10^{-34} J–s and ν is the frequency of the light in s^{-1}. Because the frequency of light is given by the relationship $\nu = c\lambda^{-1}$, equation 1.16 becomes

$$\Delta E = hc\lambda^{-1} \tag{1.17}$$

where c is the speed of light and is equal to 2.998×10^{10} cm s^{-1} and λ is the wavelength of light in cm. The energy absorbed per mole will be Avogadro's number (N) times $hc\lambda^{-1}$, where N is 6.022×10^{23} molecules per mole. Thus, equation 1.17 becomes

$$\Delta E = Nhc\lambda^{-1} \tag{1.18}$$

The energy per mole defined by equation 1.18 is representative of an einstein of photons of wavelength λ and its magnitude is a function of the wavelength of light. One einstein is a mole of quanta or photons. For the wavelength in nanometers (nm), equation 1.18 becomes

$$\Delta E \text{ (kcal/einstein)} = 2.86 \times 10^4 \lambda^{-1}\text{(nm)} \qquad (1.19)$$

$$\Delta E \text{ (kJ/einstein)} = 1.20 \times 10^5 \lambda^{-1}\text{(nm)} \qquad (1.20)$$

Energies for various wavelengths of light, obtained from equations 1.19 and 1.20, are given in Table 1-1. Inspection of the data in Table 1-1 indicates that the energy per einstein varies widely with the wavelength. For environmental photochemistry, the spectral range of interest is 290–800 nm, corresponding to an energy range of 98.6–35.8 kcal/mol or 414–150 kJ/mol. This energy is sufficient to break many chemical bonds. The lower wavelength limit for sunlight photoreaction has been set at 290 nm because ozone in the earth's atmosphere absorbs sunlight strongly at wavelengths equal to or less than 290 nm (*1–4*). The upper wavelength limit has been set at 800 nm thermochemically because light at wavelengths greater than 800 nm (35.8 kcal/mol or 150 kJ/mol) does not possess sufficient energy to break the covalent bonds of ground-state molecules (*5,6*).

1.C.2. Grotthus–Draper Law of Photochemistry

The Grotthus–Draper law, the first law of photochemistry, states that only radiations that are absorbed by the reacting system can be effective in producing chemical changes. Thus, for photochemical transfor-

Table 1-1. Energies per Einstein at Various Wavelengths of Light

		Energy per Einstein	
Wavelength (nm)	*Spectral Range*	*kilocalories*	*kilojoules*
200	ultraviolet	143	600
250	ultraviolet	114	480
300	ultraviolet	95.3	400
300–350	ultraviolet	95.3–81.7	400–343
400–450	visible (violet)	71.5–63.6	300–267
450–500	visible (blue)	63.6–57.2	267–240
500–575	visible (green)	57.2–49.7	240–209
575–590	visible (yellow)	49.7–48.5	209–203
590–650	visible (orange)	48.5–44.0	203–185
650–750	visible (red)	44.0–38.1	185–160
750–800	near infrared	38.1–35.8	160–150

mations, it is necessary to know the amount of light absorbed by the reacting chemical. Consider a photochemical reaction cell containing a chemical in a medium such as air or water. To describe the light absorbed by the chemical in the medium, let $I_{o\lambda}$ be the intensity of light entering a cell and $I_{t\lambda}$ the intensity of light transmitted through the cell. The intensity of light absorbed by the system, $I^*_{a\lambda}$, is then

$$I^*_{a\lambda} = I_{o\lambda} - I_{t\lambda} \tag{1.21}$$

Using the Beer–Lambert law, equation 1.12, the intensity of light absorbed by the system is

$$I^*_{a\lambda} = I_{o\lambda} - I_{o\lambda}\{10^{-(\alpha_\lambda + \epsilon_\lambda[C])l}\} \tag{1.22a}$$

$$I^*_{a\lambda} = I_{o\lambda}\{1 - 10^{-(\alpha_\lambda + \epsilon_\lambda[C])l}\} \tag{1.22b}$$

The fraction of the light absorbed by the system, $F_{s\lambda}$, is then given by

$$F_{s\lambda} = (I^*_{a\lambda}/I_{o\lambda}) = \{1 - 10^{-(\alpha_\lambda + \epsilon_\lambda[C])l}\} \tag{1.23}$$

1.C.3. Stark–Einstein Law of Photochemistry

The Stark–Einstein law, the second law of photochemistry, states that each molecule directly taking part in a chemical reaction induced by light absorbs one quantum or photon of light. With extremely high intensity light sources, such as lasers, biphotonic processes have been observed; however, these processes are excluded from all discussions in this text because they do not occur with conventional light sources in typical photochemical experiments.

The second law of photochemistry applies to processes in which the absorption of a quantum of light leads to the disappearance of the molecule or the deactivation of the molecule to a nonreactive state. These processes are designated as primary processes. When, as a result of the primary absorption, only one molecule transforms per quantum absorbed, the number of molecules reacting will equal the number of quanta absorbed. Thus, under these conditions there is a one-to-one correlation between the number of quanta absorbed and the number of molecules transformed. In some cases, a molecule activated by the absorption of light initiates a series of thermal reactions, designated as secondary reactions, so that many reactant molecules undergo transformation. Under these conditions, there will not be a one-to-one correlation between the reacting molecules and the number of quanta absorbed, as more than one molecule reacts per quantum absorbed. For other processes in which deactivation is significant, such as fluorescence, phosphorescence, or radiationless deactivation, less than one molecule may react per quantum absorbed.

To express the relationship between the number of molecules undergoing reaction and the number of photons absorbed, the reaction quantum yield (or disappearance quantum yield), at wavelength λ, is introduced. This quantity, ϕ_λ, is defined as

$$\phi_\lambda = \frac{\text{number of molecules that photoreact}}{\text{number of photons absorbed}} \tag{1.24}$$

$$\phi_\lambda = \frac{\text{number of moles that photoreact}}{\text{number of einsteins absorbed}} \tag{1.25}$$

where the numerator and denominator refer to the same time and volume units. The reaction quantum yield gives the number of molecules observed to undergo chemical transformation per photon of light absorbed. The reaction quantum yield can vary from a value near zero for inefficient photochemical processes (e.g., when a significant number of molecules fluoresce) to about 10^6 for the photochemical reaction of H_2 and Cl_2 in which a secondary chain reaction occurs. In environmental systems, ϕ_λ rarely exceeds one.

In the determination of the reaction quantum yield of a chemical, it is necessary to determine the intensity of the light source. This determination is conveniently done by the use of chemical actinometers, which will be discussed in more detail in Chapters 6–11.

1.C.4. Kinetic Aspects of Photochemical Processes

Consider the photoreaction of a chemical in a reaction vessel with an irradiated surface area A and volume V. It is assumed that the transformation takes place with monochromatic light of wavelength λ, is unimolecular, and follows the reaction

$$C \xrightarrow{I_{a\lambda}} \text{products} \tag{1.26}$$

without other thermal or photochemical processes taking place. The term $I_{a\lambda}$ is the average number of photons (or einsteins) absorbed by the reactant per unit volume and per unit time.

The rate of photoreaction of a chemical is a function of the rate at which light is absorbed (the Grotthus–Draper law) and the efficiency with which the absorbed light transforms the chemical into products (the Stark–Einstein law); this can be represented by $\phi_\lambda I_{a\lambda}$, where $I_{a\lambda}$ is the rate of light absorption by the chemical and ϕ_λ is the reaction quantum yield of the chemical at wavelength λ. The average rate of disappearance of a chemical in a well-mixed system is then given by

$$\text{average rate} = (-d[C]/dt)_\lambda = \phi_\lambda I_{a\lambda} \tag{1.27}$$

where $[C]$ is the molar concentration.

According to Balzani and Carassiti (7), $I_{a\lambda}$ is proportional to both the fraction of light absorbed by the system ($F_{s\lambda}$) and the fraction of light absorbed by the chemical ($F_{c\lambda}$). Thus,

$$I_{a\lambda} = I_{o\lambda}(A/V)F_{s\lambda}F_{c\lambda} \tag{1.28}$$

where $I_{o\lambda}$ is the intensity of incident light, A is the area of the irradiated cell, and V is the volume. The fraction of light absorbed by the system, $F_{s\lambda}$, is defined by equation 1.23 and is given by

$$F_{s\lambda} = \{1 - 10^{-(\alpha_\lambda + \epsilon_\lambda[C])l}\} \tag{1.23}$$

The fraction of light absorbed by the chemical, $F_{c\lambda}$, is defined by the equation

$$F_{c\lambda} = \epsilon_\lambda[C]/(\alpha_\lambda + \epsilon_\lambda[C]) \tag{1.29}$$

where ϵ_λ is the molar absorptivity of the reactant at wavelength λ in the units of M^{-1} cm^{-1}, $[C]$ is the molar concentration of the reactant, α_λ is the attenuation coefficient of the medium or solvent in cm^{-1}, and l is the reaction cell path length in cm. Equations 1.23, 1.27, 1.28, and 1.29 will be used extensively in later chapters when discussing rates of photoreaction in aqueous media.

It is important to consider the dimensions in equations 1.27 and 1.28. In equation 1.27, ϕ_λ is dimensionless so that the units of $I_{a\lambda}$ must be compatible with the units of $d[C]/dt$ (i.e., mol L^{-1} s^{-1}). Thus, $I_{a\lambda}$ must be in the units of einsteins L^{-1} s^{-1}. (An einstein is a mole of quanta or 6.022×10^{23} quanta.) In equation 1.28, $I_{o\lambda}$, as measured spectrophotometrically, and used in the Beer–Lambert law, is in the units photons (or einsteins) cm^{-2} s^{-1}. Thus, a conversion factor (j) must be introduced to make the units of $I_{a\lambda}$ compatible with molar concentration.

Because $F_{s\lambda}$ and $F_{c\lambda}$ are dimensionless quantities in equation 1.28, we need only consider the dimensions of $I_{o\lambda}(A/V)$. When $I_{o\lambda}$ is in photons cm^{-2} s^{-1}, A is in cm^2, and V is in cm^3, then the units of $I_{a\lambda}$ and $I_{o\lambda}(A/V)$ are as follows:

$$I_{a\lambda} = I_{o\lambda}(A/V) = \text{photons cm}^{-2}\text{s}^{-1}(\text{cm}^2/\text{cm}^3) \tag{1.30}$$

$$I_{a\lambda} = \frac{\text{photons s}^{-1}}{\text{cm}^3} \tag{1.31}$$

$$I_{a\lambda} = \frac{\text{photons s}^{-1}}{6.022 \times 10^{23}(\text{photons/mol})} = \frac{1}{\text{cm}^3[10^{-3}(\text{L/cm}^3)]} \tag{1.32}$$

$$I_{a\lambda} = \frac{\text{mol L}^{-1}\text{s}^{-1}}{6.022 \times 10^{20}} \tag{1.33}$$

$$I_{a\lambda} = \text{mol L}^{-1}\text{s}^{-1}/j \tag{1.34}$$

where $j = 6.022 \times 10^{20}$ and makes intensity units of $I_{a\lambda}$ compatible with the units of $d[C]/dt$.

When $I_{o\lambda}$ is in millieinsteins $\text{cm}^{-2}\text{s}^{-1}$, A is in cm^2, and V is in cm^3, then the units of $I_{a\lambda}$ and $I_{o\lambda}$ (A/V) are as follows:

$$I_{a\lambda} = I_{o\lambda}(A/V) = \text{millieinsteins cm}^{-2}\text{s}^{-1}(\text{cm}^2/\text{cm}^3) \tag{1.35}$$

$$I_{a\lambda} = \frac{10^{-3}\text{ einsteins s}^{-1}}{\text{cm}^3} \tag{1.36}$$

$$I_{a\lambda} = \frac{\text{einsteins s}^{-1}}{10^3\text{cm}^3} = \frac{\text{einsteins s}^{-1}}{\text{L}} \tag{1.37}$$

$$I_{a\lambda} = \frac{\text{einsteins L}^{-1}\text{s}^{-1}}{j\ (= 1\text{ einstein/mol})} \tag{1.38}$$

$$I_{a\lambda} = \text{mol L}^{-1}\text{s}^{-1}/j \tag{1.39}$$

In this case, the conversion factor $j = 1$ einstein/mol and makes the intensity units of $I_{a\lambda}$ compatible with the units of $d[C]/dt$.

When using equations 1.27 and 1.28, the conversion factor j will be introduced at appropriate places in the text to make the units of $I_{a\lambda}$ compatible with the units of $[C]$, the molar concentration, and ϵ_λ, the molar absorptivity, in the units of $\text{M}^{-1}\text{ cm}^{-1}$.

References

1. Demerjian, K. L.; Schere, K. L.; Peterson, J. T. *Adv. Sci. Technol.* **1980,** *10*, 389.
2. Leighton, P. A. *Photochemistry of Air Pollution*; Academic: New York, 1961; Chapter II.
3. Peterson, J. T. *Calculated Actinic Fluxes (290–700 nm) for Air Pollution Photochemistry Applications;* U.S. Environmental Protection Agency. U.S. Government Printing Office: Washington, DC, 1976; EPA 600/4–76–0025.

4. Zepp, R. G.; Cline, D. M. *Environ. Sci. Technol.* **1977**, *11*, 359.
5. Calvert, J. G.; Pitts, J. N., Jr. *Photochemistry*; Wiley: New York, 1966; Chapter 7.
6. Benson, S. W. *Thermochemical Kinetics*, 2nd ed.; Wiley–Interscience: New York, 1976.
7. Balzani, V.; Carassiti, V. *Photochemistry of Coordination Compounds*; Academic: New York, 1970; Chapter 2.

Chapter 2

Absorption and Transmission of Solar Radiation in the Atmosphere

Before discussing the photoreaction of chemicals in aqueous media in sunlight, it is necessary to consider the wavelength distribution of the light emanating from the sun, the absorption and transmission of sunlight through the atmosphere to the earth's surface, and the absorption and transmission of sunlight as it traverses through a water body. This chapter will be concerned with the nature of radiation emanating from the sun and the absorption and transmission of sunlight through the atmosphere. Chapter 3 will be concerned with the absorption and transmission of sunlight in a water body.

During the day, the atmosphere surrounding the earth receives radiation directly from the sun, from the sky, and from reflections from the earth's surface. Sky radiation results from sunlight scattered by dust, aerosols, water droplets, and gaseous molecules in the atmosphere. Leighton (*1, 2*) did the pioneering research on the application of solar radiation to photochemical atmospheric pollution problems. In his book (*2*) a simple, but effective, radiative transfer model is described to calculate the solar irradiance, or solar flux, at the earth's surface at various zenith angles over the ultraviolet (UV) and visible solar spectra. These solar irradiance data were first published by Leighton in 1956 (*1*). Though these data are over 30 years old, they are surprisingly good, considering the lack of knowledge and experimental data at the time on the absorption and transmission characteristics of sunlight through the atmosphere. Because Leighton's radiative transfer model forms the basis of solar irradiance, or solar flux, calculations, this model will be discussed in detail in this chapter. At the conclusion of this chapter, the solar irradiance data will be updated by using the best available theoretical model and the best available solar data.

1464–6/88/0013$06.00/1

2.A. Radiation Emanating from the Sun

The radiant energy emitted by the sun is responsible for the photochemical reactions (and thus smog) in the earth's atmosphere; therefore, it is necessary to understand the nature of the solar spectrum. To a good approximation, the sun can be treated as an incandescent mass having a surface temperature of approximately 6000 K. Its central temperature has been estimated to be approximately 20×10^6 K and its central pressure to be 160×10^9 atmospheres (*3*). The source of this tremendous quantity of heat and light is believed to be a nuclear reaction in which hydrogen atoms are fused to form helium atoms. When this reaction occurs, the excess mass is converted into an enormous amount of energy in the form of heat and light.

The approximate irradiance of the near UV–vis solar spectrum outside the earth's atmosphere is depicted in Figure 2.1. It is essentially that of a black body radiator at 6000 K. Superimposed on this curve is the absorption spectrum of the gases in the solar atmosphere, the Fraunhaufer absorption lines. To eliminate the detailed Fraunhaufer structure, the solar irradiances, or solar flux data, $I_{o\lambda}$, have been combined into a mean solar irradiance averaged over 10 nm intervals. The units of $I_{o\lambda}$ are photons cm^{-2} s^{-1} (10 nm)$^{-1}$ or microwatts cm^{-2} (10 nm)$^{-1}$. Because of the presence of ozone in a layer in the upper atmosphere of the earth, no radiation reaches the earth's surface below 290 nm.

The solar spectral region below 140 nm is composed of emission lines rather than a continuum. In the region 181.7 to 120.6 nm, the Lyman α-line of hydrogen atoms dominates the solar spectrum.

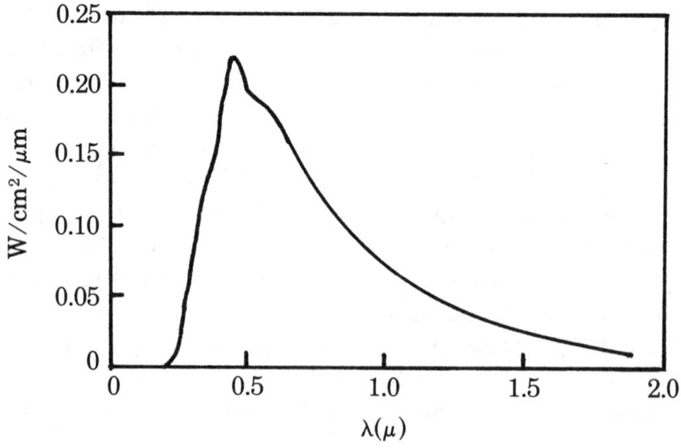

Figure 2.1. Irradiance of the near UV–vis solar spectrum outside the earth's atmosphere.

The solar UV–vis irradiance at the earth's surface is a function of the time of day, time of year, latitude, elevation above the earth's surface, thickness of the ozone layer, and turbidity in the atmosphere. The air mass, the thickness of the layer of the atmosphere through which the sunlight traverses, is a function of the time of day and year, latitude, and elevation.

2.B. Air Mass and Solar Zenith Angle

The solar radiation that reaches the earth's surface must pass through the earth's atmosphere (Figure 2.2). As sunlight passes through this atmosphere, it is modified by absorption and scattering. The solar irradiance is a function of the thickness of the atmosphere, the air mass. The air mass, *m*, is defined as the ratio of the path length of the solar radiation through the atmosphere (*L*) relative to the vertical, or nor-

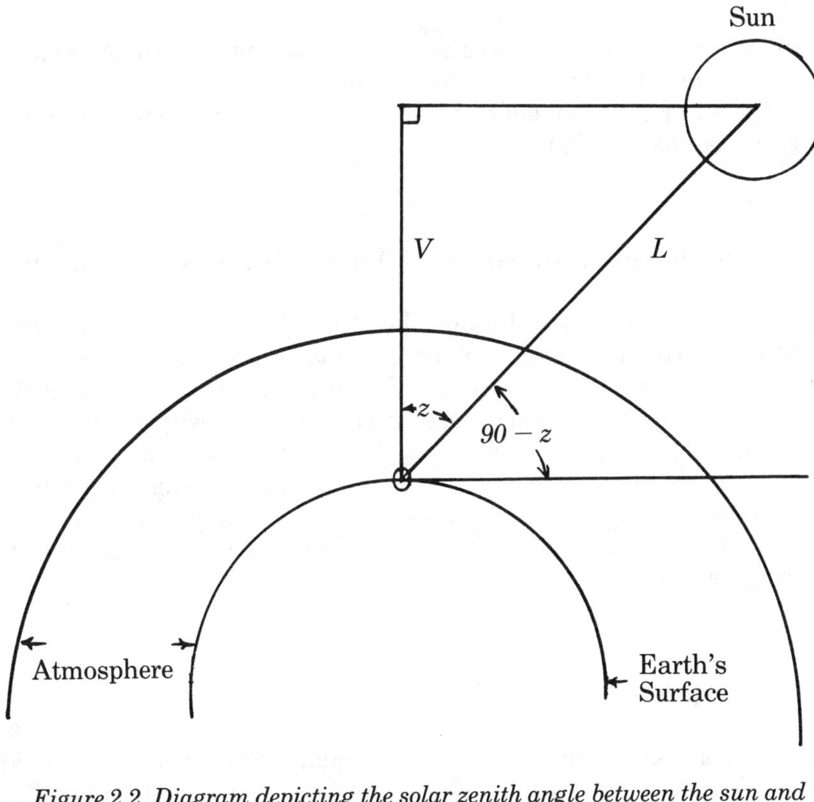

Figure 2.2. Diagram depicting the solar zenith angle between the sun and the zenith.

mal, path (V). The air mass is a function of the solar zenith angle (z), the angle formed from a point O on the earth's surface between the sun and the zenith (i.e., the normal to the earth's surface). These parameters are depicted in Figure 2.2. For solar zenith angles z less than 60°, the mathematical relationship governing z, L, and V is given by

$$m = (L/V) = \sec z \tag{2.1}$$

When the sun is directly overhead, $L = V$, $m = 1$, and $z = 0°$. Thus, the path length of solar radiation through the atmosphere is a minimum. For larger zenith angles ($z > 60°$), corrections are needed because of the curvature of the atmosphere as a result of the curvature of the surface of the earth and for reflection.

The solar zenith angle for any specified location and time may be calculated from the latitude of the location (lat), the local hour angle (lha), and the declination angle (dec) by using equation 2.2.

$$\cos z = \cos(\text{lat}) \times \cos(\text{dec}) \times \cos(\text{lha}) + \sin(\text{lat}) \times \sin(\text{dec}) \tag{2.2}$$

Declination and local hour angles can be obtained from the *Astronomical Almanac* (4), which is published annually.

The solar altitude, defined by the angle the sun makes with the horizon, is $(90-z)°$ (Figure 2.2).

2.C. Transmission of Sunlight Through the Atmosphere

Sunlight, as it traverses through the atmosphere, is attenuated, particularly in the wavelength region equal to or less than 300 nm. The attenuation of sunlight is a result of absorption and scattering in the atmosphere. Part of the decrease in sunlight intensity is due to its absorption by gases and results in the liberation of heat. The other part of the attenuation is a direct result of the scattering of light from particulates and gaseous molecules in the atmosphere. This scattered light is not changed into another form of energy but is diverted from its original path.

2.C.1. Bouger–Lambert Law

The transmission of light through a medium is governed by Lambert's law, equation 1.3 in Section 1.A. If the path length (l) is equated to the mass of air (m) through which the transmitted light passes, then equation 1.3 takes the form

$$I_{t\lambda} = I_{o\lambda} \exp(-\sigma_\lambda m) \tag{2.3}$$

where $I_{t\lambda}$ and $I_{o\lambda}$ are the transmitted and incident solar intensities, respectively, at wavelength λ, and σ_λ is the attenuation coefficient of the atmosphere at wavelength λ. Equation 2.3 is called the Bouger–Lambert law and can be used to estimate the intensity of sunlight transmitted through the atmosphere.

As just indicated, the attenuation of sunlight in the atmosphere is a function of its absorption by gases, particulate diffusion (i.e., particulate scattering), and molecular scattering. Thus, σ_λ can be defined by

$$\sigma_\lambda = \sigma_{m\lambda} + \sigma_{p\lambda} + \sigma_{a\lambda} \tag{2.4}$$

where $\sigma_{m\lambda}$ is the attenuation coefficient due to molecular scattering, $\sigma_{p\lambda}$ is the attenuation coefficient for particulate diffusion (i.e., scattering from particulates), and $\sigma_{a\lambda}$ is the attenuation coefficient for absorption by gases. Substituting equation 2.4 in equation 2.3 and taking the natural logarithm of the resulting equation yield

$$\ln(I_{t\lambda}/I_{o\lambda}) = -(\sigma_{m\lambda} + \sigma_{p\lambda} + \sigma_{a\lambda})m \tag{2.5}$$

Letting $T_\lambda = (I_{t\lambda}/I_{o\lambda})$, the fraction of light transmitted or the transmissivity and converting to decadic logarithms yield

$$\log_{10} T_\lambda = -(\sigma_{m\lambda} + \sigma_{p\lambda} + \sigma_{a\lambda})m/2.303 \tag{2.6a}$$

$$\log_{10} T_\lambda = -(S_{m\lambda} + S_{p\lambda} + S_{a\lambda})m \tag{2.6b}$$

where $S_{m\lambda}$, $S_{p\lambda}$, $S_{a\lambda}$ are now the decadic attenuation coefficients for molecular scattering, particulate diffusion, and absorption, respectively.

The transmissivity for each of these attenuation effects can be written separately as follows:

$$\log_{10} T_{m\lambda} = -S_{m\lambda}m \tag{2.7}$$

$$\log_{10} T_{p\lambda} = -S_{p\lambda}m \tag{2.8}$$

$$\log_{10} T_{a\lambda} = -S_{a\lambda}m \tag{2.9}$$

It is convenient to combine the molecular scattering and particulate diffusion into a single transmissivity scattering term $T_{s\lambda}$. This term is then the product of $T_{m\lambda}$ and $T_{p\lambda}$.

$$T_{s\lambda} = T_{m\lambda} T_{p\lambda} \tag{2.10}$$

The fraction of $I_{o\lambda}$ that is scattered and diffused is given by

$$F_{\text{scat}} = (1 - T_{s\lambda}) T_{a\lambda} \tag{2.11}$$

and the fraction of $I_{o\lambda}$ that is transmitted directly is given by

$$F_d = T_{s\lambda} T_{a\lambda} \tag{2.12}$$

2.C.2. Molecular Scattering and Particulate Diffusion

The theoretical attenuation coefficient for molecular scattering, $(S_{m\lambda})_o$, was derived by Raleigh (*5*). He showed that for standard conditions (i.e., $T = 273$ K and $P = 1$ atm), $(S_{m\lambda})_o$ is given by

$$(S_{m\lambda})_o = 4.535 \times 10^4 (\eta_{o\lambda} - 1)^2 \lambda^{-4} \tag{2.13}$$

where $\eta_{o\lambda}$ is the index of refraction of air at wavelength λ, and λ is in micrometers. Leighton (*2*) tabulated values of $(S_{m\lambda})_o$ from equation 2.13; however, in the real atmosphere, $(S_{m\lambda})_o$ is only an approximation. To obtain $S_{m\lambda}$ in the real atmosphere, the vertical temperature and density profiles in the atmosphere must be considered. For a normal variety of temperatures, Leighton indicated that $S_{m\lambda}$ can be obtained by multiplying $(S_{m\lambda})_o$ by (P/P_o), where P is the pressure at the point of observation and P_o is a reference pressure. The transmissivity for molecular scattering, $T_{p\lambda}$, is then given by

$$\log_{10} T_{m\lambda} = -(S_{m\lambda})_o (P/P_o) m \tag{2.14}$$

Particulate matter, in the form of water droplets and dust, attenuates, directs solar radiation by scattering, refraction, and diffraction. Leighton (*2*) refers to this combination of effects as particulate diffusion. Rather than resorting to Mie theory, Leighton used empirical equations to describe particulate diffusion. The transmissivity of air is a function of the molecular scattering of dry air, the scattering due to precipitable water vapor in the form of water droplets, and to the scattering from dust. The particulate diffusion transmissivity, $T_{p\lambda}$, is given by

$$\log_{10} T_{p\lambda} = -(3.75 \times 10^{-3} \lambda^{-2} w + 3.5 \times 10^{-2} \lambda^{-0.75} d) m \tag{2.15}$$

where w is a function of the precipitable water vapor in the atmosphere and d is a function of the dust content in the atmosphere.

2.C.3. Absorption by Atmospheric Gases

Ozone in the upper atmosphere is the only naturally occurring gaseous species that affects the transmission of solar radiation in the spectral region 290 to 680 nm. The spectrum of ozone shows two series of absorption bands in this region. One series of bands, the strong Huggins bands, commence absorption at approximately 350 nm and the absorption increases to a maximum at 255 nm. The other series of bands, the weak Chappius bands, absorb over the region 450 to over 700 nm with λ_{max} at approximately 600 nm.

If (O_3) is the amount of ozone in the atmosphere in terms of the thickness of an equivalent layer of pure ozone (in cm), at standard temperature and pressure, and α_λ is the decadic absorption coefficient in cm^{-1}, then the ozone attenuation coefficient of ozone is given by

$$S_{a\lambda} = \alpha_\lambda(O_3) \tag{2.16}$$

Substituting this result in equation 2.9 yields

$$\log_{10} T_{a\lambda}(O_3) = -\alpha_\lambda(O_3)m \tag{2.17}$$

Leighton (*2*) analyzed the spectral data of ozone in the region 290–350 nm and concluded that the best data for the attenuation coefficient, α_λ, can be described by

$$\log_{10} \alpha_\lambda = 16.58 - (5.64 \times 10^3)\lambda \tag{2.18}$$

where λ is in angstroms.

The transmissivity of ozone, $T_{a\lambda}(O_3)$, in the region 280–350 nm can be calculated by using the values of α_λ obtained from equation 2.18 in equation 2.17. The quantity of ozone, (O_3), is a function of latitude. According to Leighton (*2*), (O_3) can vary from 0.25 to 0.20 cm in the United States. Leighton chose a representative value of 0.22 cm and calculated the transmissivity of sunlight in the air due to ozone. With $(O_3) = 0.22$ cm and $z = 45°$, $T_{a\lambda}(O_3)$ was found to be 7×10^{-6} at 290 nm and 1×10^{-10} at 285 nm. These results demonstrate the sharp cutoff of solar radiation below 290 nm as it passes through the ozone layer.

To calculate atmospheric transmission in the visible (red) beyond 680 nm and in the near infrared (IR) regions, the absorption by O_2,

CO_2, and gaseous H_2O must be considered. Leighton (*2*) considered the weak absorption bands of O_2 in the spectral regions 766–759 and 692–686 nm, which generate the Fraunhaufer lines A and B. The oxygen atmospheric transmissivity, $T_{a\lambda}(O_2)$, averaged over these two bands is given by

$$\log_{10} T_{a\lambda}(O_2) = -k_\lambda Pm \tag{2.19}$$

where P is the atmospheric pressure in mbar and $k_\lambda = 1.5 \times 10^{-4}$ mbar^{-1} for the A band and $k_\lambda = 6 \times 10^{-5}$ mbar^{-1} for the B band.

2.D. Direct and Sky Solar Radiation

As seen in Section 2.B., $I_{o\lambda}$ represents the solar irradiance in photons cm^{-2} s^{-1} averaged over a wavelength interval $\Delta\lambda$ (10 nm), centered at λ, outside the atmosphere. On a clear day, the direct solar irradiance, $I_{d\lambda}$, over the same wavelength interval $\Delta\lambda$ per unit horizontal cross section in the lower atmosphere can be obtained as follows: the fraction of the direct solar irradiance is given by $T_{s\lambda}T_{a\lambda}$ (equation 2.12), so that the direct solar irradiance is $T_{s\lambda}T_{a\lambda}I_{o\lambda}$. From Figure 2.2, the Z-component (or vertical component) on a horizontal surface in the lower atmosphere is then given by

$$I_{d\lambda} = T_{s\lambda}T_{a\lambda}I_{o\lambda} \cos z \tag{2.20}$$

The fraction of the sky irradiance, which is the fraction of $I_{o\lambda}$ that is scattered, is given by $(1 - T_{s\lambda})\, T_{a\lambda}$ (equation 2.11), so that the sky irradiance is $(1 - T_{s\lambda})\, T_{a\lambda}I_{o\lambda}$. From Figure 2.2, the Z-component (or vertical component) of the irradiance on a horizontal surface in the lower atmosphere is then given by the equation

$$I_{s\lambda} = g(1 - T_{s\lambda})T_{a\lambda}I_{o\lambda} \cos z \tag{2.21}$$

where $I_{s\lambda}$ is the scattered solar irradiance over the interval $\Delta\lambda$, centered at λ, and received by each unit horizontal cross section in the lower layer and g is a fractional constant. The value of g is a function of molecular scattering and particulate diffusion. Leighton (*2*) chose a value of $g = 0.5$ as a useful approximation for all atmospheric conditions.

Before treating the absorption of solar radiation by a chemical in a polluted layer in the atmosphere by using the Beer–Lambert law, the path length of direct and sky radiation in that layer must be known.

2.E. Path Length of Solar Radiation in the Atmosphere

2.E.1. Path Length of Direct Solar Radiation

To describe mathematically the path length of direct solar radiation in the atmosphere, let h represent the thickness of the polluted layer and z the solar zenith angle, Figure 2.3. From triangle OAB, the path length of direct solar radiation (L_d) is given by

$$L_d = h \sec z \tag{2.22}$$

2.E.2. Path Length of Sky Solar Radiation

The derivation of the path length of sky radiation is much more complicated. If light only reached the earth in a straight line (i.e., by direct

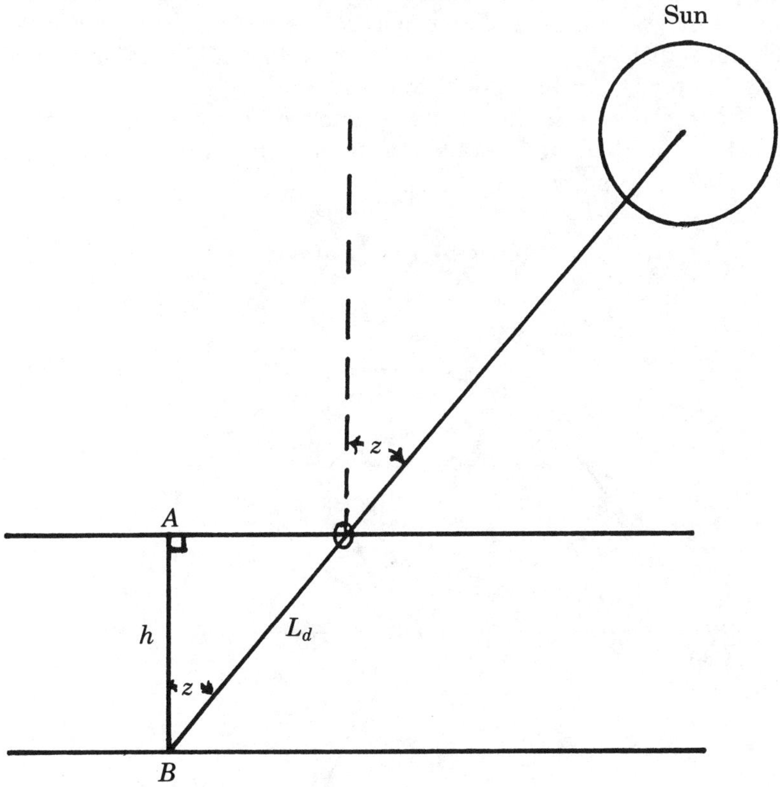

Figure 2.3. Geometry of direct solar radiation.

sunlight), the sky would appear black because no light could come from it. However, particulate scattering diverts some of the direct sunlight from its straight-line path; by repeated scattering, radiation appears to come from all parts of the sky. Scattering is most pronounced for short wavelengths in the blue and UV portions of sunlight. This scattering of the blue and UV is responsible for the blue color of the sky. Therefore, the path length of the sunlight scattered must be considered. The path length of sky radiation will vary from a minimum for scattered light entering the layer vertically to a maximum for the scattered light entering the layer horizontally. Consequently, the average path length will depend on the distribution of sky radiation from $z = 0°$ to $z = 90°$.

In the derivation of the path length of sky radiation, Leighton (*2*), to a first approximation, assumed that the sky was a luminous hemisphere of uniform brightness. Consider the sky as a hemisphere (*abc*) as depicted in Figure 2.4. Viewed from the center of the hemisphere, O, the amount of radiation from any portion of the hemisphere is proportional to the area of that portion. However, for equal areas at different zenith angles, the solar irradiance on a horizontal unit surface at the center O is proportional to cos *z*.

If I_s represents the total sky irradiance on a horizontal unit surface at O and the fraction of sky area is θ, then the incremental sky irradiance, dI_s, is given by

$$dI_s = kI_s(\cos z)\, d\theta_z \qquad (2.23)$$

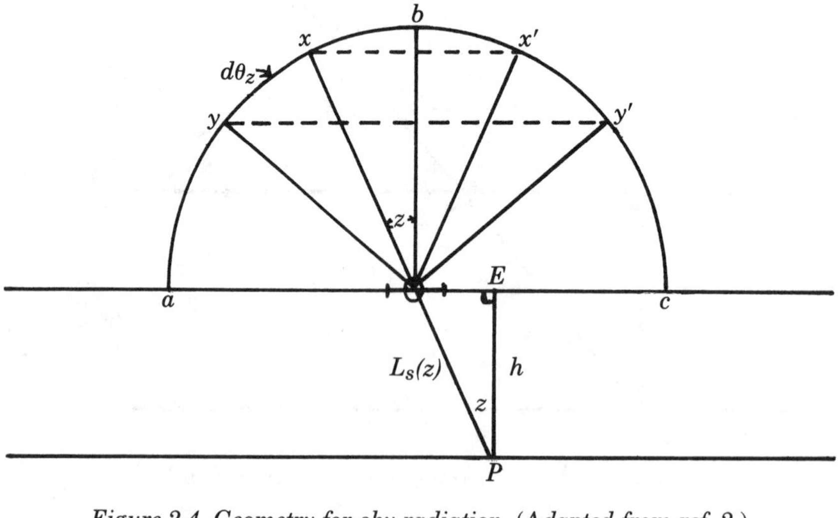

Figure 2.4. Geometry for sky radiation. (Adapted from ref. 2.)

where k is a constant and $d\theta_z$ is the incremental area on the surface of the sky.

From Figure 2.4, the area of the hemisphere (abc) is $2\pi r^2$ (where r is the radius) so that the area of the spherical segment $axx'c$ is $2\pi r^2 \cos z$. Therefore, the surface area of the spherical segment xbx', is

$$2\pi r^2 - 2\pi r^2 \cos z = 2\pi r^2(1 - \cos z) \tag{2.24a}$$

Hence,

$$\theta = \frac{\text{segment area}}{\text{hemispherical area}} = \frac{2\pi r^2(1 - \cos z)}{2\pi r^2} \tag{2.24b}$$

$$\theta = (1 - \cos z) \tag{2.24c}$$

By using equation 2.24c for annular increments of sky area, one obtains

$$d\theta_z = d(1 - \cos z) = (\sin z)\, dz \tag{2.25}$$

Substituting equation 2.25 in equation 2.23 yields

$$dI_s(z) = kI_s(\cos z \sin z)\, dz \tag{2.26}$$

Integrating equation 2.26 from $z = 0$ to $z = z$, one obtains the irradiance from the spherical segment xbx' received by a horizontal unit surface at point O in the center of the hemisphere (Figure 2.4). Thus,

$$I_s(z) = \int_0^z kI_s(\cos z \sin z)\, dz = kI_s \int_0^z \sin z\, [(\cos z)\, dz] \tag{2.27a}$$

$$I_s(z) = (kI_s/2)\sin^2 z \tag{2.27b}$$

when $z = 90°$, $\sin z = 1$, and $\sin^2 z = 1$ and for these boundary conditions, $I_s(z)$ represents the sky irradiance for the entire sky (i.e., $I_s(z) = I_s$). Using these results in equation 2.27b yields $I_s = (kI_s/2)$ or $k = 2$ and substituting this result in equation 2.27b yields

$$I_s(z) = I_s \sin^2 z \tag{2.28}$$

With equation 2.28, the irradiance from an annular segment ($xx' - yy'$) bounded by the angles z_1 and z_2 is

$$I_s(z_1 z_2) = I_s(\sin^2 z_2 - \sin^2 z_1) \tag{2.29}$$

From triangle *OEP* in Figure 2.4, it is seen that $\sec z = (L_s(z)/h)$, so that

$$L_s(z) = h \sec z \tag{2.30}$$

From equation 2.28, $I_s(z)$ is proportional to $\sin^2 z$. Therefore, the fraction of total radiation with path length $L_s(z)$ is $d(\sin^2 z)$. From equation 2.29, the fraction of radiation from the annular segment z_1z_2 is $(\sin^2 z_2 - \sin^2 z_1)$; so that over the annular segment z_1z_2, the average path length $\overline{L}_s(z_1z_2)$ is

$$\overline{L}_s(z_1z_2) = \frac{\int_{z1}^{z2} L_s(z)[d(\sin^2 z)]}{(\sin^2 z_2 - \sin^2 z_1)} \tag{2.31}$$

Substituting equation 2.30 in 2.31 yields

$$\overline{L}_s(z_1z_2) = \frac{\int_{z1}^{z2} (h \sec z)[d(\sin^2 z)]}{(\sin^2 z_2 - \sin^2 z_1)} \tag{2.32}$$

Consider the integral in the numerator of equation 2.32:

$$\int_{z1}^{z2} (h \sec z)[d \sin^2 z)] = \int_{z1}^{z2} h(\cos z)^{-1}[2(\sin z \cos z)\, dz] \tag{2.33a}$$

$$= 2h \int_{z1}^{z2} (\sin z)\, dz \tag{2.33b}$$

$$= -2h \int_{z1}^{z2} -(\sin z)\, dz \tag{2.33c}$$

$$= 2h(\cos z_1 - \cos z_2) \tag{2.33d}$$

Substituting equation 2.33d in 2.32 yields

$$\overline{L}_s(z_1z_2) = \frac{2h(\cos z_1 - \cos z_2)}{\sin^2 z_2 - \sin^2 z_1} \tag{2.34}$$

For the entire hemisphere, $z_1 = 0°$ and $z_2 = 90°$ and $\overline{L}(z_1z_2) = \overline{L}_s$, the average path length of the entire hemisphere. Using these boundary conditions in equation 2.34 yields

$$\overline{L}_s = \frac{2h(\cos 0° - \cos 90°)}{\sin^2 90° - \sin^2 0°} = \frac{2h(1 - 0)}{1 - 0} \tag{2.35a}$$

$$\overline{L}_s = 2h \tag{2.35b}$$

The validity of the derivation is dependent on the assumption that the sky is a hemisphere of uniform brightness. As Leighton indicated, this assumption is only an approximation. Leighton (*2*) evaluated some experimental data in the literature and found that $\bar{L}_s = 1.9$ h. Thus, Leighton indicated that

$$\bar{L}_s = ih \tag{2.36}$$

and $i = 2$ is a useful approximation.

2.F. Specific Light Absorption Rates and Solar Irradiance

Consider a well-mixed system of a gaseous chemical at a molar concentration $[C]$ in a layer of air of height h. From the first law of photochemistry, only light that is absorbed can cause chemical change. The average rate of disappearance of a chemical is given by equation 1.27 where $I_{a\lambda}$ represents the average rate of absorption of light. The term $I_{a\lambda}$ is given by equation 1.28 and is

$$I_{a\lambda} = I_{o\lambda}(A/V)F_{s\lambda}F_{c\lambda} \tag{2.37}$$

where $I_{o\lambda}$ is the solar irradiance of sunlight, A is the area exposed, V is the volume, and $F_{s\lambda}$ and $F_{c\lambda}$ are defined by equations 1.23 and 1.29. For this system, $F_{s\lambda}$ and $F_{c\lambda}$ are given by

$$F_{s\lambda} = \{1 - 10^{-(\alpha_\lambda + \epsilon_\lambda[C])l}\} \tag{2.38}$$

$$F_{c\lambda} = \epsilon_\lambda[C]/(\alpha_\lambda + \epsilon_\lambda[C]) \tag{2.39}$$

where α_λ is the attenuation coefficient of air, ϵ_λ is the molar absorptivity of the chemical, and l is the path length of sunlight in the layer h. For this layer, $(A/V) = h^{-1}$ so that equation 2.37 becomes

$$I_{a\lambda} = I_{o\lambda}h^{-1}F_{s\lambda}F_{c\lambda} \tag{2.40}$$

In general, the concentration of a chemical in the polluted layer h is low so that the absorbance is also low. This condition introduces a simplification in the derivation. Consider $F_{s\lambda}$ from equation 2.38 in the form

$$F_{s\lambda} = 1 - \exp\{-2.303(\alpha_\lambda + \epsilon_\lambda[C])l\} \tag{2.41}$$

$$F_{s\lambda} = 1 - \exp(-x) \tag{2.42}$$

where

$$x = 2.303(\alpha_\lambda + \epsilon_\lambda[C])l \qquad (2.43)$$

Expansion of exp(–x) in a series (*6*) yields

$$\exp(-x) = 1 - x + x^2/2! - x^3/3! + \ldots \qquad (2.44)$$

For weak absorbance, x [which is equal to $(\alpha_\lambda + \epsilon_\lambda[C])l$] is less than 0.02 and thus the terms $x^2/2!$, $x^3/3!$, ... are small relative to x and can be neglected. Hence, equation 2.44 becomes

$$x = 1 - \exp(-x) \qquad (2.45)$$

Substituting equation 2.45 in equation 2.42 and using equation 2.43 yield

$$F_{s\lambda} = x = 2.303(\alpha_\lambda + \epsilon_\lambda[C])l \qquad (2.46)$$

Finally, substituting equation 2.39 and 2.46 in equation 2.40 and canceling common terms yield

$$I_{a\lambda} = 2.303(h^{-1}\epsilon_\lambda[C])[I_{o\lambda}l] \qquad (2.47)$$

Because sunlight is composed of direct and sky radiation, $[I_{o\lambda}]$ in equation 2.47 must be the sum of direct and sky radiation. Thus, equation 2.47 must take the form

$$I_{a\lambda} = 2.303(h^{-1}\epsilon_\lambda[C])[I_{d\lambda}L_d + I_{s\lambda}\overline{L}_s] \qquad (2.48)$$

where $I_{d\lambda}$ is the intensity of direct solar radiation, L_d is the path length of the direct solar radiation, $I_{s\lambda}$ is the intensity of the sky radiation, and $\overline{L}_s$ is the average path length of the sky radiation. Introducing the conversion factor j in equation 2.48 to make the units of intensity compatible with the units of ϵ_λ and $[C]$, as described in Section 1.C.4., yields

$$I_{a\lambda} = 2.303(hj)^{-1}[C]\epsilon_\lambda[I_{d\lambda}L_d + I_{s\lambda}\overline{L}_s] \qquad (2.49)$$

If $I_{d\lambda}$ and $I_{s\lambda}$ are expressed in photons cm^{-2} s^{-1}, $[C]$ in molar concentration, and ϵ_λ is in M^{-1} cm^{-1}, then $j = 6.022 \times 10^{20}$ and when $I_{d\lambda}$ and $I_{s\lambda}$ are expressed in millieinsteins cm^{-2} s^{-1}, and $[C]$ and ϵ_λ are as previously defined, then $j = 1$.

Summing equation 2.49 over all wavelengths at which the chemical absorbs light yields

$$I_a = \sum_{\lambda} I_{a\lambda} \tag{2.50}$$

$$I_a = 2.303[C](hj)^{-1} \sum_{\lambda} \epsilon_{\lambda}[I_{d\lambda}L_d + I_{s\lambda}\overline{L}_s] \tag{2.51}$$

Substituting equations 2.22 and 2.36 in equation 2.51 yields

$$I_a = 2.303[C]j^{-1} \sum_{\lambda} \epsilon_{\lambda}[I_{d\lambda} \sec z + iI_{s\lambda}] \tag{2.52}$$

Equation 2.52 can be written in the form

$$I_a = k_a[C] \tag{2.53}$$

where

$$k_a = 2.303j^{-1} \sum_{\lambda} \epsilon_{\lambda}[I_{d\lambda} \sec z + iI_{s\lambda}] \tag{2.54}$$

and k_a is called the specific light absorption rate constant.

Equations 2.49 and 2.54 can be reduced to more useful forms by using equations 2.20 and 2.21 as follows:

$$I_{a\lambda} = 2.303[C]\epsilon_{\lambda}[I_{d\lambda}\sec z + iI_{s\lambda}] \tag{2.55a}$$

$$I_{a\lambda} = 2.303[C]\epsilon_{\lambda}[I_{d\lambda}(\cos z)^{-1} + iI_{s\lambda}] \tag{2.55b}$$

$$I_{a\lambda} = 2.303[C]\epsilon_{\lambda}[T_{s\lambda}T_{a\lambda}I_{o\lambda} + ig(1 - T_{s\lambda})T_{a\lambda}I_{o\lambda}\cos z] \tag{2.55c}$$

$$I_{a\lambda} = 2.303[C]\epsilon_{\lambda}J_{\lambda} \tag{2.55d}$$

where

$$J_{\lambda} = T_{a\lambda}I_{o\lambda}[T_{s\lambda} + ig(1 - T_{s\lambda})\cos z] \tag{2.56}$$

and

$$k_a = 2.303j^{-1} \sum_{\lambda} \epsilon_{\lambda}[I_{d\lambda}(\cos z)^{-1} + iI_{s\lambda}] \tag{2.57a}$$

$$k_a = 2.303j^{-1} \sum_{\lambda} \epsilon_{\lambda}\{T_{a\lambda}I_{o\lambda}[T_{s\lambda} + ig(1 - T_{s\lambda})\cos z]\} \tag{2.57b}$$

$$k_a = 2.303j^{-1} \sum_{\lambda} \epsilon_{\lambda}J_{\lambda} \tag{2.57c}$$

The parameter J_λ is an extremely useful term. According to Leighton (*2*), J_λ represents the irradiance that would be measured by a weakly absorbing chemical actinometer in a *flat* horizontal surface exposed to sun and sky. J_λ represents the actinic irradiance to which the polluted layer, *h*, is exposed. Peterson (*7*) pointed out that J_λ really refers to the irradiance of a *spherical* chemical actinometer because J_λ is the solar flux on a spherical surface.

The rate of disappearance of chemical in the polluted layer, *h*, in the atmosphere is proportional to the rate of absorption of light, k_a. From equation 2.57, k_a is proportional to $\sum_\lambda \epsilon_\lambda J_\lambda$, so that the rate of disappearance of chemical is proportional to the overlap of the absorption of light ϵ_λ and the solar actinic irradiance J_λ summed for all wavelengths for which the chemical absorbs light. Inspection of equation 2.56 indicates that J_λ contains all the parameters that are functions of the solar zenith angle (*z*). Because J_λ exhibits all diurnal variations, the rate of disappearance of a chemical exhibits the same diurnal variations.

2.G. Surface Reflection and Diffusion Within the Polluted Layer

Because the earth's surface can reflect light, the solar actinic irradiance will increase in the polluted layer; consequently, the rate of disappearance of the gaseous chemical in the polluted layer will increase. Surface reflections are characterized by a parameter called the *albedo*. Albedos have been reported for a variety of the earth's surfaces–for example, snow, oceans, forests, fields and meadows, bare ground, and dune sand. Snow will have a relatively high albedo value, and the bare ground will have a relatively low albedo value. Leighton (*2*) did not incorporate surface reflections into his calculations. In addition, Leighton's early calculations (*2*) did not account for the effects of diffusion within the polluted layer *h*.

2.H. Best Available Solar Irradiance Data

Because Leighton's work had become outdated, Peterson (*7*) carried out research to (1) update the radiative transfer model of Leighton; (2) develop a computer program to calculate solar irradiance values (J_λ) from 290 to 700 nm as a function of the solar zenith angle (*z*) from 0° to 86°; and (3) input into the computer program the best available atmospheric data on molecular scattering; ozone absorption; aerosol scattering and absorption; absorption by water vapor, oxygen, and carbon dioxide; and the earth's surface albedo as a function of wavelength.

For all of the calculations, Peterson used the radiative transfer model first developed by Dave (*8*) and improved by Braslau and Dave (*9,10*). This model was used to calculate the optical properties of aerosols. References 9 and 10 present details of this model.

The Dave model was designed to calculate radiative fluxes from a flat, horizontal, surface (F_λ). However, for photochemical applications, the irradiance (J_λ) is spherical because the incident radiation enters the polluted layer from all directions. Peterson showed that $J_\lambda = 2F_\lambda$ so that the spherical flux is composed of the sum of the upward and downward components. Thus, Peterson modified the Dave model to include the upward and downward components of the actinic flux.

The ozone absorption coefficients used by Peterson were similar to those used by Leighton. However, Peterson used the value of $(O_3) =$ 0.295 cm, based on the latest ozone measurements, rather than the value of 0.22 cm used by Leighton (*see* Section 2.C.3).

The atmospheric aerosols were assumed (1) to be spherical and homogeneous; (2) to have a given size distribution and known refractive index, both of which were assumed to be independent of the height and wavelength; (3) to have density (i.e., the number of aerosols per unit volume) that could be varied with the height; and (4) to be partly absorbing.

Peterson used the best available data (*5*) on molecular scattering and included absorption by water vapor, oxygen, and carbon dioxide. Finally, the best available surface albedo factors were used to account for reflection from the earth's surface as a function of wavelength.

All of these data were used in the computer program to calculate the solar irradiance, J_λ, in photons cm^{-2} s^{-1} for clear sky conditions, for 10 solar zenith angles (0°, 10°, 20°, ... , 70°, 78°, and 86°), and for 48 spectral wavelength intervals from 290 to 700 nm to give the best values of J_λ.

Schere and Demerjian (*11*) improved the work of Peterson by calculating J_λ over 10-nm intervals from 290 to 700 nm and expanded the wavelength region from 700 to 800 nm.

The J_λ values obtained by Peterson are instantaneous values in photons cm^{-2} s^{-1}. Because many chemicals require more than 1 day to photoreact, Hendry (*12*) modified the J_λ values as follows: the J_λ values from the computer program of Schere and Demerjian (*11*) were integrated over a 24-h day to obtain values in photons cm^{-2} day^{-1}. Hendry prepared tables of J_λ values as a function of λ in the spectral region 290 to 800 nm in 31 intervals at 10°, 30°, and 50° N. latitude for the winter and summer solstices and the equinox. In addition, these tables included averaged spring–summer and fall–winter J_λ values. These tables of solar irradiance have been incorporated into a test method for determining rate constants and half-lives for the gas-phase photoreaction of chemicals in sunlight in the environment (*13*).

Recently, Davenport (*14*) updated the J_λ values of Hendry for the wavelength region 290 to 800 nm in 10-nm intervals for 20°, 30°, 40°, and 50° N. latitude for the winter and summer solstices, the equinox, and the averaged spring–summer and fall–winter values (*15*). In addition, tables of J_λ have been prepared for 0°, 10°, 60°, and 70° N. latitude (*14*). The tables corresponding to 20°, 30°, 40°, and 50° N. latitude have now been incorporated into the method for determining rate constants and half-lives for the gas-phase photoreaction of chemicals in sunlight in the environment (*16*).

References

1. Leighton, P. A.; Perkins, W. A. "Solar Radiation Absorption Rates and Photochemical Primary Processes in Urban Air"; Report No. 14; Air Pollution Foundation: Los Angeles, CA, 1956.
2. Leighton, P. A. *Photochemistry of Air Pollution*; Academic: New York, 1961; Chapter II.
3. Koller, L. P. *Ultraviolet Radiation*; 2nd ed., Wiley: NY, 1965.
4. *Astronomical Almanac*; Nautical Almanac Office. U.S. Naval Observatory: Washington, DC, 1987.
5. Raleigh, Lord. *Philos. Mag.* **1899**, *47*, 375.
6. Peirce, B. O. *A Short Table of Integrals;* Ginn: New York, 1929; Integral No. 26, p 5; Series Expansion No. 759, p 89.
7. Peterson, J. T. *Calculated Actinic Fluxes (290–700 nm) for Air Pollution Photochemistry Applications;* U.S. Environmental Protection Agency. U.S. Government Printing Office: Washington, DC, 1976; EPA 600/4–76–0025.
8. Dave, J. V. "Development Programs for Computing Characteristics of Ultraviolet Radiation"; National Aeronautics and Space Administration Final Report under Contract NAS5–21680; National Aeronautics and Space Administration: Greenbelt, MD, 1972; NASA CR–139134; NTIS Report No. N75–10746/6SL. 1972.
9. Braslau, N.; Dave, J. V. *J. Appl. Meteorol.* **1973**, *12*, 601.
10. Ibid, p 616.
11. Schere, K. L.; Demerjian, K. L. *Calculation of Selected Photolytic Rate Constants over a Diurnal Range;* U.S. Environmental Protection Agency. U.S. Government Printing Office: Washington, DC, 1977; EPA 600/ 4–77–015.
12. Hendry, D. G.; Kenley, R. A. *Atmospheric Reaction Products of Organic Compounds;* U.S. Environmental Protection Agency. U.S. Government Printing Office: Washington, DC, 1979; EPA–500/12–79–001.
13. U.S. Environmental Protection Agency. *Chemical Fate Test Guidelines and Support Documents*; "Gas Phase Absorption Spectra and Photolysis, CG, CS–7000"; EPA 560/6–83–003; NTIS PB 83–257717; National Technical Information Service: Springfield, VA, 1983.
14. Davenport, J. E., SRI International, Menlo Park, CA, personal communication, 1985.
15. Mill, T., SRI International, Menlo Park, CA, personal communication, 1984.
16. *Fed. Regist.* **1985**, *50*, 39311, 40 CFR 796.3800.

Chapter 3

Absorption and Transmission of Solar Radiation in a Natural Water Body

The rates of all photochemical processes in a natural water body are affected by solar irradiance at the water surface, radiative transfer from air to water, and the transmission of sunlight in the water body. Therefore, as a prelude to the discussion of rates of photochemical processes in a water body, it is necessary to describe in detail solar irradiance at the water surface, radiative transfer of sunlight from the atmosphere into a water body, the path length of sunlight in water, and the absorption and transmission of solar radiation in a water body. This chapter will be devoted to a description of these processes, the development of the theory of absorption and transmission of solar irradiance in water, and the calculation of solar irradiance in a natural water body as developed by Zepp and Cline (*1*). At the conclusion of this chapter, solar irradiance computations will be revised by using the best available solar data in the UV spectral region 280–380 nm (*2*).

The theory of absorption and transmission of solar radiation in a natural water body will be used in Chapter 4 to derive equations that will translate readily available laboratory data into sunlight-direct photoreaction rate constants and half-lives as a function of time of day, season, latitude, depth in water bodies, and the ozone layer thickness.

3.A. Solar Irradiance at the Water Surface

Direct and sky irradiance data for the UV spectral region 297.5–380 nm were obtained from the research of Bener (*3*). The sky irradiance $H(\lambda, h_o, X)$ in W cm^{-2} nm^{-1}, is a function of the wavelength λ, solar

1464–6/88/0031$06.00/1

altitude h_o, and the thickness of a layer of atmospheric ozone X (in cm at standard temperature and pressure [STP]) and is given by

$$\log_{10}H(\lambda, h_o, X) = \log_{10}H(\lambda, h_o, X_o) - T(\lambda, h_o)[X - X_o] \quad (3.1)$$

where $H(\lambda, h_o, X_o)$ represents the sky irradiance at λ, h_o, and X_o (0.3375 cm of ozone) and $T(\lambda, h_o)$ is a coefficient.

The direct solar irradiance $S(\lambda, h_o, X, Z)$, in W cm^{-2} nm^{-1}, can be described by

$$S(\lambda, h_o, X, Z) = S_o(\lambda)\{10^{-[0.4343(t_R + t_a)m + kyX]}\} \quad (3.2)$$

where $S_o(\lambda)$ is the extraterrestrial solar irradiance in W cm^{-2} nm^{-1}, $t_R(\lambda, z)$ is the optical thickness for Raleigh scattering, $t_a(\lambda, z)$ is the optical thickness for aerosol scattering, $m(h_o)$ is the air mass according to Bemporad that takes into account the curvature of the earth's surface, $k(\lambda)$ is the absorption coefficient for ozone at –44 °C, $y(h_o)$ is the relative slant path through the ozone layer normalized to 1.0 for $h_o = 90°$, X is the thickness of a layer of atmospheric ozone in cm at STP, Z is the elevation above sea level in km, h_o is the solar altitude, and λ is the wavelength in nm. Equations 3.1 and 3.2 were converted from W cm^{-2} nm^{-1} to photons cm^{-2} s^{-1} nm^{-1} as a function of λ (in nm) according to

$$I(\text{photons cm}^{-2}\text{s}^{-1}\text{nm}^{-1}) = 5.035 \times 10^{15} I(\text{W cm}^{-2}\text{nm}^{-1})\lambda \quad (3.3)$$

Direct and sky irradiances for visible radiation in the spectral region 390–800 nm were obtained from the research of Leighton (*4*) as described in Chapter II.

Two solar spectral regions need to be defined because they are responsible for the direct photoreaction of most chemicals in sunlight. The short UV wavelength region 280–320 nm, called UV-B radiation, is responsible for the direct photoreaction of many chemicals in sunlight including most commonly used pesticides. This spectral region is also responsible for causing sunburn on humans and other biological effects. The long wavelength UV region 320–400 nm, called UV-A radiation, is also important for the direct photoreaction of many chemicals in sunlight. Solar irradiance data in these two spectral regions will be described in detail in Section 3.D. of this chapter and in Chapter 4.

3.B. Absorption and Transmission of Sunlight from the Atmosphere into a Water Body

3.B.1. Radiative Transfer from the Atmosphere into a Water Body

Consider Figure 3.1, which depicts the passage of sunlight through the atmosphere into a water body. The angle of incidence of the sunlight is z, which is the solar zenith angle, θ is the angle of refraction in water, h is the height of a layer of atmosphere, D is the depth in the water body, and n is the refractive index of water. According to Snell's law, the index of refraction is given by

$$n = \sin z/\sin \theta \qquad (3.4)$$

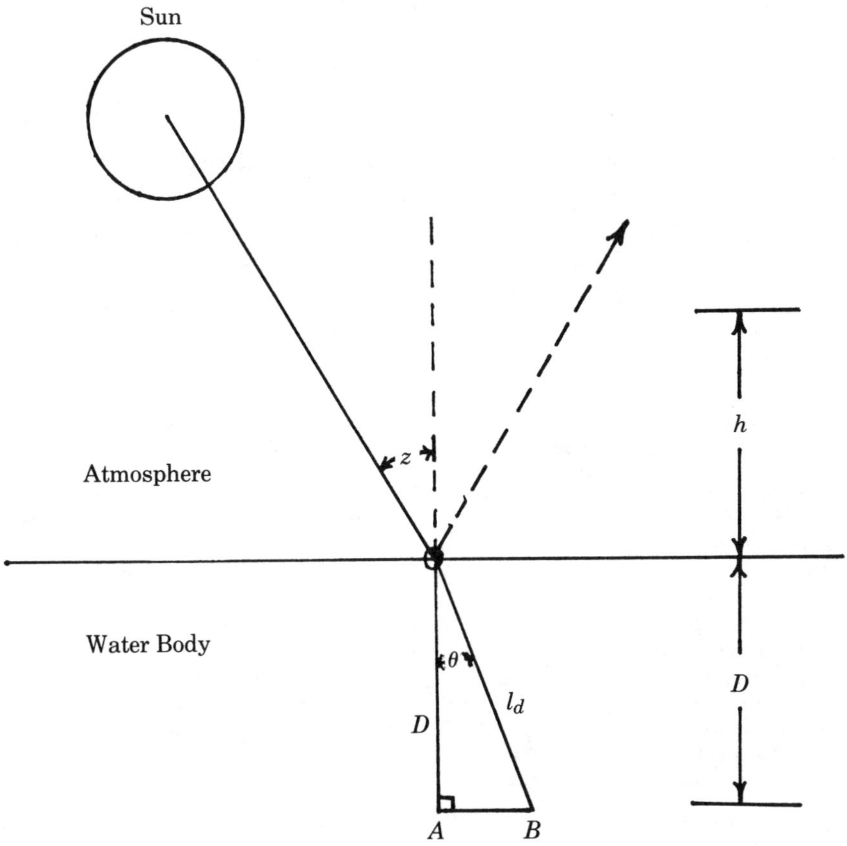

Figure 3.1. Passage of sunlight through the atmosphere and into a water body.

The fraction of direct solar radiation reflected (RFD) can be computed from Fresnel's law. RFD is less than 0.1 at all angles of z except when z approaches 90°. As an approximation, Zepp and Cline (*1*) assumed that the sky is uniformly bright and found that the fraction of sky radiation reflected (RFS) is about 0.07. The values of RFD and RFS were computed by assuming that n for water is 1.34, a value corresponding to $\lambda = 436$ nm (blue light). The refractive index of water, n, varies slightly with λ and increases slightly with decreasing λ. Thus, the computed reflectivities are slightly higher for UV light and slightly lower for longer wavelength visible light. Nevertheless, the use of $n = 1.34$ is a good approximation as the computed values of RFD and RFS used in all calculations agreed closely with the published experimental data for reflection from natural water bodies. The sunlight irradiance equations for direct and sky radiation in the atmosphere and in water are summarized in Table 3-1.

3.B.2. Path Length of Sunlight in Water

3.B.2.a. Path Length of Direct Sunlight in Water

As direct sunlight enters a water body from the atmosphere, it is bent downward and assumes a path length l_d (Figure 3.1). From triangle *OAB*,

$$l_d = D \sec \theta \tag{3.5}$$

Table 3-1. Irradiance and Path Length of Solar Radiation in the Atmosphere and in a Water Body

Radiation	*Atmosphere*	*Water Body*
Direct irradiance	S	$I_{d\lambda} = S[1 - \text{RFD}]$
Sky irradiance	H	$I_{s\lambda} = H[1 - \text{RFS}]$
Path length		$l_d = D \sec \theta$
of direct radiation	$L_d = h \sec z$	$\underline{l_d} = Dn[(n^2 - \sin^2 z)]^{-1/2}$
Average path length	$\bar{L}_s = ih$	$\bar{l}_s = 1.20D$
of sky radiation	$i \approx 2$	

NOTES: Spectral region, 297.5–380 nm: S (from equation 3.2), H (from equation 3.1) Spectral region, 390–800 nm: $S = T_{s\lambda}T_{a\lambda}I_{o\lambda} \cos z$ (from equation 2.20), $H = g(1 - T_{s\lambda})T_{a\lambda}I_{o\lambda}\cos z$ (from equation 2.21)

Path lengths for the atmosphere: $L_d = h \sec z$ (from equation 2.22), $\bar{L}_s = ih$; $i \approx 2$ (from equation 2.36)

Path lengths for a water body: $l_d = D \sec \theta$ (from equation 3.5), $\underline{l_d} = Dn[(n^2 - \sin^2 z)]^{-1/2}$ (from equation 3.8b), $\bar{l}_s = 2Dn[n - (n^2 - 1)^{1/2}]$ (from equation 3.20b), $\bar{l}_s = 1.20D$ (for $n = 1.34$ for water, equation 3.21) where θ is the angle of refraction and D is the depth in water. Equation 3.5 can be written in the form

$$l_d = D \cos^{-1} \theta \tag{3.6}$$

However, $\cos \theta = [1 - \sin^2 \theta]^{1/2}$ so that equation 3.6 becomes

$$l_d = D[1 - \sin^2 \theta]^{-1/2} \tag{3.7}$$

From equation 3.4, $\sin \theta = n^{-1} \sin z$. Thus, equation 3.7 becomes

$$l_d = D[1 - (n^{-2} \sin^2 z)]^{-1/2} \tag{3.8a}$$

$$l_d = Dn[(n^2 - \sin^2 z)]^{-1/2} \tag{3.8b}$$

The path length of direct sunlight in water is listed in Table 3-1 (equations 3.5 and 3.8) along with the path length in the atmosphere (equation 2.22). As the solar zenith angle increases, the amount of light refracted increases. When the solar zenith angle z is $\geq 85°$, and the refractive index of water $n = 1.34$, equation 3.4 yields $\theta \geq 48°$; hence, under these conditions, sunlight is strongly bent as it traverses through water.

3.B.2.b. Path Length of Sky Light in Water

The average path length for sky radiation in the atmosphere has been derived in Section 2.E.2, so this discussion will be extended to derive the average path length of sky radiation in a water body. To a good approximation, Zepp and Cline (*1*) assumed that the sky was a luminous hemisphere of uniform brightness. Consider the sky as a hemisphere (*abc*) of uniform brightness (Figure 3.2); the irradiance from an annular segment ($xx' - yy'$) bounded by the angles z_1 and z_2 is given by equation 2.29 and is

$$I_s(z_1 z_2) = I_s(\sin^2 z_2 - \sin^2 z_1) \tag{3.9}$$

From triangle *OAB* in Figure 3.2 the path length of sky radiation in the water body is given by the equation

$$l_s(\theta) = D \cos^{-1} \theta \tag{3.10}$$

However, $\cos \theta = [1 - \sin^2 \theta]^{1/2}$ so that equation 3.10 becomes

$$l_s(\theta) = D[1 - \sin^2 \theta]^{-1/2} \tag{3.11}$$

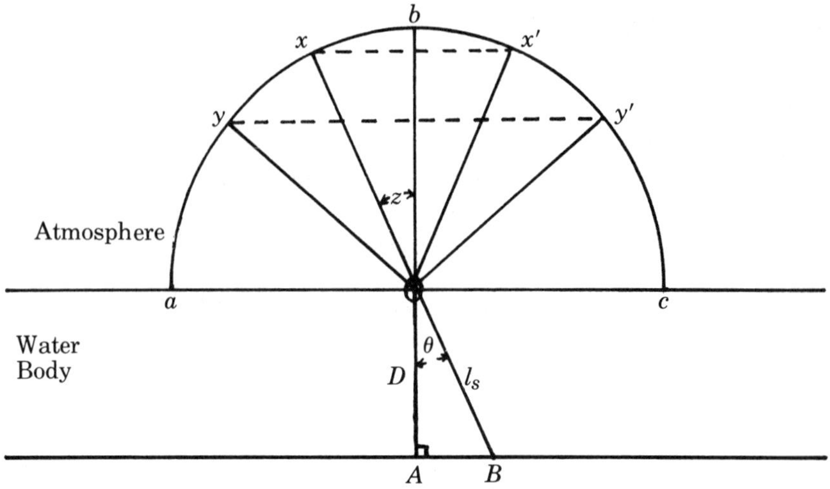

Figure 3.2. Geometry of sky radiation in the atmosphere and over a water body.

From equation 3.4, sin θ $= n^{-1}\sin z$. Thus, equation 3.11 becomes

$$l_s(z) = D[1 - n^{-2}\sin^2 z]^{-1/2} \tag{3.12}$$

$$l_s(z) = Dn[(n^2 - \sin^{-2} z)]^{-1/2} \tag{3.13}$$

From equation 2.28, $I_s(z)$ is proportional to $\sin^2 z$. Therefore, the fraction of total radiation with path length $l_s(z)$ is $d(\sin^2 z)$. From equation 3.9, the fraction of radiation from the annular segment z_1z_2 is $(\sin^2 z_2 - \sin^2 z_1)$, so that over the annular segment z_1z_2, the average path length $\bar{l}_s(z_1z_2)$ is given by the equation

$$\bar{l}_s(z_1z_2) = \frac{\int_{z_1}^{z_2} l_s(z)[d(\sin^2 z)]}{(\sin^2 z_2 - \sin^2 z_1)} \tag{3.14}$$

Substituting equation 3.13 in equation 3.14 yields

$$\bar{l}_s(z_1z_2) = \frac{Dn \int_{z_1}^{z_2} [n^2 - \sin^2 z]^{-1/2}[d(\sin^2 z)]}{(\sin^2 z_2 - \sin^2 z_1)} \tag{3.15}$$

Consider the integral in the numerator of equation 3.15.

$$\int_{z_1}^{z_2} [n^2 - \sin^2 z]^{-1/2}[d(\sin^2 z)]$$

$$= 2\int_{z_1}^{z_2} [n^2 - \sin^2 z]^{-1/2}(\sin z)[d(\sin z)] \quad (3.16)$$

From a table of integrals (*5*), one obtains

$$\int x(a^2 - x^2)^{-1/2}dx = -(a^2 - x^2)^{1/2} \quad (3.17)$$

By making the substitution $a = n$, $x = \sin z$, $dx = d(\sin z)$ and by using equation 3.17, the integral in equation 3.16 becomes

$$2\int_{z_1}^{z_2} [n^2 - \sin^2 z]^{-1/2}(\sin z)[d(\sin) z] = 2\{-[n^2 - \sin^2]^{1/2}\}_{z_1}^{z_2}$$

$$= 2\{(n^2 - \sin^2 z_1)^{1/2} - (n^2 - \sin^2 z_2)^{1/2}\} \quad (3.18)$$

Substituting equation 3.18 in equation 3.15 yields

$$\bar{l}_s(z_1z_2) = \frac{2Dn\{(n^2 - \sin^2 z_1)^{1/2} - (n^2 - \sin^2 z_2)^{1/2}\}}{(\sin^2 z_2 - \sin^2 z_1)} \quad (3.19)$$

For the entire hemisphere $z_1 = 0°$, $z_2 = 90°$, and $\bar{l}_s(z_1z_2) = \bar{l}_s$, the average path length for sky radiation under water for the entire hemisphere. Substituting these boundary conditions in equation 3.19 yields

$$\bar{l}_s = \frac{2Dn[n - (n^2 - 1)^{1/2}]}{(1 - 0)} \quad (3.20a)$$

$$\bar{l}_s = 2Dn[n - (n^2 - 1)^{1/2}] \quad (3.20b)$$

For the refractive index of water equal to 1.34, $\bar{l}_s$ becomes

$$\bar{l}_s = 1.20D \quad (3.21)$$

The average path length of sky radiation in water is listed in Table 3-1 (equations 3.20b and 3.21) along with the path length of sky radiation in the atmosphere (equation 2.36).

Reflection from the surface of the water body has little effect on the computed value of $\bar{l}_s$. Taking reflection into account, the computed path length was found to be 1.19D (*1*). Reflection has little effect on the computed value because the most strongly reflected part of the sky radiation near the horizon contributes only a small fraction of the total sky intensity on the surface.

3.B.3. Light Attenuation in Natural Water Bodies

In Chapter 2, it was shown that sunlight in the atmosphere is attenuated by scattering from water droplets and particulates; from gaseous molecules; and from absorption, principally by ozone. However, for natural water bodies, the attenuation of sunlight is quite different. For the ocean, attenuation is primarily due to absorption by water. For inland natural waters, attenuation is primarily due to dissolved natural organic matter such as humic acids. Thus, $(\alpha_\lambda)_{inland} > (\alpha_\lambda)_{ocean}$. The attenuation coefficient is wavelength-dependent and increases as λ decreases from the visible to the UV. Average attenuation coefficients for 10 river water samples collected in the Southeastern United States are presented in Figure 3.3.

Attenuation due to light scattering is, in general, less important than attenuation by absorption for most natural inland waters, espe-

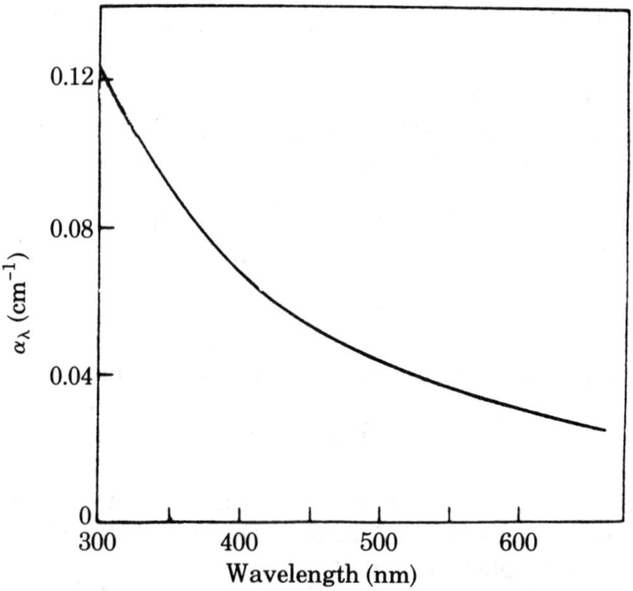

Figure 3.3. Average attenuation coefficients for 10 river water samples collected in southeastern United States. (Reproduced from ref. 1. Copyright 1977 American Chemical Society.)

cially in the UV spectral region. As a first approximation, Zepp and Cline (*1*) ignored scattering when making calculations for direct and sky radiation in natural water bodies.

3.C. Absorption and Transmission of Solar Radiation in a Natural Water Body

3.C.1. Very Weak Light Absorption by a Chemical

Consider a well-mixed natural water body with attenuation coefficient α_λ containing a chemical with molar absorptivity ϵ_λ and molar concentration $[C]$. According to equation 1.22, the rate of light absorption ($I^*_{a\lambda}$) by this system at wavelength λ is given by

$$I^*_{a\lambda} = I_{o\lambda}\{1 - 10^{-(\alpha_\lambda + \epsilon_\lambda[C])l}\} \tag{3.22}$$

where $I_{o\lambda}$ is the incident light intensity (i.e., irradiance) and l is the path length in the water body. The intensities $I^*_{a\lambda}$ and $I_{o\lambda}$ are in photons cm^{-2} s^{-1}. In general, chemicals are present in natural water bodies in the environment at very low concentrations. Hence, $\alpha_\lambda > \epsilon_\lambda[C]$ so that equation 3.22 reduces to

$$I^*_{a\lambda} = I_{o\lambda}(1 - 10^{-\alpha_\lambda l}) \tag{3.23}$$

Equation 3.23 is a form of the Lambert law describing the rate of light absorption by the water.

Because sunlight is composed of direct radiation ($I_{d\lambda}$) and sky radiation ($I_{s\lambda}$), equation 3.23 takes the form

$$I'_{a\lambda} = I^*_{a\lambda}D^{-1} = [I_{d\lambda}(1 - 10^{-\alpha_\lambda l_d}) + I_{s\lambda}(1 - 10^{-\alpha_\lambda \bar{l}_s})]D^{-1} \tag{3.24}$$

where $I'_{a\lambda}$ is the rate of absorption of sunlight by the system, which is essentially the rate of light absorption by the water, in photons cm^{-3} s^{-1} in a layer of depth D in cm; l_d is the path length of direct radiation in water in cm; and $\bar{l}_s$ is the average path length of sky radiation in water in cm.

The term $I'_{a\lambda}$ must be modified by the fraction of light absorbed by the chemical. According to equation 1.29, the fraction of light absorbed by the chemical is $\epsilon_\lambda[C]/(\alpha_\lambda + \epsilon_\lambda[C])$. Under normal environmental conditions $\alpha_\lambda > \epsilon_\lambda[C]$; therefore, the fraction of the light absorbed by the chemical is $\epsilon_\lambda[C]/\alpha_\lambda$. When this result is combined with $I'_{a\lambda}$, the rate of light absorption by the chemical in this natural water body ($I_{a\lambda}$) is given by

$$I_{a\lambda} = I'_{a\lambda}\epsilon_\lambda[C]\alpha_\lambda^{-1} \tag{3.25}$$

Introducing the conversion factor j in equation 3.25 to make the units of $I'_{a\lambda}$ compatible with ϵ_λ and $[C]$, as described in Section 1.C.4., yields

$$I_{a\lambda} = I'_{a\lambda}\epsilon_\lambda[C](\alpha_\lambda j)^{-1} \tag{3.26}$$

When $[C]$ is the molar concentration, ϵ_λ is in $M^{-1}\ cm^{-1}$, and $I'_{a\lambda}$ is in photons $cm^{-3}s^{-1}$, $j = 6.022 \times 10^{20}$.

From equation 3.26, the average rate of light absorption by the chemical can then be expressed as

$$I_{a\lambda} = k_{a\lambda}[C] \tag{3.27}$$

where

$$k_{a\lambda} = I'_{a\lambda}\epsilon_\lambda(\alpha_\lambda j)^{-1} \tag{3.28}$$

and $k_{a\lambda}$ is called the specific light absorption rate constant at wavelength λ.

The equation for $k_{a\lambda}$ can be simplified further under either of two conditions: (1) very weak light absorption by the chemical but very strong light absorption by the water; or (2) very weak light absorption by the chemical and the water.

Condition 1. If the absorbance $A_\lambda = \alpha_\lambda l_d = \alpha_\lambda \bar{l}_s > 2$, then essentially all the light is absorbed by the water. Under these conditions $10^{-\alpha_\lambda l_d}$ and $10^{-\alpha_\lambda \bar{l}_s}$ are essentially 0. Using this result in equation 3.24 yields

$$I'_{a\lambda} = [I_{d\lambda} + I_{s\lambda}]D^{-1} = W_\lambda D^{-1} \tag{3.29}$$

where

$$W_\lambda = I_{d\lambda} + I_{s\lambda} \tag{3.30}$$

Substituting equation 3.29 in equation 3.28 yields

$$k_{a\lambda} = W_\lambda\epsilon_\lambda(D\alpha_\lambda j)^{-1} \tag{3.31}$$

The specific light absorption rate constant $k_{a\lambda}$ is inversely proportional to the depth D so that at great depths $k_{a\lambda}$ approaches zero (and direct photoreaction also approaches zero as described in Chapter 4).

Condition 2. If the absorbance $A_\lambda = \alpha_\lambda l_d = \alpha_\lambda \bar{l}_s < 0.02$, then the light is weakly absorbed by both the chemical and the water. Equation 3.24 simplifies as follows: consider the term $(1 - 10^{-\alpha_\lambda l_d})$; conversion of this term to $[1 - \exp(-2.303\alpha_\lambda l_d)]$; expansion of the exponential term in a power series, neglecting all higher power terms (quadratic, cubic, etc.), because $\alpha_\lambda l_d$ is < 0.02; and simplification yields

$$(1 - 10^{-\alpha_\lambda l_d}) = 2.303\alpha_\lambda l_d \tag{3.32}$$

The steps in this derivation are the same as those in Chapter 2, equations 2.41 through 2.46. In the same manner

$$(1 - 10^{-\alpha_\lambda \bar{l}_s}) = 2.303\alpha_\lambda \bar{l}_s \tag{3.33}$$

Substituting equations 3.32 and 3.33 in equation 3.24 yields

$$I'_{a\lambda} = 2.303\alpha_\lambda[I_{d\lambda} l_d + I_{s\lambda}\bar{l}_s]D^{-1} \tag{3.34}$$

and substituting equation 3.34 in equation 3.28 yields

$$k_{a\lambda} = 2.303\epsilon_\lambda(Dj)^{-1}[I_{d\lambda} l_d + I_{s\lambda}\bar{l}_s] \tag{3.35}$$

Finally, substituting equations 3.5 and 3.21 in equation 3.35 yields

$$k_{a\lambda} = 2.303\epsilon_\lambda j^{-1}[I_{d\lambda}\sec\theta + 1.20I_{s\lambda}] \tag{3.36}$$

$$k_{a\lambda} = 2.303\epsilon_\lambda j^{-1}Z_\lambda \tag{3.37}$$

where

$$Z_\lambda = [I_{d\lambda}\sec\theta + 1.20I_{s\lambda}] \tag{3.38}$$

Equation 3.38 is approximately applicable to shallow depths in any natural water body and to depths up to 0.5 m in pure (e.g., distilled) water.

3.C.2. Very Strong Light Absorption by a Chemical

Another scenario should be considered when discussing environmental photochemistry. For example, when an accidental chemical spill occurs, a large amount of chemical could enter a small water body. Thus, the molar concentration of a chemical could become large so that the chemical absorbs most of the sunlight, that is, $\sum_{\lambda} \epsilon_\lambda[C] \gg \sum_{\lambda} \alpha_\lambda$. Under these conditions, essentially all the sunlight is absorbed by the chemical and the kinetics of direct photoreaction change dramatically, as described in Chapter 4. Some sunlight photoreaction rate constants reported in the literature have been measured under experimental conditions in which the chemical absorbs essentially all the light.

For this scenario, consider a well-mixed natural water body with attenuation coefficient α_λ containing a chemical at a high molar concentration $[C]$ with molar absorptivity ϵ_λ. Under the condition $\epsilon_\lambda[C] \gg \alpha_\lambda$, equation 3.22 becomes

$$I^*_{a\lambda} = I_{o\lambda}[1 - 10^{-\epsilon_\lambda[C]l}] \qquad (3.39)$$

Because essentially all the light is absorbed by the chemical, $\epsilon_\lambda[C]l > 2$, and $10^{-\epsilon_\lambda[C]l}$ is <0.01 and is essentially 0. Equation 3.39 reduces to

$$I^*_{a\lambda} = I_{o\lambda} \qquad (3.40)$$

Furthermore, because sunlight is composed of direct radiation ($I_{d\lambda}$) and sky radiation ($I_{s\lambda}$) equation 3.40 takes the form

$$I'_{a\lambda} = I^*_{a\lambda}D^{-1} = I_{o\lambda}D^{-1} = [I_{d\lambda} + I_{s\lambda}]D^{-1} \qquad (3.41)$$

where $I'_{a\lambda}$ is the rate of absorption of sunlight by the system in photons $cm^{-3}\ s^{-1}$ and D is the depth in the water body in cm. Using equation 3.30 in 3.41 yields

$$I'_{a\lambda} = W_\lambda D^{-1} \qquad (3.42)$$

The term I'_λ must be modified by the fraction of light absorbed by the chemical. According to equation 1.29 for the condition $\epsilon_\lambda[C] \gg \alpha_\lambda$, the fraction of the light absorbed by the chemical is $\epsilon_\lambda[C]/\epsilon_\lambda[C]$, which is equal to 1. By combining this result with equation 3.42, the rate of absorption of light by the chemical ($I_{a\lambda}$) is given by

$$I_{a\lambda} = I'_{a\lambda}(1) = W_\lambda D^{-1} \qquad (3.43)$$

Introducing the conversion factor j in equation 3.43 to make the units of $I'_{a\lambda}$ compatible with ϵ_λ and $[C]$ yields

$$I_{a\lambda} = W_\lambda(jD)^{-1} \qquad (3.44)$$

From equation 3.44, the average rate of light absorption by the chemical can be expressed as

$$I_{a\lambda} = k_{a\lambda} \quad (3.45)$$

where

$$k_{a\lambda} = W_{\lambda}(jD)^{-1} \quad (3.46)$$

$k_{a\lambda}$ is the specific light absorption rate constant at wavelength λ.

3.D. Computer Calculation of Solar Irradiance

Zepp and Cline (*1*) wrote a computer program in Fortran IV (called SOLAR) that uses all the equations just derived to compute solar irradiance values W_{λ} and Z_{λ} [in photons cm^{-2} s^{-1} $\Delta\lambda$ (nm)] for 39 wavelength intervals $\Delta\lambda$ (nm) from 297.5 to 800 nm as a function of latitude and midseason dates for winter, spring, summer, and fall. The values of solar irradiance correspond to wavelength intervals $\Delta\lambda$ (nm) centered at a specific wavelength λ (Appendix A, Table A-1). For example, the solar irradiance value for λ = 400 nm corresponds to the $\Delta\lambda$ interval of 10 nm from 395 to 405 nm. Values of W_{λ} and Z_{λ} have been tabulated for midday, midseason (spring, summer, fall, and winter) at 40° N. latitude (*1*). The difference between W_{λ} and Z_{λ} must be emphasized. The value of W_{λ} defined by equation 3.30, is valid only when essentially all the sunlight is absorbed. The value of Z_{λ} defined by equation 3.38 is valid only when the sunlight is very weakly absorbed. Molar absorptivities, ϵ_{λ}, used in all calculations must be averaged over the wavelength intervals corresponding to those listed in Table A-1, Appendix A. Data for rates of photoreaction and Z_{λ} will be discussed in more detail in Chapter 4 and in subsequent chapters.

Some interesting observations can be made concerning the solar irradiance data obtained from the SOLAR computer program. Many chemicals, including pesticides, absorb light strongly in the UV spectral region 297.5–320 nm (the short-wavelength UV region in the UV-B region) and thus have the potential to undergo rapid direct photoreaction. Solar irradiance is strongly affected by the solar zenith angle (z) and the atmospheric ozone content (O_3). Ozone absorption causes a sharp decrease in the solar irradiance at wavelengths < 320 nm (Figure 3.4). At wavelengths 300 nm, the solar irradiance is very low and below 290 nm it is essentially zero. This conclusion is in agreement with the research of Leighton (*4*), Chapter II, and other research workers (*7,8*). Both Z_{λ} and (O_3) are functions of season and latitude.

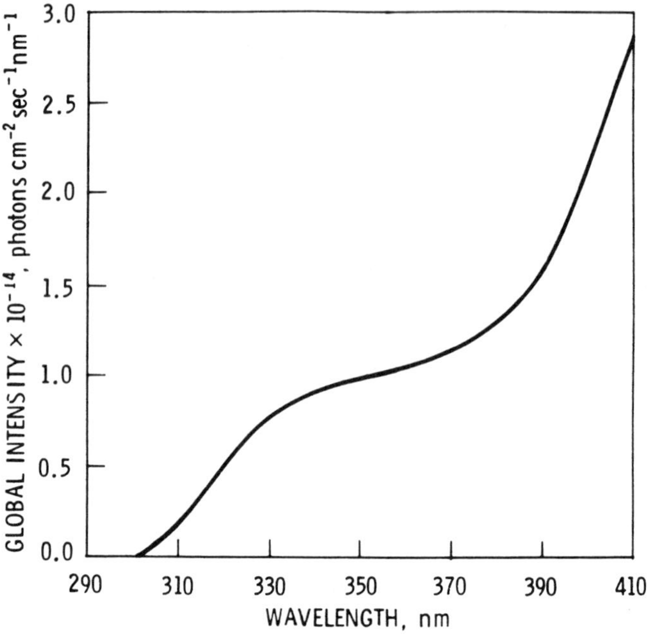

Figure 3.4. Midday solar irradiance for summer, latitude 40° N. (Reproduced from ref. 6.)

Seasonal variations of Z_λ for several latitudes are depicted in Figure 3.5 for the short-wavelength region 297.5–330 nm (UV-B) and in Figure 3.6 for the long-wavelength region 330–390 nm (essentially all the UV-A. There are large seasonal variations of Z_λ in the short-wavelength region because Z_λ is a function of the solar zenith angle z and (O_3), the ozone content. (O_3) is a function of season and latitude. By contrast, the seasonal variation in Z_λ in the long-wavelength region is relatively small because Z_λ is a function of the solar zenith angle z and is not a function of (O_3). Inspection of these two figures indicates that (1) the Z_λ values are a maximum in the summer months and decrease to a minimum in the winter months; (2) at 0° N. latitude (i.e., at the equator), the Z_λ values are at a maximum and seasonal variations are at a minimum; and (3) as the latitude increases, the Z_λ values decrease and the seasonal variations in Z_λ become larger, especially in the spectral region <320 nm.

Because direct photoreaction rates are directly proportional to Z_λ (equation 3.37), these rates are at a maximum in the summer months and decrease to a minimum in the winter months. Furthermore, direct photoreaction rates are at a maximum at 0° N. latitude and decrease to a minimum in the winter months. For chemicals that absorb light

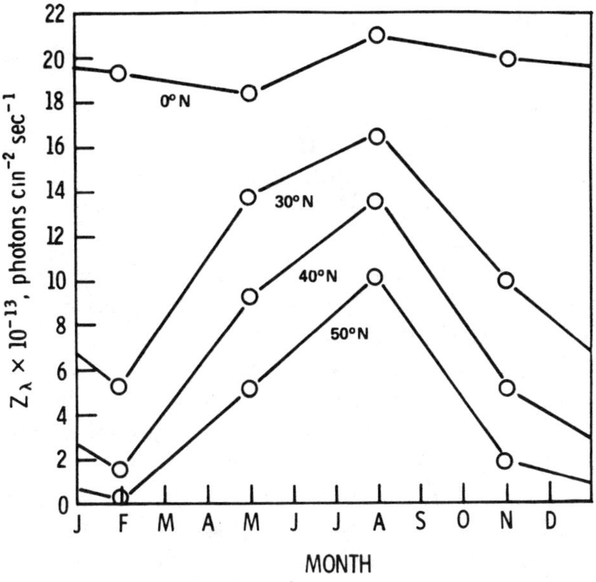

Figure 3.5. Dependence of Z_λ in the short-wavelength UV spectral region (297.5–330 nm) on season and latitude. (Reproduced from ref. 6.)

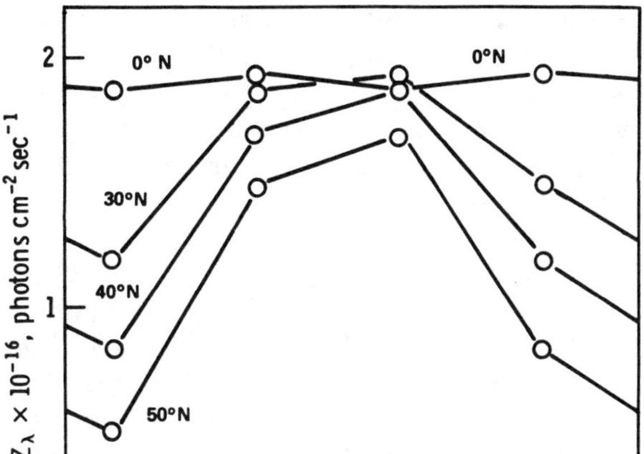

Figure 3-6. Dependence of Z_λ in the long-wavelength UV spectral region (330–390 nm) on season and latitude. (Reproduced from ref. 6.)

strongly at wavelengths <320 nm, direct photoreaction rates vary considerably with season. These variations in direct photoreaction rates will be discussed in detail in Chapter 4.

All sunlight intensity data apply to clear sky conditions. Cloud cover will reduce the solar irradiance. Zepp (*9*) describes procedures for estimating the variation in solar irradiance with cloud cover. On the average, for cloudy days, solar irradiance is reduced by 30% and $Z'_\lambda = Z_\lambda$ (*10*).

The Z_λ values obtained by Zepp and Cline (*1*) are instantaneous values in photons cm^{-2} s^{-1}. Because many chemicals require more than 1 day to photoreact, Mill et al. (*11*) calculated L_λ values in millieinsteins cm^{-2} day^{-1} (i.e., averaged over a 24-h day) by using the Zepp and Cline SOLAR computer program. These L_λ values correspond to solar irradiance in natural water bodies, shallow depths, and for chemicals with absorbance less than 0.02 in pure water. These L_λ values also correspond to values over a specified wavelength interval $\Delta\lambda$ centered at λ (Table A-1, Appendix A) for winter, spring, summer, and fall midseason dates as a function of latitude. A more detailed discussion of L_λ data is in Section 7.B.1.i.

3.E. Best Available Solar Irradiance Data

Zepp (*2*) improved the SOLAR computer program for calculating solar irradiance in water in the middle UV spectral region. The improvement involved the replacement of the solar irradiance data of Bener (*3*) with the more recent data of Green et al. (*12*) in the middle UV region 280–380 nm. The new computer program is called GCSOLAR. The GCSOLAR program now computes solar irradiance in 39 wavelength intervals from 297.5 to 800 nm and can calculate solar irradiance down to 280 nm if desired.

Green et al. (*12*) developed analytical expressions describing sky radiation in the spectral region 280–380 nm. They used the more precise radiative transfer model of Braslau and Dave (*13*) and Dave and Halpern (*14*) and indicated that their sky radiation data is now accurate to the 5% level. These papers (*12–14*) give details of the analytical expressions for describing sky radiation in the region 280–380 nm.

Other changes were made in the new GCSOLAR computer program that are applicable to rates of photoreaction in natural water bodies; these changes will be discussed in Section 4.D.

By using the new GCSOLAR program, Mill et al. (*15*) calculated more reliable L_λ values for the midseason dates corresponding to winter, spring, summer, and fall seasons for 39 wavelength intervals (Table

A-1, Appendix A) as a function of latitude from 0° to 70° N. These data are summarized in Tables A-4 to A-11 of Appendix A. More detailed discussions of these L_λ values will be given in Section 7.B.1.i.

References

1. Zepp, R. G.; Cline, D. M. *Environ. Sci. Technol.* **1977**, *11*, 359.
2. Zepp, R. G., U.S. Environmental Protection Agency, Office of Research and Development, Athens, GA, personal communication, 1982.
3. Bener, P. "Approximate Values of Intensity of Natural Ultraviolet Radiation for Different Amounts of Atmospheric Ozone"; U.S. Army Report DASA 37–68–C–1017; Davos Platz, Switzerland, 1972; National Technical Information Service, U.S. Department of Commerce: Springfield, VA, 1972.
4. Leighton, P. A. *Photochemistry of Air Pollution*; Academic: New York, 1961; Chapter II.
5. Peirce, B. O. *A Short Table of Integrals;* Ginn: New York, 1929; Integral No. 132, p 20.
6. Wolfe, N. L.; Zepp, R. G.; Baughman, G. L.; Fincher, R. C.; Gordon, J. A. *Chemical and Photochemical Transformation of Selected Pesticides in Aquatic Systems;* U.S. Environmental Protection Agency. U.S. Government Printing Office: Washington, DC, 1976; EPA 600/3–76–067.
7. Peterson, J. T. *Calculated Actinic Fluxes (290–700 nm) for Air Pollution Photochemistry Applications;* U.S. Environmental Protection Agency. U.S. Government Printing Office: Washington, DC, 1976; EPA 600/4–76–0025.
8. Demerjian, K. L.; Schere, K. L.; Peterson, J. T. *Adv. Sci. Technol.* **1980,** *10*, 389.
9. Zepp, R. G. In *The Handbook of Environmental Photochemistry*; Hutzinger, O., Ed.; Springer–Verlag: New York, 1981; Vol. 2, Part B, p 19.
10. Zepp, R. G., U.S. Environmental Protection Agency, Office of Research and Development, Athens, GA, personal communication, 1985.
11. Mill, T.; Mabey, W. R.; Bomberger, D. C.; Chou, T–W.; Hendry, D. G.; Smith, J. H. *Laboratory Protocols for Evaluation of the Fate of Organic Chemicals in Air and Water;* U.S. Environmental Protection Agency. U.S. Government Printing Office: Washington, DC, 1982; EPA 600/3–82–002.
12. Green, A. E. S.; Cross, K. R.; Smith, L. A. *Photochem. Photobiol.* **1980,** *31*, 59.
13. Braslau, N.; Dave, J. V. "Effect of Aerosols on the Transfer of Solar Energy Through Realistic Model Atmospheres." "Part III. Ground Level Fluxes in the Biologically Active Bands, 0.2850–0.23700)m". IBM Research Report RC 4308; Thomas Watson Laboratories: Yorktown Heights, NY, 1973.
14. Dave, J. V.; Halpern, P. *Atmos. Environ.* **1976,** *10*, 547.
15. Mill, T., SRI International, Menlo Park, CA, personal communication, 1985.

Chapter 4

Direct Photoreaction Rates and Half-Lives in Natural Water Bodies in the Environment

The theory and practice for determining direct sunlight photoreaction rates of a chemical in a natural water body in the environment are described in this chapter, particularly the research of Zepp and Cline (*1*). Photochemical studies are presented to illustrate how laboratory experimental data and solar data are used in a computer program (called SOLAR) to calculate direct photoreaction rate constants and half-lives in natural water bodies as a function of time of day, season, depth in water bodies, and ozone layer thickness. Several photochemical studies that verify the method of Zepp and Cline are presented. The chapter concludes with a discussion of the latest version of the computer program (called GCSOLAR) and the calculation of rates of direct photoreaction using this computer program.

4.A. Theoretical Discussion

4.A.1. Very Weak Light Absorption by a Chemical

Consider a well-mixed pure natural water body containing a chemical at a molar concentration $[C]$ with molar absorptivity ϵ_λ, and with α_λ as the attenuation coefficient of the water. The rate of direct photoreaction of the chemical is a function of the rate of absorption of sunlight (the Grotthus–Draper law) and the efficiency with which the absorbed light transforms the chemical into products (the Stark–Einstein law). Therefore, the average rate of disappearance of

1464–6/88/0049$07.75/1

the chemical undergoing direct photoreaction, at a fixed wavelength λ, is given by equation 1.27 and is

$$\text{average rate} = -(d[C]/dt)_{d\lambda} = \phi_{d\lambda} I^{d}_{a\lambda} \quad (4.1)$$

where $I^{d}_{a\lambda}$ is the average rate of light absorption by the chemical and $\phi_{d\lambda}$ is the reaction quantum yield of the chemical. The subscript and superscript d is used to denote that the chemical is undergoing direct photoreaction.

In general, chemicals are present in the environment at low concentrations. Thus, for a chemical at a low molar concentration $\alpha_{\lambda} > \epsilon_{\lambda}[C]$. Under these environmental conditions, the average rate of light absorption, $I^{d}_{a\lambda}$ is given by equation 3.27 and is

$$I^{d}_{a\lambda} = k_{a\lambda}[C] \quad (4.2)$$

where

$$k_{a\lambda} = I'_{a\lambda}\epsilon_{\lambda}(\alpha_{\lambda} j)^{-1} \quad (4.3)$$

and $k_{a\lambda}$ is the specific light absorption rate constant at wavelength λ and $I'_{a\lambda}$ is the rate of absorption of sunlight by the water. Under the conditions of weak sunlight absorbance by the system (i.e., the absorbance of the chemical and water is <0.02), $k_{a\lambda}$ reduces to equations 3.36, 3.37, and 3.38 which are

$$k_{a\lambda} = 2.303\epsilon_{\lambda} j^{-1}[I_{d\lambda}\sec\theta + 1.20 I_{s\lambda}] \quad (4.4)$$

$$k_{a\lambda} = 2.303\epsilon_{\lambda} j^{-1} Z_{\lambda} \quad (4.5)$$

where

$$Z_{\lambda} = [I_{d\lambda}\sec\theta + 1.20 I_{s\lambda}] \quad (4.6)$$

Substituting equation 4.2 in equation 4.1 yields

$$-(d[C]/dt)_{d\lambda} = \phi_{d\lambda} k_{a\lambda}[C] \quad (4.7)$$

Taking the sum of equation 4.7 over all wavelengths at which the chemical absorbs sunlight yields

$$-\sum_{\lambda} (d[C]/dt)_{d\lambda} = \sum_{\lambda} \phi_{d\lambda} k_{a\lambda}[C] \quad (4.8)$$

For most organic chemicals in aqueous solution, the reaction quantum yield is independent of λ and $\phi_{d\lambda} = \phi_{d}$. This assumption is valid for complex molecules in solution because the photoreaction from second or higher electronic states usually cannot compete with the rapid radiationless decay of excited molecules to the first excited state (2). Thus, the complete rate equation can be written in the form

$$-(d[C]/dt)_d = \phi_d k_a [C] \tag{4.9}$$

where

$$k_a = \sum_{\lambda} k_{a\lambda} \tag{4.10}$$

and k_a represents the sum of the $k_{a\lambda}$ values for all wavelengths λ that are absorbed by the chemical.

Equation 4.9 is a first-order differential equation in the form

$$-(d[C]/dt)_d = k_{dE}[C] \tag{4.11}$$

where

$$k_{dE} = \phi_d k_a \tag{4.12}$$

and k_{dE} is the first-order direct photoreaction rate constant in a natural water body in the environment. The subscript E is used to denote that direct photoreaction is taking place in a natural water body in the environment.

Integrating equation 4.11 under the boundary conditions ($t = 0$, $[C_o]$) and (t, $[C_t]$) yields

$$\ln([C_o]/[C_t]) = k_{dE}t \tag{4.13}$$

where k_{dE} is in the units of reciprocal time.

The half-life, $t_{(1/2)E}$, is defined as the time required for the chemical to reach one-half its initial concentration. With the boundary condition ($t = t_{(1/2)E}$, $[C_t] = [C_o]/2$), equation 4.13 yields

$$t_{(1/2)E} = \ln 2/k_{dE} = 0.693/k_{dE} \tag{4.14}$$

For this system, the sunlight is absorbed weakly (i.e., $A_\lambda < 0.02$). Hence, using equation 4.5 in equation 4.10 yields

$$k_a = 2.303 j^{-1} \sum_{\lambda} \epsilon_\lambda Z_\lambda \tag{4.15}$$

and substituting equation 4.15 in equation 4.12 yields

$$k_{dE} = 2.303 j^{-1} \phi_d \sum_{\lambda} \epsilon_\lambda Z_\lambda \tag{4.16}$$

Therefore, the first-order direct photoreaction rate constant in a natural water body in the environment is proportional to the reaction quantum yield of the chemical ϕ_d and the overlap between the molar absorptivity ϵ_λ and the solar irradiance Z_λ, summed over all wavelengths at which light is absorbed by the chemical.

4.A.2. Very Strong Light Absorption by a Chemical

Consider another environmental scenario of a well-mixed pure natural water body with attenuation coefficient α_λ, containing a chemical with molar absorptivity ϵ_λ, where the molar concentration $[C]$ is large. Under these conditions $\epsilon_\lambda[C] \gg \alpha_\lambda$ and essentially all the light is absorbed by the chemical. This condition could occur as a result of an accidental spill in which a large quantity of a chemical enters a small pure water body. For this scenario, the average rate of light absorption, $I^d_{a\lambda}$, is given by equations 3.45, 3.46, and 3.30.

$$I^d_{a\lambda} = k_{a\lambda} \tag{4.17}$$

where

$$k_{a\lambda} = W_\lambda (jD)^{-1} \tag{4.18}$$

$$W_\lambda = I_{d\lambda} + I_{s\lambda} \tag{4.19}$$

Substituting equation 4.17 in equation 4.1 yields

$$-(d[C]/dt)_{d\lambda} = \phi_{d\lambda} k_{a\lambda} \tag{4.20}$$

Taking the sum of equation 4.20 over all wavelengths at which the chemical absorbs sunlight yields

$$-\sum_\lambda (d[C]/dt)_{d\lambda} = \sum_\lambda \phi_{d\lambda} k_{a\lambda} \tag{4.21}$$

For most organic chemicals in aqueous solution, the reaction quantum yield is independent of λ and $\phi_{d\lambda} = \phi_d$ (2). Thus, the complete rate equation can be written in the form

$$-(d[C]/dt)_d = \phi_d k_a \tag{4.22}$$

where

$$k_a = \sum_\lambda k_{a\lambda} \tag{4.23}$$

and k_a represents the sum of the $k_{a\lambda}$ values for all wavelengths that are absorbed by the chemical. Substituting equation 4.18 in equation 4.23 yields

$$k_a = (jD)^{-1} \sum_{\lambda} W_\lambda \tag{4.24}$$

Equation 4.22 is a zero-order differential equation in the form

$$-(d[C]/dt)_d = k_{dE} \tag{4.25}$$

where

$$k_{dE} = \phi_d k_a \tag{4.26}$$

and k_{dE} is the zero-order direct photoreaction rate constant in a natural water body. Substituting equation 4.24 in equation 4.26 yields

$$k_{dE} = \phi_d (jD)^{-1} \sum_{\lambda} W_\lambda \tag{4.27}$$

Integrating equation 4.25 under the boundary conditions ($t = 0$, $[C_o]$) and (t, $[C_t]$) yields

$$[C_t] = -k_{dE}t + [C_o] \tag{4.28}$$

where k_{dE} is in the units of molar time^{-1}.

With the boundary condition ($t = t_{(1/2)E}$, $[C_t] = [C_o]/2$), equation 4.28 yields

$$t_{(1/2)E} = [C_o]/2k_{dE} \tag{4.29}$$

Substituting equation 4.27 in equation 4.29 yields

$$t_{(1/2)E} = jD[C_o]/(2\phi_d \sum_{\lambda} W_\lambda) \tag{4.30}$$

Therefore, under conditions in which essentially all the sunlight is absorbed by the chemical, the direct photoreaction half-life in a natural water body in the environment is directly proportional to the initial concentration of the chemical $[C_o]$. Consequently, half-lives measured under high molar concentration conditions can be much longer than those observed at very low concentration conditions normally found in a natural water body in the environment.

4.B. Computation of Direct Photoreaction Rate Constants and Half-Lives

On the basis of the equations derived in Chapters 3 and 4, Zepp and Cline (*1*) wrote a computer program (called SOLAR) for 39 specific wavelength intervals from 297.5 to 800 nm (Appendix A, Table A-1) that calculates instantaneous rate constants at specific times of day for a given latitude and date, or midseason date, and takes their sum to obtain k_{dE}. The term "midseason date" does not refer to a day midway in the season, but rather to a day during the season when the angle of declination of the sun is at specified values (listed in a subsequent paragraph). The program also calculates a photoreaction rate constant, k_{dE}, that is averaged over a 24-h calendar day in the units of days^{-1}. Required data inputs are

1. molar absorptivities of the chemical (ϵ_λ) at specified wavelengths equal to or greater than 297.5 nm where the chemical absorbs light
2. the refractive index of the medium (e.g., for water n = 1.34);
3. the attenuation coefficient (α_λ) of the medium (e.g., α_λ for distilled water or for natural water bodies containing dissolved humic acids);
4. the reaction quantum yield of the chemical (ϕ_d);
5. the solar declination, solar right ascension, and siderial time for the date of interest obtained from the *Astronomical Almanac* (*4*), which is published annually;
6. the latitude and longitude;
7. the depth in the water body; and
8. the average ozone layer thickness for the season of the year.

The computer program can calculate rate constants and half-lives at shallow depths or an average rate constant and half-life over a depth D in a natural water body.

For routine computations, the computer program calculates rate constants, k_{dE}, and half-lives, $t_{(1/2)E}$, for midseason dates corresponding to solar declinations of +10° for spring, +20° for summer, −10° for fall, and −20° for winter as a function of local time of day, longitude, and latitude from 0° to 90° in 10° increments.

The program can also compute representative effects of the variation in the ozone layer thickness (O_3) upon the photoreaction rates for three latitudes 0°, 40°, and 73° N.

4.C. Photochemical Studies: Experimental Verification of the Theory of Zepp and Cline and the Calculation of Direct Photoreaction Rate Constants and Half-Lives

A few photochemical studies will be presented in this section to describe (1) the experimental verification of the theory of Zepp and Cline (*1*) and (2) the calculation of direct photoreaction rate constants, k_{dE}, and half-lives, $t_{(1/2)E}$, in natural water bodies with respect to the electronic absorption spectroscopic properties and reaction quantum yield of a chemical as a function of some environmental variables (e.g., time of day, season, latitude, attenuation coefficient of a water body, depth in the water body, and ozone layer thickness). All calculations are for clear sky conditions. Refs. 2 and 5–12 describe additional photochemical studies and give details.

4.C.1. Direct Photoreaction Studies of Carbaryl and Trifluralin

Direct photoreaction studies and calculations were carried out by Zepp and Cline (*1*) for carbaryl (labeled A), a carbamate insecticide, and trifluralin (labeled B), a dinitroaniline type herbicide. The compounds were chosen because they absorb light in different spectral regions and, therefore, could be expected to exhibit quite different photolytic behaviors. On the basis of the electronic absorption spectra of carbaryl and trifluralin in a mixture of water and acetonitrile, it was found that carbaryl absorbed light strongly in the spectral region 300–340 nm (and thus absorbed sunlight strongly in the UV-B spectral region) while trifluralin absorbed light strongly at approximately 300 nm and well into the visible region beyond 500 nm. Molar absorptivities, ϵ_λ, were calculated for these two compounds from the spectra. By using the molar absorptivities in equation 4.5, the computer program calculated $k_{a\lambda}$ for shallow depths at midday, midsummer, at 40° N. latitude; these results are depicted in Figure 4.1. For carbaryl, the rate of absorption of sunlight is high in the UV-B region while the rate of absorption of sunlight by trifluralin is higher in the visible region. By using equation 4.10, the computer program then calculated k_a (which represents the sum of the $k_{a\lambda}$ values over all wavelengths of sunlight absorbed by the chemical). The results were $k_a(\text{A}) = 6.21 \times 10^{-4}\ \text{s}^{-1}$ and $k_a(\text{B}) = 0.258\ \text{s}^{-1}$. Thus, $k_a(\text{B}) = 416\ k_a(\text{A})$, and the sunlight absorption rate constant of trifluralin is over 400 times larger than the sunlight absorption rate constant of carbaryl.

The reaction quantum yield for carbaryl and trifluralin was determined in air-saturated distilled water by the methods outlined in

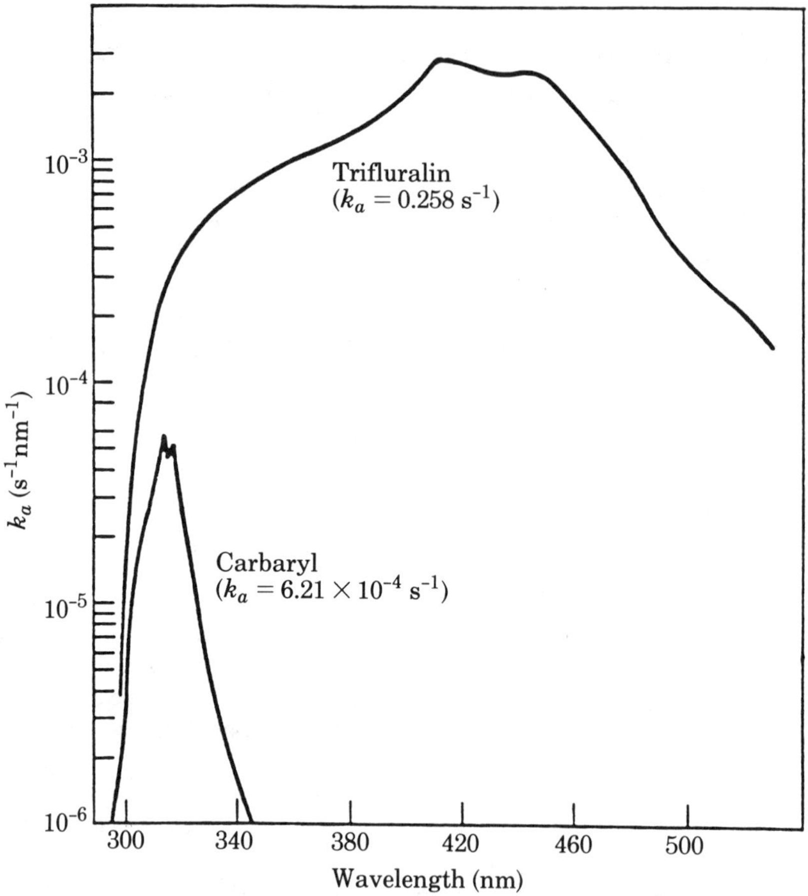

Figure 4.1. Specific sunlight absorption rates of carbaryl and trifluralin as a function of wavelength at midday and midsummer, latitude 40° N. (Reproduced from ref. 1. Copyright 1977 American Chemical Society.)

Chapters 9–11. The results were, for carbaryl, ϕ_d = 0.006 at 313 nm and, for trifluralin, ϕ_d = 0.002 at 366 nm. Using these quantum yields and the rates of sunlight absorption in equation 4.12 yields k_{dE}(A) = 3.7×10^{-6} s^{-1} and k_{dE}(B) = 5.2×10^{-4} s^{-1}. Thus, k_{dE}(B) = 141 k_{dE}(A), and the direct photoreaction rate constant of trifluralin is over 2 orders of magnitude larger than that of carbaryl at midday, midsummer, and 40° N. at shallow depths in a natural water body.

Midday direct photoreaction half-lives were computed for shallow depths as a function of time of year and latitude in the northern hemisphere. These results are depicted in Figure 4.2 in which the relative midday half-lives for each pesticide were computed on the basis of the

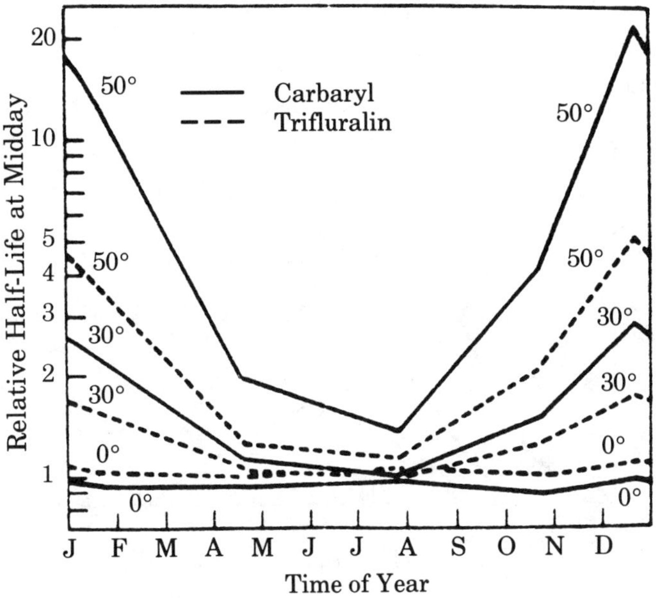

Figure 4.2. Relative midday half-lives for direct photoreaction of carbaryl and trifluralin at midseason for several northern latitudes. (Reproduced from ref. 1. Copyright 1977 American Chemical Society.)

half-life of 1 for midsummer at 30° N. latitude. For the tropical zone, the half-lives are approximately the same throughout the year. For the midlatitudes, the half-lives are at a minimum during the summer months and at a maximum during the winter months. As the latitude increases, both the half-lives and the amplitudes of their seasonal variations increase. The data in Figure 4.2 indicate that at the midlatitudes 30° and 50° N., carbaryl is more sensitive to seasonal changes. This sensitivity is a direct result of the large decrease in UV-B radiation with increasing latitude and change in season from summer to winter and the fact that carbaryl absorbs sunlight strongly only in the UV-B region.

The intensity of sunlight increases from sunrise to noon and then decreases throughout the remainder of the day. Direct photoreaction rate constants parallel this change in sunlight intensity (Figure 4.3). The computed values in Figure 4.3 are relative to the photoreaction rates at midday, midsummer, at 40° N. latitude and apply to shallow depths. The calculations were made for trifluralin and carbaryl on July 24, 1975 and January 21, 1975 at 40° N. As expected, the relative rates are much larger in the summer than in the winter. Furthermore, because UV-B "sunrise" and "sunset" lag and precede visible sunrise and

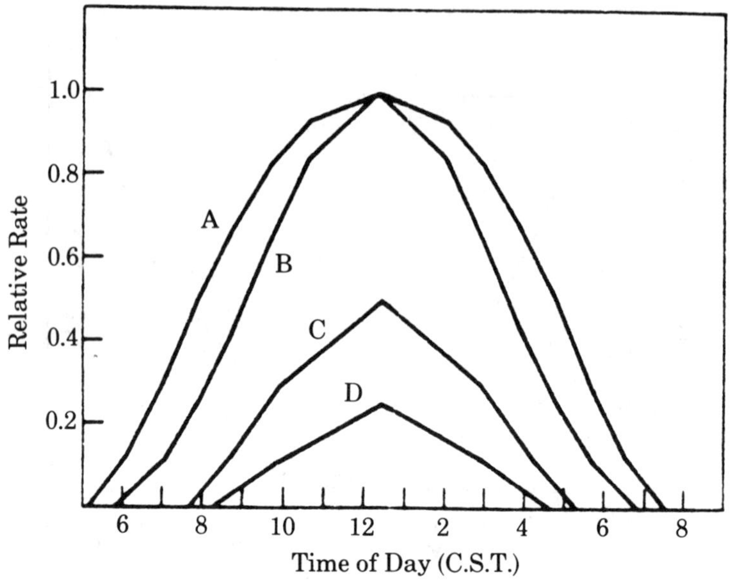

Figure 4.3. Diurnal variation of direct photoreaction rates of trifluralin and carbaryl at latitude 40° N., longitude 90° W. A. trifluralin, July 24, 1975; B. carbaryl, July 24, 1975; C. trifluralin, January 21, 1975; D. carbaryl January 21, 1975. (Reproduced from ref. 1. Copyright 1977 American Chemical Society.)

sunset, respectively, the period of photoreaction for carbaryl is shorter than the period for photoreaction of trifluralin.

As sunlight passes through a natural water body, its intensity decreases and its spectral distribution changes considerably. The decrease in photoreaction rate with depth is a function of the attenuation coefficient of the water body (α_λ), the molar absorptivity of the chemical (ϵ_λ), and the solar irradiance. The computed depth dependence for carbaryl and trifluralin at midday and midsummer at 40° N. latitude is shown in Figure 4.4. These data were calculated by using α_λ for pure water and α_λ for average river water (Figure 3.3). For a lake or ocean, the depth dependence is similar to that shown for pure water. Figure 4.4 shows that direct photoreaction rates drop off more rapidly in river water than in ocean or lake water.

As indicated, carbaryl absorbs strongly only in the UV-B region, but trifluralin absorbs strongly in the visible region. Because UV-B radiation is absorbed more strongly than visible radiation in both pure water and river water (Figure 3.3), the photoreaction rate of carbaryl decreases more rapidly with depth than that of trifluralin. In addition, Figure 4.4 shows that as depth increases, the rate of direct photoreaction of both compounds levels off and both rates become approximately

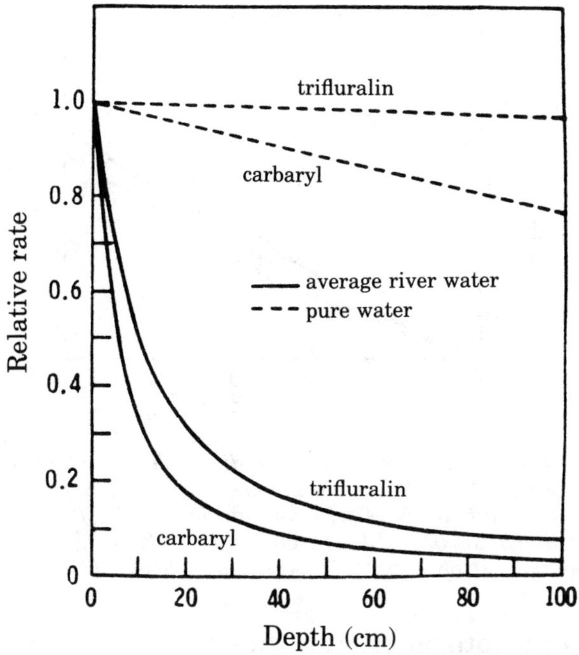

Figure 4.4. Computed depth dependence of direct photoreaction of carbaryl and trifluralin at midday and midsummer, latitude 40° N. (Reproduced from ref. 1. Copyright 1977 American Chemical Society.)

the same. At these large depths, the photic zone is exceeded and essentially all the sunlight is absorbed by the water (i.e., the absorbance is >2). This effect can be predicted from equation 3.31 because $k_{a\lambda}$, and thus k_a, is inversely proportional to the depth.

As the solar zenith angle increases and the sun is lower in the sky, the underwater path length of sunlight increases; therefore, the depth dependence becomes more pronounced with increasing north latitude and from summer to winter in the midlatitudes.

All calculations were based on average UV-B sunlight intensities for each season and latitude. Natural variations in the ozone layer thickness (O_3) can occur over a period of a few years and (O_3) varies longitudinally within a given latitude (*13*). Variations in the ozone layer thickness can cause significant changes in the amount of UV-B radiation that reaches earth. The effect of ozone reduction on direct photoreaction rates of carbaryl at 0°, 40°, and 73° N. latitude is shown in Figure 4.5. The direct photoreaction rate of trifluralin is essentially unaffected by ozone thickness variations because trifluralin absorbs most strongly in the visible region where ozone has no effect on solar radiation.

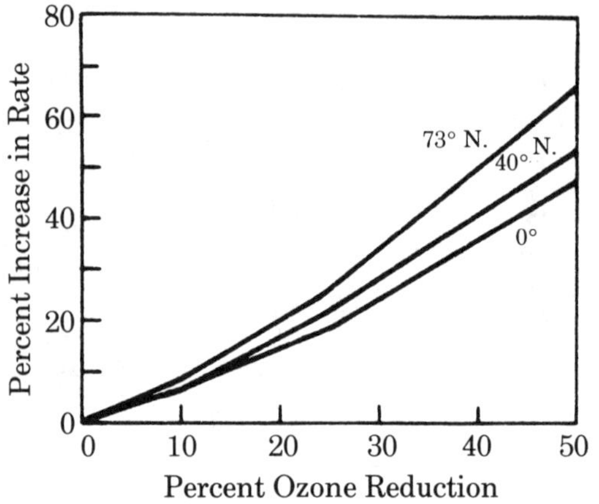

Figure 4.5. Effect of ozone reduction upon direct photoreaction of carbaryl at selected northern latitudes. Ozone thickness assumed for zero reduction: lat. 0°, 0.248 cm; lat. 40° N., 0.319 cm; and lat. 73° N., 0.366 cm. Solar altitudes assumed: lat. 0° and 40° N., 60°; and lat. 73° N., 40°. (Reproduced from ref. 1. Copyright 1977 American Chemical Society.)

4.C.2. Direct Photoreaction Studies of Hexacyanoferrate(II) and Hexacyanoferrate(III) Complexes

Detailed studies were carried out to determine direct photoreaction rates of hexacyanoferrate(II), $[Fe(CN)_6]^{4-}$, and hexacyanoferrate(III), $[Fe(CN)_6]^{3-}$, complexes in aquatic media in sunlight (*14*). Before this research is discussed, it is necessary to summarize some background information on the photoreaction of these complexes obtained from Broderius and Smith (*14*) and Balzani and Carassiti (*15*).

The photochemical behavior of potassium hexacyanoferrate(II) and hexacyanoferrate(III) in aqueous solution has been studied extensively in the laboratory. Solutions of both complexes are stable in the dark or in the presence of diffuse light but are transformed when exposed to light of 254 nm and >300 nm. The irradiation of aqueous $[Fe(CN)_6]^{4-}$ with light >300 nm involves ligand substitution yielding aquopentacyanoferrate(II), $[Fe(CN)_5(H_2O)]^{3-}$, and cyanide ion. The overall photochemical reaction is

$$[Fe(CN)_6]^{4-} + 2H_2O \xrightarrow{h\nu} [Fe(CN)_5(H_2O)]^{3-} + HCN + OH^- \quad (4.31)$$

After prolonged exposure, both thermal and photochemical reactions occur. The photochemical reaction involves progressive photosubstitution, which results in an increase in pH and the release of Fe^{2+} and CN^-. The cyanide ion hydrolyzes to form HCN.

The photochemical transformation of the hexacyanoferrate(III) complex is much more complicated, and the details of the reaction are not completely known. The photochemical reaction involves first the formation of the aquopentacyanoferrate(III) complex plus cyanide ion followed by the thermal and photochemical reduction of Fe(III) to Fe(II). The overall reaction proposed is

$$2[Fe(CN)_6]^{3-} + 2H_2O + OH^- \xrightarrow[\Delta]{h\nu} 2[Fe(CN)_5(H_2O)]^{3-} + CNO^- + HCN \quad (4.32)$$

The electronic absorption spectra of freshly prepared dilute aqueous solutions of $[Fe(CN)_6]^{4-}$ and $[Fe(CN)_6]^{3-}$ were measured spectrophotometrically by Broderius and Smith (*14*) and the results are shown in Figure 4.6. Inspection of Figure 4.6 indicates that the hexacyanoferrate(II) complex absorbs light at $\lambda < 420$ nm and the hexacyanoferrate(III) complex absorbs light at $\lambda < 480$ nm. Therefore, photoreaction rates would be expected to reflect the fluctuations in intensity of the shorter wavelengths of sunlight. The molar absorptivities were determined from the spectra at the specified wavelengths (Appendix A, Table A-1) of the SOLAR computer program of Zepp and Cline (*1*).

By using the molar absorptivity data in the computer program, $k_{a\lambda}$ was computed as a function of wavelength for shallow depths for midday, midsummer, at 40° N. latitude. The results are shown in Figure 4.7. Only wavelengths <480 nm are important. The maximum overlap of ϵ_λ with Z_λ (which is $k_{a\lambda}$) occurs at approximately 330 nm for the hexacyanoferrate(II) solution and at approximately 420 nm for the hexacyanoferrate(III) solution (Figure 4.7). The k_a values for each complex (which represents the sum of the $k_{a\lambda}$ values for all wavelengths in which each complex absorbs light) were computed: $k_a(\text{III}) = 7.79 \times 10^{-2}\ s^{-1}$ for $[Fe(CN_6)]^{3-}$ and $k_a(\text{II}) = 4.66 \times 10^{-3}\ s^{-1}$ for $[Fe(CN)_6]^{4-}$. Thus, $k_a(\text{III}) = 16.7\ k_a(\text{II})$, and the sunlight absorption rate constant for hexacyanoferrate(III) is about 17 times larger than that for hexacyanoferrate(II).

The direct photoreaction of these complexes is first-order in the concentration range 25–100 ppb of total cyanide and follows equation 4.13. The half-life can then be computed with equation 4.14. Experimentally, the disappearance of each of these complexes was followed by measuring the formation of HCN as a function of time. Reactions 4.31 and 4.32 show that for every mole of complex that photoreacts, 1 mol of HCN is formed. Thus, the concentration of each complex is given by the relationship $[C_t] = \{[C_o] - [HCN]_t\}$, and k_{dE} can be computed by measuring $[HCN]_t$ as a function of time at midday with equation 4.13. HCN can volatilize, but the rate of volatilization is very slow relative

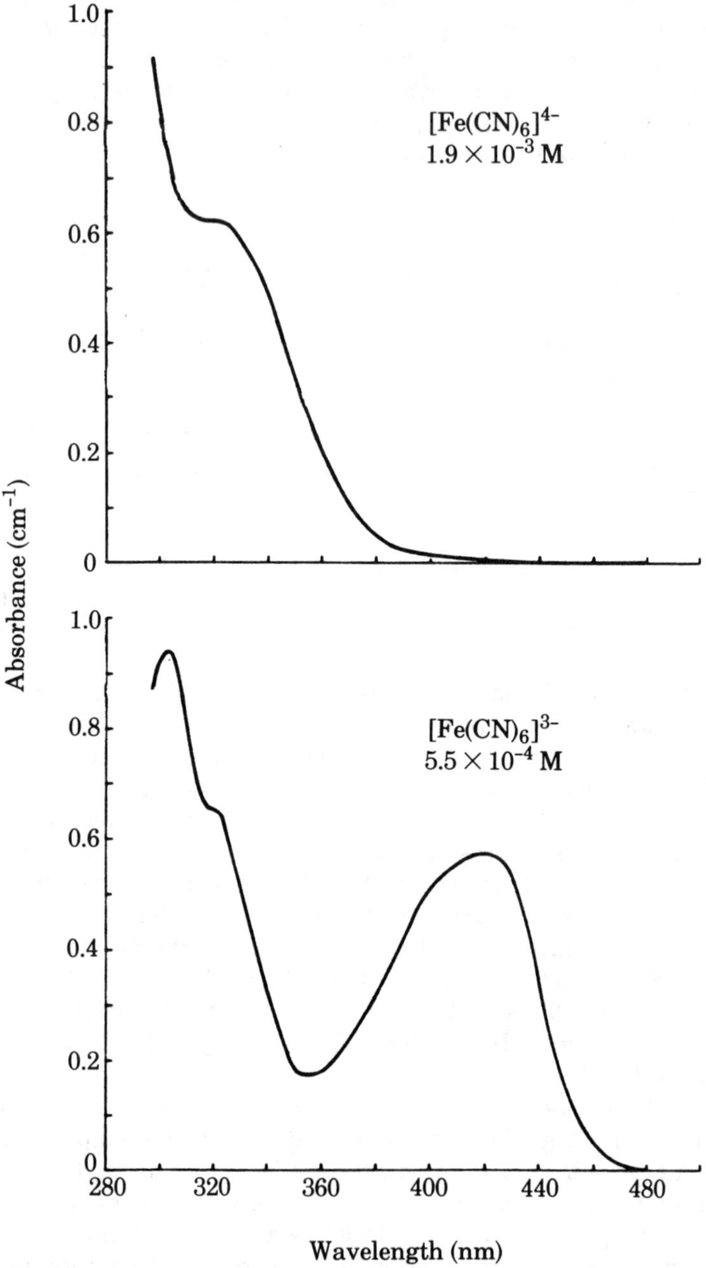

Figure 4.6. Electronic absorption spectra of hexacyanoferrate(II) and hexacyanoferrate(III) in dilute aqueous solution. (Reproduced from ref. 14.)

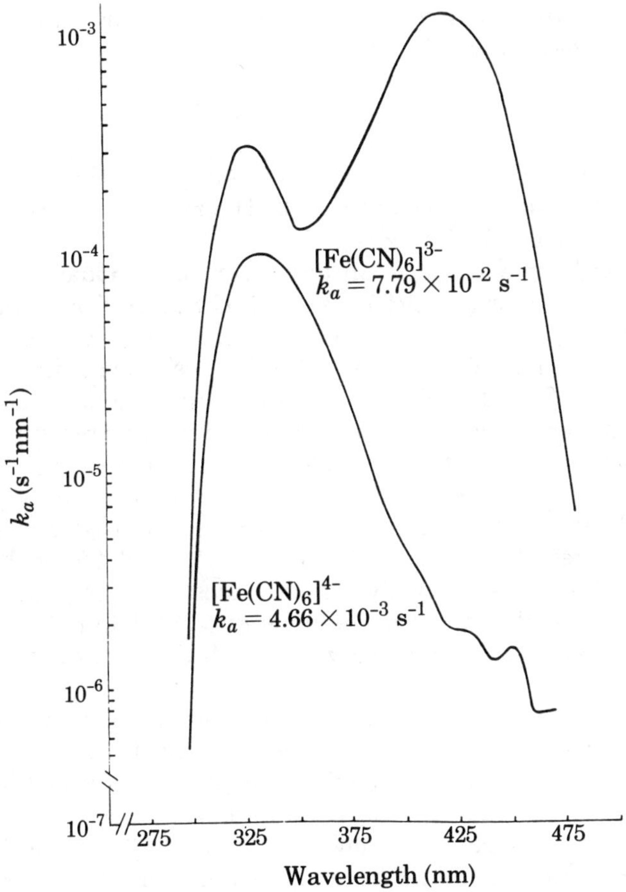

Figure 4.7. Specific sunlight absorption rates of hexacyanoferrate(II) and hexacyanoferrate(III) complexes as a function of wavelength at midday and midsummer, latitude 40° N. (Reproduced from ref. 14.)

to the photoreaction time (which is on the order of 1 h). The half-life can be calculated by using k_{dE} in equation 4.14. The irradiated solutions were buffered at pH 7.8 with phosphate or borate buffer at 10^{-3} M.

The reaction quantum yield of each complex was measured by irradiating aqueous solutions at 20 °C for approximately 1 h under clear sky conditions at midday as just described. These experiments were repeated many times over several months from spring to fall at St. Paul, MN (45° N. latitude). The rate constants k_{dE} and the corresponding half-lives $t_{(1/2)E}$ were calculated for each date. In addition, the values of k_a were computed for each date with the computer program of Zepp and Cline (*1*). According to equations 4.12 and 4.14, the reaction quantum yield is given by

$$\phi_d = 0.693/[t_{(1/2)E}]k_a \tag{4.33}$$

The reaction quantum yield was determined by the visual best fit analysis of the data to equation 4.33. The results were $\phi_d = 0.14$ for $[Fe(CN)_6]^{4-}$ and 0.0023 for $[Fe(CN)_6]^{3-}$.

The experimental near-surface, clear sky, midday half-lives for $[Fe(CN)_6]^{4-}$ solutions at pH 7.8 and 20 °C ranged from about 50 min in late fall to a minimum of about 18 min in midsummer. For comparable conditions for $[Fe(CN)_6]^{3-}$ solutions, the midday half-lives ranged from about 160 min in late fall to a minimum of about 64 min in midsummer. By using the molar absorptivity and reaction quantum yield data for each of these complexes in the computer program, the near-surface, clear sky, midday, midsummer half-lives at St. Paul, MN (45° N. latitude) were calculated as a function of time of year. The theoretical results are depicted in Figures 4.8 and 4.9 for $[Fe(CN)_6]^{4-}$ and $[Fe(CN)_6]^{3-}$, respectively (smooth lines). The experimental data (solid dots) are also shown in these figures. The data indicate that the experimental values agreed very well with the theoretical calculated results.

The direct photoreaction rate of both complexes changes diurnally as the intensity of sunlight increases from sunrise to a maximum at noon and then decreases throughout the day. This variation in computed rates for shallow depths, relative to photoreaction rates at midday at 45° N. latitude, is shown in Figure 4.10 for $[Fe(CN)_6]^{4-}$ on October 20, 1977, and in Figure 4.11 for $[Fe(CN)_6]^{3-}$ on October 21, 1977 (smooth lines).

Experimental photoreaction rates relative to midday values were determined at different times of the day for $[Fe(CN)_6]^{4-}$ on October 20, 1977, and for $[Fe(CN)_6]^{3-}$ on October 21, 1977, at St. Paul, MN, and the results are depicted in Figures 4.10 and 4.11, respectively (solid dots). Figures 4.10 and 4.11 indicate that the experimental results were close to the theoretical calculated results.

By using the reaction quantum yield and molar absorptivity data for both complexes, midday near-surface half-lives were calculated as a function of time of the year and latitude (Figure 4.12). The results were computed as relative values with the half-life of each complex on July 1 at 45° N. latitude assigned a value of 1. For both complexes, the half-lives at the midlatitudes are minimal during the summer months and maximal during the winter months. Both the half-lives and the amplitude of the seasonal variation of half-lives increase with an increase in latitude. In the tropical zone, half-lives are relatively

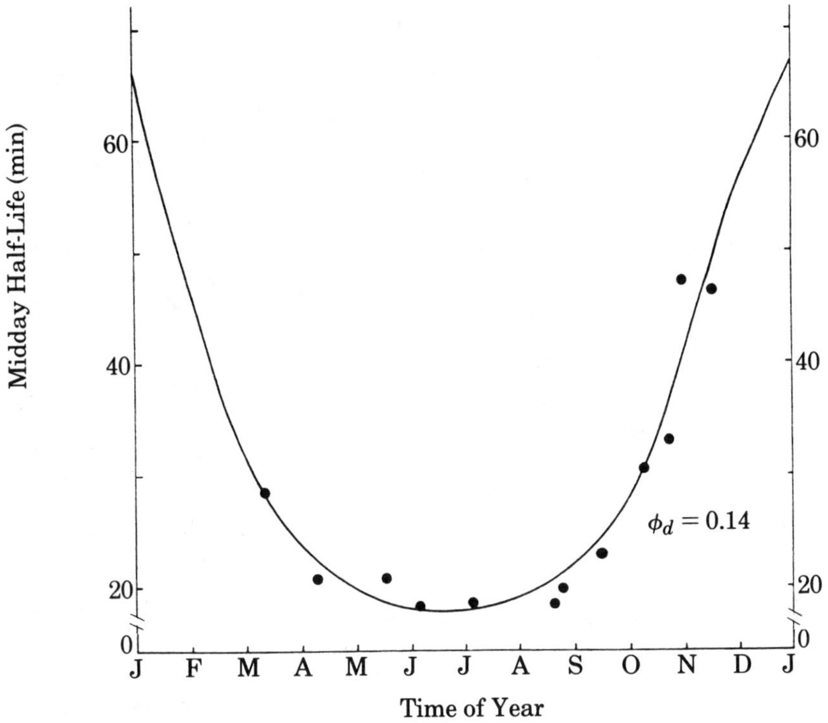

Figure 4.8. Midday half-lives for $[Fe(CN)_6]^{4-}$ solutions at near-surface depths for various times of the the year under full sunlight at St. Paul, MN. Theoretical data are indicated by the smooth line; experimental are data indicated by the solid dots. (Reproduced from ref. 14.)

constant throughout the year. During the summer months, the half-lives are at a minimum with less than a 1.3-fold increase from 0° to 60° N. latitude.

4.C.3. Direct Photoreaction Studies of 1,1-Bis(*p*-methoxyphenyl)-2,2-dichloroethylene

To check the results of the computer calculations, Zepp and Cline (*1*) measured the direct photoreaction rate constant of 1,1-bis(*p*-methoxyphenyl)-2,2-dichloroethylene (DMDE) at shallow depths in sunlight in Athens, GA (34° N. latitude). DMDE was chosen because it absorbs UV-B solar radiation very strongly with a high reaction quantum yield and, therefore, photoreacts rapidly in sunlight. The electronic absorption spectrum of DMDE was obtained in a solvent mixture of aceto-

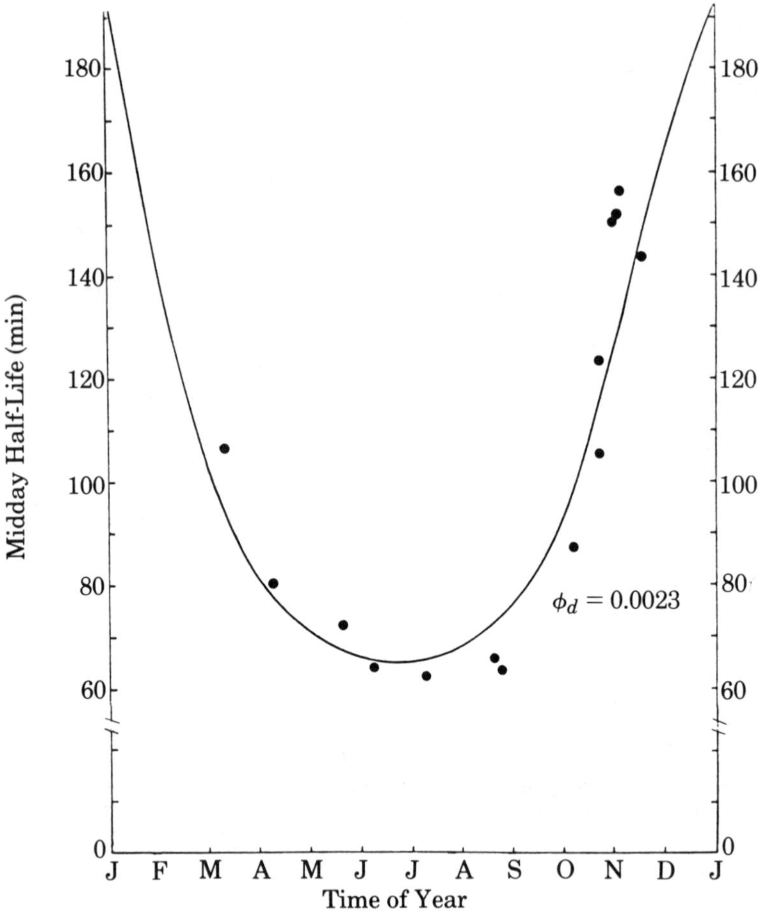

Figure 4.9. Midday half-lives for $[Fe(CN)_6]^{3-}$ solutions at near-surface depths for various times of the year under full sunlight at St. Paul, MN. Theoretical data are indicated by the smooth line; experimental data are indicated by the solid dots. (Reproduced from ref. 14.)

nitrile and water, and ϵ_λ was determined at the specified wavelengths in the computer program (Appendix A, Table A-1). The reaction quantum yield of 0.30 was determined in water at 313 nm by the methods outlined in Chapters 9–11 (*1*).

The computer program calculated $k_{a\lambda}$ and the results are given in Figure 4.13. The data clearly show that the sunlight absorption rate constant appears in the UV-B solar region with $\lambda_{max} \approx 310$ nm. Direct photoreaction rate constants k_{dE} ($k_{dE} = \phi_d k_a$ equation 4.12) were calculated by the computer at specific times of day on June 20, 1975 and

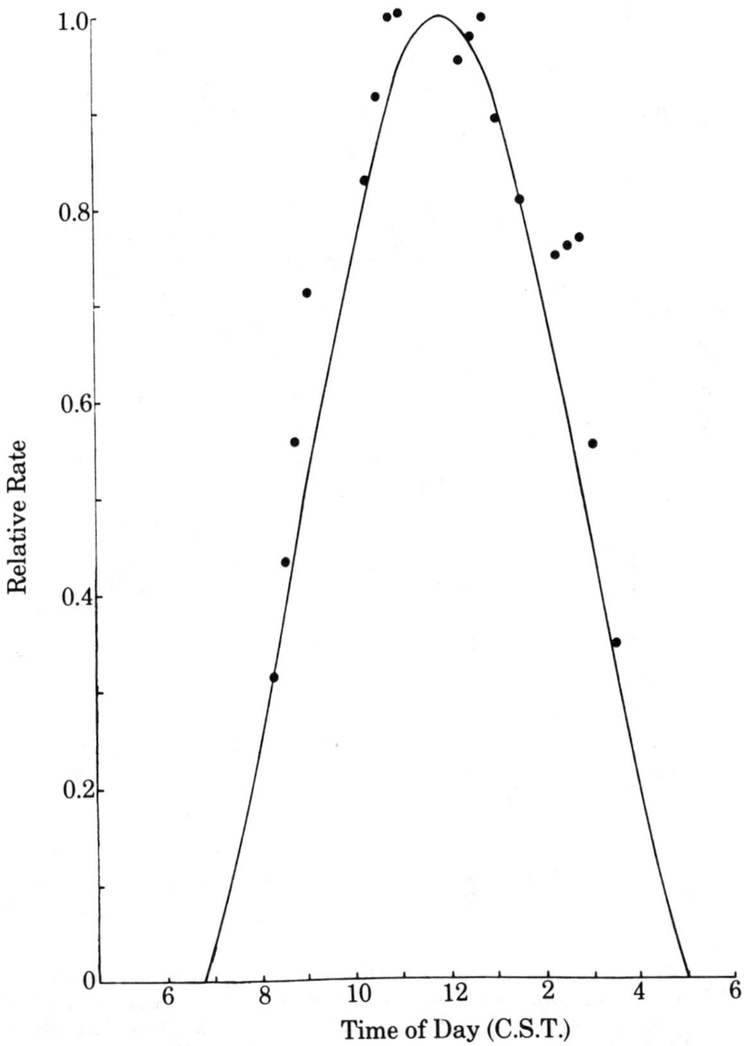

Figure 4.10. Time-of-day dependence of direct photoreaction rates of $[Fe(CN)_4]^{4-}$ *solutions (near-surface) relative to photoreaction rates for midday at St. Paul, MN on October 20, 1977. Theoretical data are indicated by the smooth line; experimental data are indicated by the solid dots. (Reproduced from ref. 14.)*

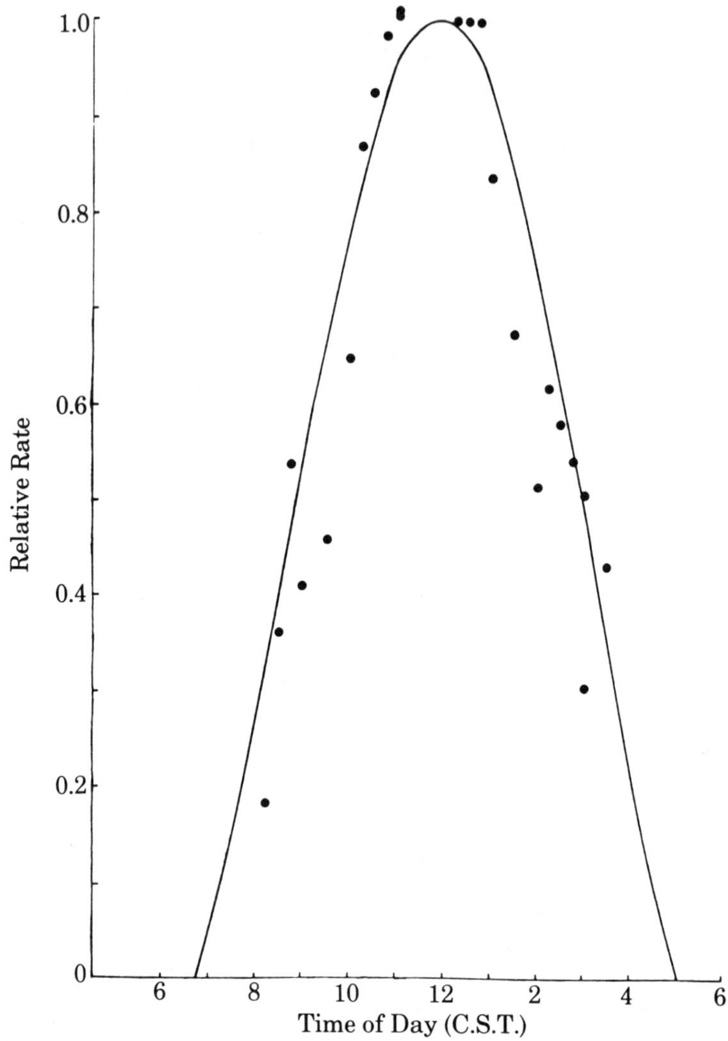

Figure 4.11. Time-of-day dependence of direct photoreaction rates of $[Fe(CN)_6]^{3-}$ solutions (near-surface) relative to photoreaction rates for midday at St. Paul, MN on October 21, 1977. Theoretical data are indicated by the smooth line; experimental data are indicated by the solid dots. (Reproduced from ref. 14.)

June 23, 1975 at Athens, GA, and the computed results are depicted in Figure 4.14.

Dilute aqueous solutions of DMDE (1.00×10^{-7} M) were irradiated in uncovered dishes in sunlight for 1–1.5 h at various times of the day on June 20, 1975 and June 23, 1975 in an open field. The solutions

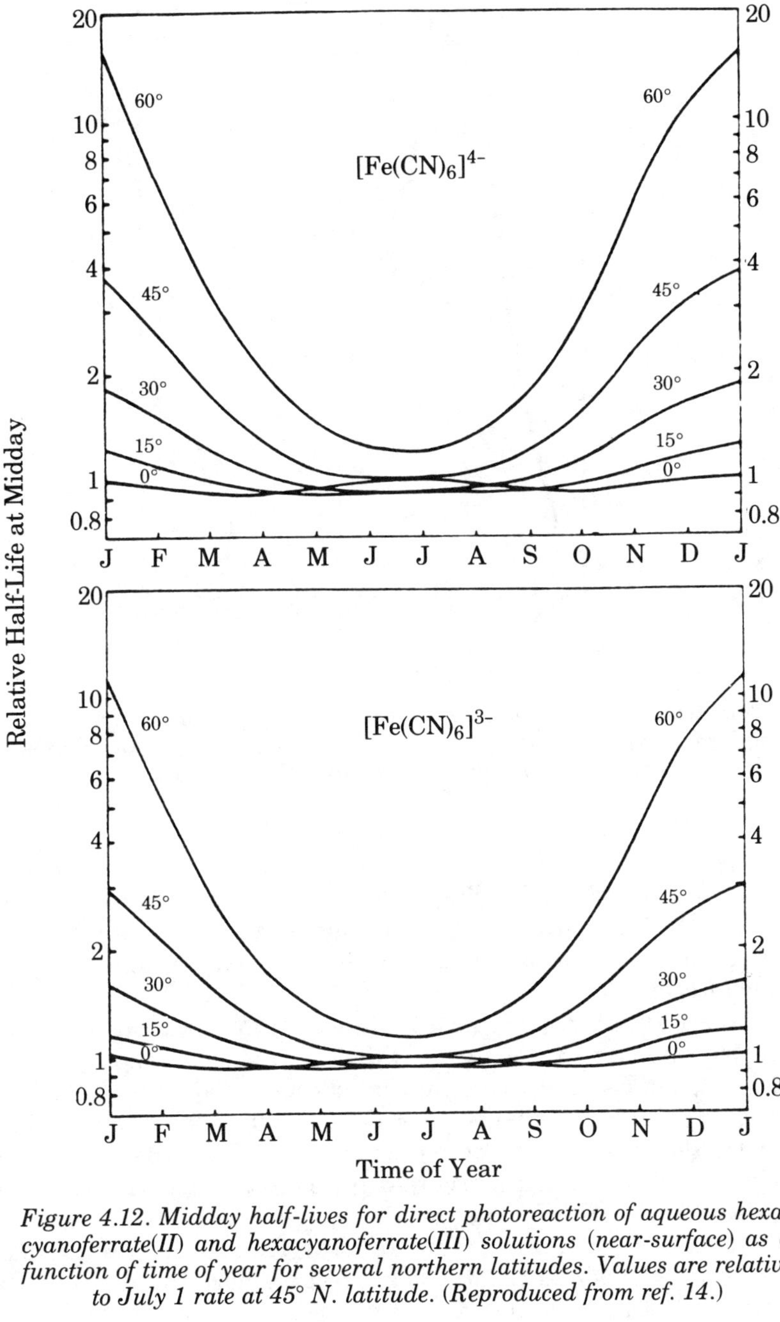

Figure 4.12. Midday half-lives for direct photoreaction of aqueous hexacyanoferrate(II) and hexacyanoferrate(III) solutions (near-surface) as a function of time of year for several northern latitudes. Values are relative to July 1 rate at 45° N. latitude. (Reproduced from ref. 14.)

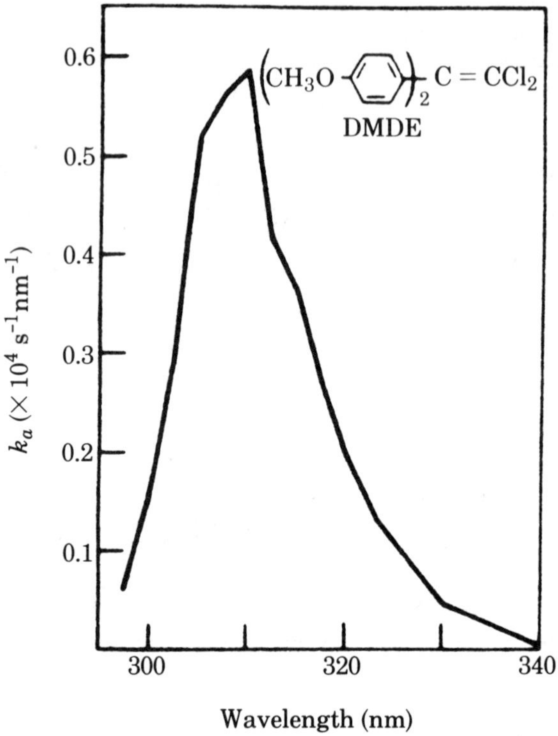

Figure 4.13. Solar absorption rate constant, $k_{a\lambda}$*, as a function of wavelength for DMDE at midday, midsummer, latitude 40° N.* (*Reproduced from ref. 1. Copyright 1977 American Chemical Society.*)

were analyzed for DMDE, and k_{dE} was calculated from equation 4.13. The results are depicted in Figure 4.14. The experimental values for the 2 days in June agreed closely with the theoretical computed values. During both days, the mornings were clear with increasing cloudiness in the afternoon.

Experimental midday, near-surface photoreaction rate constants were determined periodically for DMDE during the year 1975 as just described, and half-lives were calculated from equation 4.14. With the computer program, the half-lives were calculated for the same dates the photoreaction experiments were carried out. The experimental and computer-calculated half-lives are depicted as a function of season in Figure 4.15. The experimental results agreed reasonably well with the theoretical calculated results (Figure 4.15). The rapid increase in half-lives during the winter months is a direct result of the sharp decline in the UV-B solar radiation during this period.

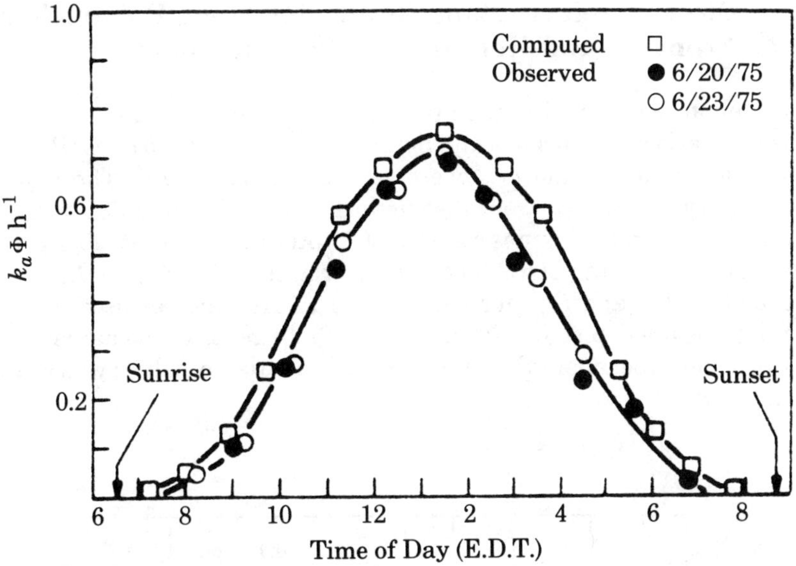

Figure 4.14. Time-of-day dependence of direct photoreaction of DMDE at Athens, Georgia. (Reproduced from ref. 1. Copyright 1977 American Chemical Society.)

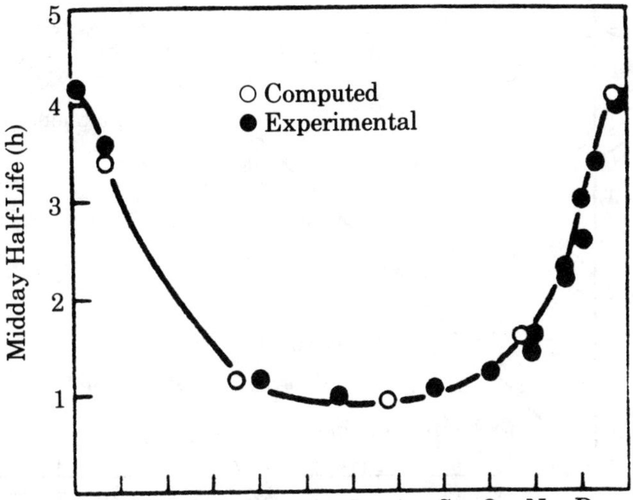

Figure 4.15. Seasonal variation of direct photoreaction of DMDE at Athens, Georgia during 1975. (Reproduced from ref. 1. Copyright 1977 American Chemical Society.)

4.C.4. Direct Photoreaction Studies for Some Polynuclear Aromatic and Heteroaromatic Compounds

Direct photoreaction experiments were carried out for eight polynuclear aromatic hydrocarbons and heteroaromatic hydrocarbons (PAHs) in sunlight and k_{dE} and $t_{(1/2)E}$ were calculated (*7, 16, 17*). The molar absorptivity, ϵ_λ, and ϕ_d were also determined at 313 and 366 nm according to the methods described in Chapters 9–11. The laboratory data were used in the SOLAR computer program of Zepp and Cline (*1*) to calculate k_{dE} and $t_{(1/2)E}$ for each of the PAHs. The results for $t_{(1/2)E}$ calculations are depicted in Figure 4.16 (*16*). In general, the agreement between the experimental and theoretical calculated values was quite good.

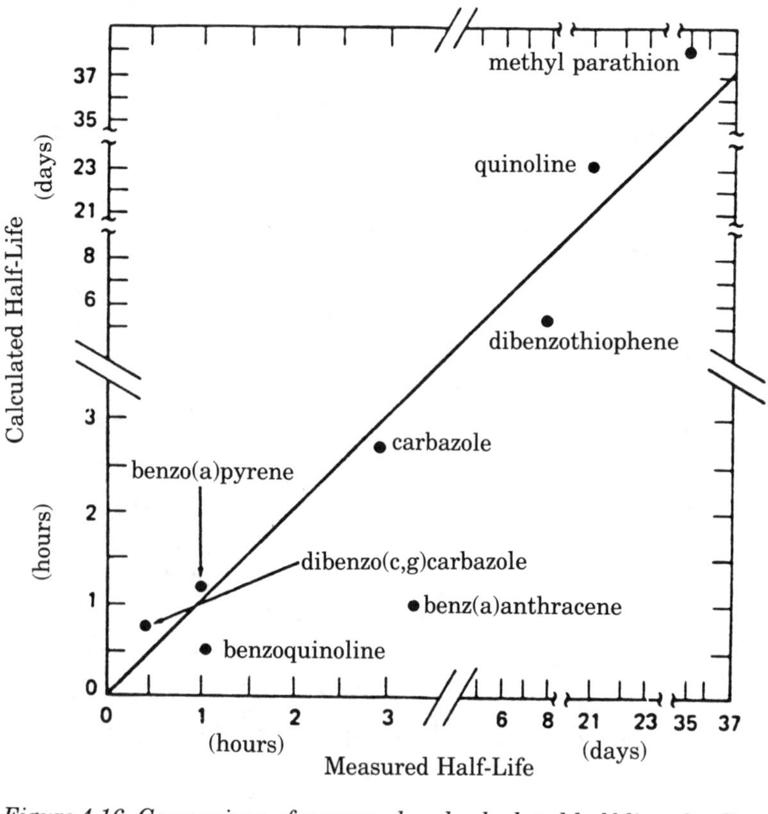

Figure 4.16. Comparison of measured and calculated half-lives for direct photoreaction of PAHs. (Reproduced with permission from ref. 16. Copyright 1981 Pergamon.)

4.C.5. Direct Photoreaction Studies for Several Pesticides and Organomercury Compounds

Zepp and Cline (*1*) carried out photoreaction studies for several pesticides and organomercury compounds at shallow depths in sunlight at Athens, GA (34° N. latitude). The experimental half-lives ranged from 3 min to 168 h; all the results are depicted in Figure 4.17 (*1*). By using the experimentally determined reaction quantum yield for each of these compounds in the computer program, the half-lives were computed, and the results are also depicted in Figure 4.17. Figure 4.17 indicates that the theoretical computed values were within reasonable agreement with the experimental values (i.e., within 30% of the experimental values).

4.D. Revision of the Computer Program and Direct Photoreaction Rates

As indicated in Chapter 3, Zepp (*18*) improved the original SOLAR computer program of Zepp and Cline (*1*) by incorporating the best

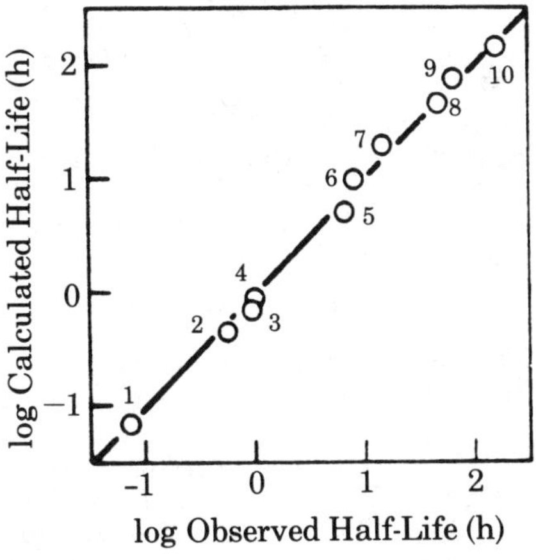

Figure 4.17. Comparison of calculated and observed half-lives for direct photoreaction of several pesticides and pesticide derivatives (reaction quantum yield in parentheses). 1, N*-nitrosoatrazine in water* (ϕ_d = *0.30*)*; 2, trifluralin in water* (ϕ_d = *0.0020*)*; 3, DMDE in water* (ϕ_d = *0.30*)*; 4, DMDE in hexadecane* (ϕ_d = *0.20*)*; 5, DDE in hexadecane* (ϕ_d = *0.26*)*; 6, diphenylmercury in water* (ϕ_d = *0.27*)*; 7, phenylmercuric acetate in water* (ϕ_d = *0.25*)*; 8, 2,4-D-butoxyethyl ester in hexadecane* (ϕ_d = *0.17*)*; 9, carbaryl in water* (ϕ_d = *0.0060*)*; and 10, 2,4-D-butoxyethyl ester in water* (ϕ_d = *0.056*). (*Reproduced from ref. 1. Copyright 1977 American Chemical Society.*)

available irradiance data in the middle UV spectral region 280–380 nm. The new GCSOLAR computer program can now calculate rate constants and half-lives at a specific depth in a water body as well as the average values over the entire depth. In addition, rate constants and half-lives can now be calculated for any water body at any distance above sea level as well as at sea level.

Direct photoreaction studies were conducted for the compounds (2,4,5-trichlorophenoxy)acetic acid [2,4,5-T] and 4-amino-3,5,6-trichloropicolinic acid (picloram) in distilled water (*19*). When the reaction quantum yields determined in this study were used, near-surface half-lives for direct photoreaction were computed (by GCSOLAR) to be 15 days for 2,4,5-T and 2.2 days for picloram during late summer at 40° N. latitude. These values are in close agreement with the experimentally observed values (*20, 21*). These results again verified the methods of Zepp and Cline (*1*) and the new GCSOLAR computer program.

References

1. Zepp, R. G.; Cline, D. M. *Environ. Sci. Technol.* **1977,** *11*, 359.
2. Zepp, R. G. *Environ. Sci. Technol.* **1978,** *12*, 327.
3. Zepp, R. G.; Baughman, G. L. In *Aquatic Pollutants: Transformation and Biological Effects*; Hutzinger, O.; Van Lelyweld, I. H.; Zoetman, B. C. J., Eds.; Pergamon: New York, 1978; p 236.
4. *Astronomical Almanac*; Nautical Almanac Office. U.S. Naval Observatory: Washington, DC, 1987.
5. Zepp, R. G.; Wolfe, N. L.; Gordon, J. A.; Baughman, G. L. *Environ. Sci. Technol.* **1975,** *9*, 1144.
6. Zepp, R. G.; Wolfe, N. L.; Gordon, J. A.; Fincher, R. C. *J. Agric. Food Chem.* **1976,** *24*, 727.
7. Zepp, R. G.; Schlotzhauer, P. F. In *Polynuclear Aromatic Hydrocarbons*; Jones, P. W.; Leber, P., Eds.; Ann Arbor Science: Ann Arbor, MI, 1979; p 141.
8. Smith, J. H.; Mabey, W. R.; Bohonos, N.; Holt, B. R.; Lee, S. S.; Chou, T.-W.; Bomberger, D. C.; Mill, T. *Environmental Pathways of Selected Chemicals in Fresh Water Systems. Part II: Laboratory Studies;* U.S. Environmental Protection Agency. U.S. Government Printing Office: Washington, DC, 1978; EPA 600/7–78–074.
9. Zepp, R. G. In *The Role of Solar Ultraviolet Radiation in Marine Ecosystems*; Calkins, J., Ed.; Plenum: New York, 1982; p 283.
10. Zepp, R. G. In *The Handbook of Environmental Photochemistry*; Hutzinger, O., Ed.; Springer–Verlag: New York, 1982; Vol. 2, Part B, p 19.
11. Zepp, R. G. In *Dynamics, Exposure, and Hazard Assessment of Toxic Chemicals*; Haque, R., Ed.; Ann Arbor Science: Ann Arbor, MI, 1980; p 69.
12. Zafiriou, O. C.; Joussot-Dubien, J.; Zepp, R. G.; Zika, R. G. *Environ. Sci. Technol.* **1984,** *18*, 358A.
13. London, J. *Beitr. Phys. Atmos.* **1963,** *36*, 254.
14. Broderius, S. J.; Smith, L. L., Jr. *Direct Photolysis of Hexacyanoferrate*

Complexes; U.S. Environmental Protection Agency. U.S. Government Printing Office: Washington, DC, 1980; EPA–600/3–80–003.

15. Balzani, V.; Carassiti, V. *Photochemistry of Coordination Compounds*; Academic: New York, 1970; Chapter 2.
16. Mill, T.; Mabey, W. R.; Lau, B. Y.; Baraze, A. *Chemosphere* **1981,** *10*, 1281.
17. Smith, J. H.; Mabey, W. R.; Bohonos, N.; Holt, B. R.; Lee, S. S.; Chou, T.-W.; Bomberger, D. C.; Mill, T. *Environmental Pathways of Selected Chemicals in Fresh Water Systems. Part I: Background and Experimental Procedures;* U.S. Environmental Protection Agency. U.S. Government Printing Office: Washington, DC, 1977; EPA 600/7–77–113.
18. Zepp, R. G., U.S. Environmental Protection Agency, Office of Research and Development, Athens, GA, personal communication, 1982.
19. Skurlatov, Y. I.; Zepp, R. G.; Baughman, G. L. *J. Agric. Food Chem.* **1983,** *31*, 1065.
20. Crosby, D. G.; Wong, A. S. *J. Agric. Food Chem.* **1973**, *21*, 1652.
21. Hedlund, R. T.; Youngson, C. R. In *Fate of Organic Pesticides in the Aquatic Environment;* Faust, S., Ed.; Advances in Chemistry 209; American Chemical Society: Washington, DC, 1972; pp 159–172.

Chapter 5

Theory of the Kinetics of Photochemical Reactions in Aquatic Media in the Environment

A generalized theory of the kinetics of photochemical reactions in aquatic media with any light source, particularly when the light source is sunlight and the aqueous medium is a natural water body (e.g., a lake or a river), is developed in this chapter. This generalized theory is developed by applying the fundamental principles outlined in Chapter 1. Specifically, these photochemical principles are used to

1. Develop the kinetics of direct photoreaction
   a. in pure water with any light source
   b. in a pure natural water body in the environment with sunlight as the light source
2. Study the attenuation of the kinetics of direct photoreaction in natural waters containing humic substances
3. Formulate the theory of the kinetics of indirect (or sensitized) photoreaction
   a. in aqueous media containing dissolved humic acids with any light source with wavelength distribution $\lambda \geq 290$ nm
   b. with sunlight as the light source in a natural water body containing dissolved humic acids, and
3. Develop a generalized theory of the kinetics of direct and indirect (sensitized) sunlight photoreaction
   a. in a natural water containing dissolved humic acids, and
   b. in a natural humic water body in the environment.

1464–6/88/0077$06.00/1

The theory of the kinetics of photochemical reactions in aquatic media in sunlight in the environment is developed systematically and with mathematical rigor; all mathematical derivations are developed step by step so that the environmental scientist can follow the theory and grasp the concepts easily. All key parameters are defined in the course of the derivations. At appropriate places in the derivations, the photochemical units, and other pertinent units, are discussed to emphasize their importance in the theoretical discussions.

5.A. Theory of the Kinetics of Direct Photoreaction in Natural Waters

5.A.1. Direct Photoreaction Rates in Pure Natural Waters

Consider a well-mixed pure natural water containing a chemical at a very low concentration so that the absorbance is less than 0.02 at all wavelengths λ where the chemical absorbs light. Because this natural water is considered to be pristine (e.g., distilled water), the chemical can undergo only direct photoreaction. As a result of the first and second laws of photochemistry, the rate of direct photoreaction of the chemical in pure water, at a fixed wavelength λ, is a function of the average rate of light absorption, $I^d_{a\lambda}$, and the reaction quantum yield, $\phi_{d\lambda}$. The average rate is given by equation 1.27 and is thus

$$\text{average rate} = -(d[C]/dt)_{d\lambda} = \phi_{d\lambda} I^d_{a\lambda} \tag{5.1}$$

where $[C]$ is the molar concentration, t is the time, and the subscript and superscript d denote direct photoreaction.

5.A.1.a. Rate of Light Absorption

Let us now consider the direct photoreaction of a chemical in a sample of this pure natural water in a reaction cell containing volume V of solution (in cm^3) and exposed area A (in cm^2). The average rate of light absorption in the cell can be defined by equation 1.28 and is

$$I^d_{a\lambda} = I_{o\lambda}(A/V)F_{s\lambda}F_{c\lambda} \tag{5.2}$$

where $I_{o\lambda}$ is the intensity of the incident light in the cell at λ in the units photons (or einsteins) cm^{-2} s^{-1}, $F_{s\lambda}$ is the fraction of light absorbed by the system, and $F_{c\lambda}$ is the fraction of light absorbed by the chemical. By letting $I_{\lambda} = I_{o\lambda}(A/V)$, equation 5.2 becomes

$$I_{a\lambda}^{d} = I_{\lambda}F_{s\lambda}F_{c\lambda} \tag{5.3}$$

where I_{λ} is in the units photons (or einsteins) $cm^{-3}\ s^{-1}$. The dimensionless terms $F_{s\lambda}$ and $F_{c\lambda}$ are defined by equations 1.23 and 1.29 and are

$$F_{s\lambda} = \{1 - 10^{-(\alpha_{\lambda} + \epsilon_{\lambda}[C])l}\} \tag{5.4}$$

$$F_{c\lambda} = \epsilon_{\lambda}[C]/(\alpha_{\lambda} + \epsilon_{\lambda}[C]) \tag{5.5}$$

where α_{λ} is the attenuation coefficient of water in cm^{-1}, ϵ_{λ} is the molar absorptivity of the chemical in $M^{-1}\ cm^{-1}$, and l is the cell path length in cm.

For an aqueous solution of a chemical at low absorbance (or low optical density), the incident light intensity (now defined by I_{λ}) is weakly absorbed. That is, only a small fraction of the incident light in the solution is absorbed by the system. Under these conditions, $F_{s\lambda}$, defined by equation 5.4, can be simplified as follows. The term $F_{s\lambda}$ can be put in the form

$$F_{s\lambda} = [1 - \exp\{-2.303(\alpha_{\lambda} + \epsilon_{\lambda}[C])\}] \tag{5.6}$$

$$F_{s\lambda} = [1 - \exp(-x)] \tag{5.7}$$

where

$$x = 2.303(\alpha_{\lambda} + \epsilon_{\lambda}[C])l \tag{5.8}$$

Expansion of $\exp(-x)$ in a series (*1*) yields

$$\exp(-x) = 1 - x + x^2/2! - x^3/3! + \cdots \tag{5.9}$$

For weak absorbance, x {which is the absorbance and is equal to $(\alpha_{\lambda} + \epsilon_{\lambda}\ [C])l$} is less than 0.02 and thus the terms $x^2/2!$, $x^3/3!$, ... are small relative to x and can be neglected. Hence, equation 5.9 becomes

$$x = [1 - \exp(-x)] \tag{5.10}$$

Substituting equation 5.10 in equation 5.7 and using equation 5.8 yield

$$F_{s\lambda} = x = 2.303\{(\alpha_{\lambda} + \epsilon_{\lambda}[C])l\} \tag{5.11}$$

Substituting equations 5.11 and 5.5 in equation 5.3 yields

$$I^{d}_{a\lambda} = 2.303 I_{\lambda} l \epsilon_{\lambda} [C] \tag{5.12}$$

In equation 5.12, I_{λ} is the incident light intensity in photons (or einsteins) cm^{-3} s^{-1}, l is the cell path length in cm, ϵ_{λ} is the molar absorptivity of the chemical in M^{-1} cm^{-1}, and $[C]$ is the molar concentration. Introducing the conversion factor j in equation 5.12 to make the units of intensity compatible with the units of ϵ_{λ} and $[C]$ yields

$$I^{d}_{a\lambda} = 2.303 I_{\lambda} l \epsilon_{\lambda} [C] j^{-1} \tag{5.13}$$

The significance of j was discussed in Section 1.C.4.

The average irradiance in a reaction cell of path length l is denoted by the symbol $E^{av}_{o\lambda}$ and represents the average number of photons (or einsteins) per unit area and per unit time. The irradiance represents the number of photons (or einsteins) that flow through a unit surface area per second, so it is also called a photon flux. Zepp (*2*) called $E^{av}_{o\lambda}$ "the photon scalar irradiance." Because the average irradiance is defined by

$$E^{av}_{o\lambda} = I_{\lambda} l \tag{5.14}$$

equation 5.13 becomes

$$I^{d}_{a\lambda} = 2.303 E^{av}_{o\lambda} \epsilon_{\lambda} [C] j^{-1} \tag{5.15}$$

When $E^{av}_{o\lambda}$ is expressed in photons cm^{-2} s^{-1}, $[C]$ is the molar concentration, and ϵ_{λ} is in M^{-1} cm^{-1}, then $j = 6.022 \times 10^{20}$; and when $E^{av}_{o\lambda}$ is expressed in millieinsteins cm^{-2} s^{-1}, and $[C]$ and ϵ_{λ} are as just defined, then $j = 1$.

Equation 5.15 can be written as

$$I^{d}_{a\lambda} = k_{a\lambda} [C] \tag{5.16}$$

where

$$k_{a\lambda} = 2.303 E^{av}_{o\lambda} \epsilon_{\lambda} j^{-1} \tag{5.17}$$

and $k_{a\lambda}$ is called the specific light absorption rate constant at a specific wavelength λ. The specific light absorption rate constant represents a measure of the spectral overlap between the irradiance of the light source, $E^{av}_{o\lambda}$, and the light absorption, (ϵ_{λ}), of the photoreactive chemical at wavelength λ.

5.A.1.b. The Kinetics of Direct Photoreaction

The kinetics of direct photoreaction can be derived in the following manner. Substituting equation 5.16 in equation 5.1 yields the first-order differential equation

$$-(d[C]/dt)_{d\lambda} = \phi_{d\lambda} k_{a\lambda}[C] \qquad (5.18)$$

Taking the sum of equation 5.18 over all wavelengths for which the chemical absorbs light yields

$$-\sum_{\lambda} (d[C]/dt)_{d\lambda} = \sum_{\lambda} \phi_{d\lambda} k_{a\lambda}[C] \qquad (5.19)$$

For most organic chemicals in aqueous solution, the reaction quantum yield is independent of λ. This assumption is valid for complex molecules in solution because the photoreaction from second or higher electronic states usually cannot compete with the rapid radiationless decay of excited molecules to the first excited state (*3*). Therefore, $\phi_{d\lambda} = \phi_d$ and equation 5.19 becomes

$$-\sum_{\lambda} (d[C]/dt)_{d\lambda} = \phi_d \sum_{\lambda} k_{a\lambda}[C] \qquad (5.20)$$

By defining the relationships

$$\text{average rate} = \sum_{\lambda} (d[C]/dt)_{d\lambda} = (d[C]/dt)_d \qquad (5.21)$$

and

$$k_d = \phi_d \sum_{\lambda} k_{a\lambda} \qquad (5.22)$$

equation 5.20 becomes

$$-(d[C]/dt)_d = k_d[C] \qquad (5.23)$$

where k_d is the first-order direct photoreaction rate constant. Integrating equation 5.23 under the boundary conditions ($t = 0$, $[C_o]$) and (t, $[C_t]$) yields

$$\ln([C_o]/[C_t]) = k_d t \qquad (5.24)$$

Using the mathematical relationship

$$k_a = \sum_{\lambda} k_{a\lambda} \qquad (5.25)$$

then the specific light absorption rate constant, k_a, which is the sum of the $k_{a\lambda}$ values for all wavelengths λ that are absorbed by the chemical. Substituting equation 5.17 in equation 5.25 yields

$$k_a = 2.303j^{-1} \sum_{\lambda} \epsilon_\lambda E_{o\lambda}^{av} \tag{5.26}$$

Finally, substituting equation 5.25 in equation 5.22 yields

$$k_d = \phi_d k_a \tag{5.27}$$

and substituting equation 5.26 in equation 5.27 yields

$$k_d = 2.303j^{-1}\phi_d \sum_{\lambda} \epsilon_\lambda E_{o\lambda}^{av} \tag{5.28}$$

Equation 5.28, shows that the first-order direct photoreaction rate constant is proportional to both the reaction quantum yield ϕ_d and to the overlap between the molar absorptivity ϵ_λ and the irradiance $E_{o\lambda}^{av}$, summed for all wavelengths of light absorbed by the chemical. The units of k_d are reciprocal time.

The half-life, $t_{(1/2)}$, is defined as the time required for the chemical to reach one-half its initial concentration. Under the boundary conditions ($[C_t] = [C_o]/2$, $t = t_{(1/2)}$), equation 5.24 yields

$$t_{(1/2)} = \ln 2/k_d = 0.693/k_d \tag{5.29}$$

If $E_{o\lambda}^{av}$ represents the average solar irradiance (or solar flux), then one obtains all the relevant equations for direct photoreaction in sunlight. The rate constant k_d, defined by equation 5.28, is proportional to the reaction quantum yield and the overlap of the chemical's molar absorptivity with the average solar irradiance, summed over all wavelengths λ where the chemical absorbs sunlight.

Zepp and Cline (*4*) published a paper on the determination of direct photoreaction rates in aquatic environments. They indicated that the rates of all photochemical processes in a natural water body are affected by the solar irradiance at the water surface, radiative transfer from air to water, and the transmission of sunlight in a water body. A detailed discussion of the transmission and absorption of sunlight in a natural water body is given in Chapter 3. These researchers developed a computer program called SOLAR and an updated version called GCSOLAR to calculate solar irradiance, $E_{o\lambda}^{av}$, in a natural water body. For the conditions corresponding to weak absorbance of a chemical (i.e., $A_\lambda < 0.02$) in a natural water body, $E_{o\lambda}^{av} = Z_\lambda$ in photons cm^{-2} s^{-1} (equation 3.38). Under these conditions, $E_{o\lambda}^{av}$ corresponds to the solar irradiance in a natural water body for shallow depths and for depths up to 0.5 m in a pure water body. Furthermore, because $E_{o\lambda}^{av} = Z_\lambda$, $E_{o\lambda}^{av}$ corresponds to the solar irradiance at midday and is a function of latitude and season of the year. Under the conditions of weak absorbance of the chemical in a pure natural water body at shallow depths (<0.5 m), $E_{o\lambda}^{av} = L_\lambda$ (in millieinsteins cm^{-2} day^{-1}), and corresponds to day-averaged values of solar irradiance, which are a function of latitude and season of the year.

The theoretical discussions described have shown that for the photoreaction of a chemical at a low molar concentration with absorbance less than 0.02, the kinetics of direct photoreaction can be described by equations 5.17 to 5.29 with $E_{o\lambda}^{av}$ now representing the average solar irradiance in a natural water body, k_{dE} representing the first-order direct photoreaction rate constant in a natural water body, and $t_{(1/2)E}$ representing the corresponding half-life. The subscript E is used to indicate that the chemical is undergoing direct photoreaction in a natural water body (e.g., a river or lake) in sunlight in the environment.

The method of Zepp and Cline (*4*) for determining environmentally relevant direct photoreaction rate constants and half-lives in natural water bodies was discussed in detail in Chapter 4. The theory of direct photoreaction will be expanded in Chapters 6–8 and detailed methods will be described for determining environmentally relevant direct photoreaction rate constants and half-lives in aqueous media by carrying out photoreaction experiments with actinometers either outdoors in sunlight (Chapters 6–8) or in the laboratory (Chapters 9–11) with monochromatic light and actinometers.

5.A.2. Attenuation of Direct Photoreaction Rates in Natural Waters Containing Humic Substances

Rates of photoreaction can be affected by dissolved and suspended matter in aqueous media. For example, rates of photoreaction can be affected by the absorption or scattering of light in aqueous media via physical processes or by photochemical processes involving sensitization or quenching of excited states. This section will be devoted to the discussion of the attenuation of direct photoreaction rates by naturally occurring humic substances in aqueous media via physical processes. The next section will discuss the increase in rates of photoreaction via photosensitization processes involving dissolved humic acids in natural waters (i.e., indirect photoreaction).

Natural waters can contain humic substances that form from decayed vegetation. These humic substances are composed of complex matter containing humic and fulvic acids. The humic substances, either dissolved or suspended, can attenuate sunlight penetrating these water bodies. Consider the attenuation of direct photoreaction rates by dissolved humic acids in a natural water. Specifically, let us consider a system composed of a well-mixed natural water containing dissolved humic acids and a chemical at a very low molar concentration $[C]$. Furthermore, this system will be constrained by the condition that these humic acids do not photosensitize the chemical. The molar absorptivity of the chemical is ϵ_λ, and the attenuation coefficient of the humic acid solution is α_λ. Equations 5.4 and 5.5 can be used to describe this natural water system. Because the chemical is at a very low molar concentration, $\alpha_\lambda \gg \epsilon_\lambda[C]$, so that the equations become

$$F_{s\lambda} = (1 - 10^{-\alpha_\lambda l}) \tag{5.30}$$

$$F_{c\lambda} = \epsilon_\lambda[C]/\alpha_\lambda \tag{5.31}$$

Substituting equations 5.30 and 5.31 in equation 5.3 yields

$$I^d_{a\lambda} = (1 - 10^{-\alpha_\lambda l})(I_\lambda \epsilon_\lambda/\alpha_\lambda)[C] \tag{5.32}$$

Introducing the conversion factor j to make the units of $[C]$ and ϵ_λ compatible with $I^d_{a\lambda}$ yields

$$I^d_{a\lambda} = (1 - 10^{-\alpha_\lambda l})(I_\lambda \epsilon_\lambda/j\alpha_\lambda)[C] \tag{5.33}$$

Substituting equation 5.33 in equation 5.1 yields

$$-(d[C]/dt)_{d\lambda} = \phi_{d\lambda}(1 - 10^{-\alpha_\lambda l})(I_\lambda \epsilon_\lambda/j\alpha_\lambda)[C] \tag{5.34}$$

Equation 5.34 can be written in the form of the first-order differential equation

$$-(d[C]/dt)_{d\lambda} = k_{d\beta}[C] \tag{5.35}$$

where $k_{d\beta}$ is the first-order direct photoreaction rate constant in the humic water and is given by

$$k_{d\beta} = \phi_{d\lambda}(1 - 10^{-\alpha_\lambda l})(I_\lambda \epsilon_\lambda/j\alpha_\lambda) \tag{5.36}$$

Under the conditions of weak absorbance of a chemical at a low molar concentration $[C]$ in pure water (e.g., distilled water), equation 5.13 is valid. Thus, substituting equation 5.13 in equation 5.1 yields

$$-(d[C]/dt)_{d\lambda} = 2.303 I_\lambda l \epsilon_\lambda \phi_{d\lambda} j^{-1}[C] \tag{5.37}$$

and the first-order differential equation is given by the expression

$$-(d[C]/dt)_{d\lambda} = k_d[C] \tag{5.38}$$

where k_d, the first-order direct photoreaction rate constant in pure water, is defined by

$$k_d = 2.303\phi_{d\lambda}I_\lambda l\epsilon_\lambda j^{-1} \tag{5.39}$$

Consider the case of two identical photochemical cells, one containing a chemical in pure water at a low molar concentration $[C]$ and absorbance <0.02; and the other cell containing the chemical under the same conditions but in a natural water containing dissolved humic acids with attenuation coefficient α_λ. Both cells are then exposed to the same light source at wavelength λ. Under these conditions, a "light screening factor (S_λ)" can be defined as the ratio of the first-order rate constant in the natural water containing dissolved humic acids to the first-order rate constant in pure water. Therefore,

$$S_\lambda = k_{d\beta}/k_d \tag{5.40}$$

Substituting equations 5.39 and 5.36 in equation 5.40 yields

$$S_\lambda = (1 - 10^{-\alpha_\lambda l})/(2.303\alpha_\lambda l) \tag{5.41}$$

Figure 5.1 (*5*) depicts how S_λ varies with $\alpha_\lambda l$. According to equation 1.13, $A_\lambda = (\alpha_\lambda + \epsilon_\lambda[C])l$. Because $\alpha_\lambda \gg \epsilon_\lambda[C]$, $A_\lambda = \alpha_\lambda l$. For a cell with a path length of 1 cm, $A_\lambda = \alpha_\lambda$. Thus, the absorbance is equal to the attenuation coefficient of the natural water containing dissolved humic acids in the 1-cm cell. In a reaction cell with l equal to 1 cm, S_λ for a natural humic acid solution is >0.5 at λ > 300 nm (*2*).

Equation 5.41 is very general and is applicable to the direct sunlight photoreaction of a chemical in a natural humic water body in the environment. In effect, the rate of direct photoreaction of the chemical in the humic natural water is decreased by the attenuation coefficient of the natural water (α_λ) because most of the sunlight is absorbed by the humic acids and only a very small fraction of the sunlight is available for the direct photoreaction of the chemical.

For example, in ref. 2, equation 5.41 is used in a different form. If the attenuation coefficient, α_λ, is expressed in the Napierian logarithm system, a_λ, [i.e., $a_\lambda = 2.303\alpha_\lambda$], then it can be readily shown that

$$S_\lambda = [1 - \exp(-a_\lambda l)]/(a_\lambda l) \tag{5.42}$$

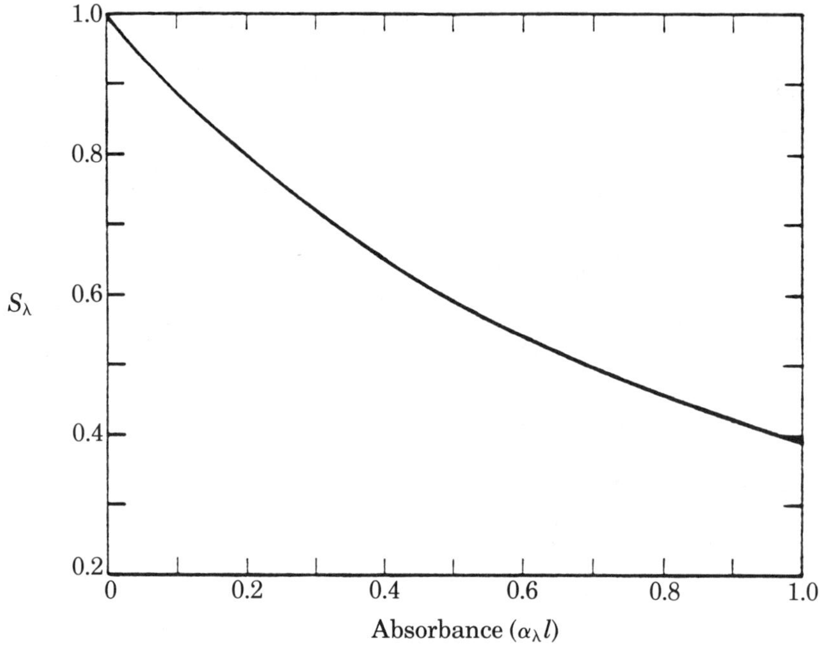

Figure 5.1. Plot of light screening factor, S_λ, *against absorbance* ($\alpha_\lambda l$). *(Reproduced from ref. 5.)*

The attenuation coefficients a_λ and α_λ cannot be used to compute S_λ when the natural water contains high concentrations of suspended matter and a large fraction of attenuation is caused by scattering. Under these conditions, light attenuation effects are overestimated when using equation 5.41 or 5.42 (*2*).

Light absorption can be determined in natural waters containing suspended matter by using Ulbricht spheres in a spectrophotometer. Ulbricht spheres collect all the light, including scattered light. Miller and Zepp (*6, 7*) demonstrated that light scattering by particulates in natural waters causes the light to become more diffuse, and thus photoreaction rates are enhanced.

5.B. Theory of the Kinetics of Indirect Photoreaction in Natural Humic Waters

Consider a well-mixed natural water system composed of dissolved humic acids and a chemical at a very low molar concentration $[C]$ with weak absorbance (i.e., the absorbance of the chemical is <0.02). The molar absorptivity of the chemical is ϵ_λ, and the attenuation coefficient of the humic acid solution is α_λ. When this system is exposed to light of wavelengths $\lambda \geq 290$ nm, only the humic acids absorb the light. It is assumed that the chemical, because of the nature of its molecular structure, does not absorb light at wavelengths λ so that direct photoreaction does not occur (i.e., $k_d = 0$). However, even though the chemical does not absorb the incident light at wavelengths λ, it does undergo phototransformation via photosensitization, or indirect photoreaction, by the humic acids present in the natural water. This section will describe indirect photoreaction, or photosensitization of a chemical, in the presence of dissolved humic acids and light, specifically in the presence of sunlight.

In the scenario just described, only indirect or sensitized photoreaction takes place so that equation 1.27 takes the form

$$\text{average rate} = (-d[C]/dt)_{s\lambda} = \phi_{s\lambda} I^s_{a\lambda} \tag{5.43}$$

where $[C]$ is the molar concentration of the chemical, $I^s_{a\lambda}$ is the average rate of light absorbed by the humic acids that results in the indirect (or sensitized) photoreaction of the chemical, and $\phi_{s\lambda}$ is the photosensitized reaction quantum yield of the chemical. The superscript and subscript *s* is used to denote that the chemical is undergoing sensitized or indirect photoreaction.

5.B.1. Rate of Light Absorption

The rate of light absorption by the humic acids in a reaction cell of area *A* and volume *V* is given by equation 1.28 and is

$$I^s_{a\lambda} = I_{o\lambda}(A/V)F_{s\lambda}F_{c\lambda} \tag{5.44}$$

where $I_{o\lambda}$ is the intensity of the incident light entering the cell in photons (or einsteins) cm^{-2} s^{-1}, $F_{s\lambda}$ is the fraction of the light absorbed by the system, and $F_{c\lambda}$ is the fraction of the light absorbed by the sensitizers present in the humic acids. Because for this system $\alpha_\lambda \gg \epsilon_\lambda[C]$, $F_{s\lambda}$, defined by equation 5.4, becomes

$$F_{s\lambda} = (1 - 10^{-\alpha_\lambda l}) \tag{5.45}$$

where α_λ is the attenuation coefficient of the natural water in cm^{-1} and l is the cell path length in cm.

Using equations 1.5 and 1.6 for a 1-cm cell yields

$$A_\lambda = \alpha_\lambda l = \alpha_\lambda = \log_{10}(I_{o\lambda}/I_{t\lambda}) \tag{5.46}$$

so that α_λ is obtained directly from the absorbance A_λ of the humic water solution measured in this cell in a spectrophotometer.

If it is assumed that the humic acids present in the natural water are composed of substances that attenuate the light and one or more sensitizers with molar absorptivity $\epsilon_{\lambda s}$ in M^{-1} cm^{-1} and molar concentration $[S]$, then

$$\alpha_\lambda = (\alpha'_\lambda + \epsilon_{\lambda s}[S]) \tag{5.47}$$

where α'_λ represents the attenuation coefficient of the nonsensitizers in cm^{-1}. Substituting equation 5.47 in equation 5.45 yields

$$F_{s\lambda} = \{1 - 10^{-(\alpha'_\lambda + \epsilon_{\lambda s}[S])l}\} \tag{5.48}$$

In addition, the fraction of the light absorbed by the sensitizers (or sensitizer) that results in photosensitization, as defined by equation 5.5, is then given by

$$F_{c\lambda} = \epsilon_{\lambda s}[S]/(\alpha'_\lambda + \epsilon_{\lambda s}[S]) \tag{5.49}$$

$$F_{c\lambda} = \epsilon_{\lambda s}[S]/\alpha_\lambda \tag{5.50}$$

By making the substitution $I_\lambda = I_{o\lambda}(A/V)$, with I_λ in the units of photons (or einsteins) cm^{-3} s^{-1}, equation 5.44 becomes

$$I^s_{a\lambda} = I_\lambda F_{s\lambda} F_{c\lambda} \tag{5.51}$$

Substituting equations 5.45 and 5.50 in equation 5.51 yields

$$I^s_{a\lambda} = I_\lambda(1 - 10^{-\alpha_\lambda l})\alpha_\lambda^{-1}\epsilon_{\lambda s}[S] \tag{5.52}$$

Multiplying and dividing equation 5.52 by 2.303 yield

$$I^s_{a\lambda} = 2.303[I_\lambda(1 - 10^{-\alpha_\lambda l})(2.303\alpha_\lambda)^{-1}]\epsilon_{\lambda s}[S] \tag{5.53}$$

The average irradiance $E^{av}_{o\lambda}$ in the reaction cell of path length l is the number of photons (or einsteins) per unit area per unit time, and is given by

$$E^{av}_{o\lambda} = I_\lambda(1 - 10^{-\alpha_\lambda l})(2.303\alpha_\lambda)^{-1} \tag{5.54}$$

Substituting equation 5.54 in equation 5.53 yields

$$I^s_{a\lambda} = 2.303E^{av}_{o\lambda}\epsilon_{\lambda s}[S] \tag{5.55}$$

The conversion factor j is now introduced in equation 5.55 to make the units of $E_{o\lambda}^{av}$ compatible with those of $\epsilon_{\lambda s}$ and $[S]$ as described previously. Thus,

$$I_{a\lambda}^{s} = 2.303j^{-1}E_{o\lambda}^{av}\epsilon_{\lambda s}[S] \quad (5.56)$$

Multiplying and dividing equation 5.54 by l yield

$$E_{o\lambda}^{av} = I_{\lambda}l(1 - 10^{-\alpha_{\lambda}l})(2.303\alpha_{\lambda}l)^{-1} \quad (5.57)$$

and using the expression for S_{λ} defined by equation 5.41 in equation 5.57 yields

$$E_{o\lambda}^{av} = I_{\lambda}lS_{\lambda} \quad (5.58)$$

Thus, the incident light intensity entering the reaction cell is attenuated by the natural substances present in the natural water (i.e., by the light screening factor S_{λ}). The average irradiance $E_{o\lambda}^{av}$ in the cell now represents the remaining light available for indirect photoreaction.

5.B.2. Photosensitized Reaction Quantum Yield and the Kinetics of Indirect Photoreaction

In photosensitization experiments, when the concentration of the chemical is low, as is generally the case under environmental conditions, the presence of the chemical has negligible effect on the lifetime of the electronically excited sensitizer. Under these conditions, the photosensitized reaction quantum yield of the chemical is given by the expression

$$\phi_{s\lambda} = Q_{\lambda}[C] \quad (5.59)$$

where Q_{λ} is a constant and $[C]$ is the molar concentration of the chemical (*2, 8–10*). The theoretical derivation of equation 5.59 is given in Appendix B.

Substituting equations 5.59 and 5.56 in equation 5.43 yields

$$\text{average rate} = -(d[C]/dt)_{s\lambda} = 2.303j^{-1}E_{o\lambda}^{av}\epsilon_{\lambda s}Q_{\lambda}[S][C] \quad (5.60)$$

Equation 5.60 shows that the rate of disappearance of the chemical is second-order; that is, the average rate is proportional to $[S][C]$. If the conditions are such that the sensitizer concentration remains constant (*9*), then equation 5.60 reduces to the first-order differential equation

$$-(d[C]/dt)_{s\lambda} = k_{s\lambda}[C] \tag{5.61}$$

where

$$k_{s\lambda} = 2.303j^{-1}E^{av}_{o\lambda}\epsilon_{\lambda s}Q_{\lambda}[S] \tag{5.62}$$

The rate constant $k_{s\lambda}$ is called the rate constant for indirect, or sensitized, photoreaction at wavelength λ and is proportional to the average irradiance $E^{av}_{o\lambda}$ of the light source and the molar absorptivity of the sensitizers (or sensitizer), $\epsilon_{\lambda s}$, at wavelength λ.

Taking the sum of equation 5.61 over all wavelengths λ absorbed by the sensitizers (or sensitizer) in the humic acids yields

$$-\sum_{\lambda} (d[C]/dt)_{s\lambda} = \sum_{\lambda} k_{s\lambda}[C] \tag{5.63}$$

By defining the relationships

$$\text{average rate} = \sum_{\lambda} (d[C]/dt)_{s\lambda} = (d[C]/dt)_s \tag{5.64}$$

$$k_s = \sum_{\lambda} k_{s\lambda} \tag{5.65}$$

equation 5.63 becomes

$$-(d[C]/dt)_s = k_s[C] \tag{5.66}$$

In equation 5.66, k_s is the first-order indirect, or sensitized, photoreaction rate constant. Substituting equation 5.62 in equation 5.65 yields

$$k_s = 2.303j^{-1}[S] \sum_{\lambda} Q_{\lambda}\epsilon_{\lambda s}E^{av}_{o\lambda} \tag{5.67}$$

Thus, the first-order indirect photoreaction rate constant, k_s, is proportional to the overlap of the molar absorptivity of the sensitizers (or sensitizer), $\epsilon_{\lambda s}$, and the irradiance of the light source, $E^{av}_{o\lambda}$, summed for all wavelengths λ absorbed by the humic sensitizers (or sensitizer).

Because equation 5.66 is a first-order differential equation, integration in the standard manner yields

$$\ln([C_o]/[C_t]) = k_s t \tag{5.68}$$

where k_s is the first-order indirect photoreaction rate constant in the units of reciprocal time. The associated half-life $t_{(1/2)}$ (obtained from equation 5.68) is given by

$$t_{(1/2)} = \ln 2/k_s = 0.693/k_s \tag{5.69}$$

If $E_{o\lambda}^{av}$ is the average solar irradiance, then all the equations just derived are relevant to indirect (or sensitized) photoreaction in sunlight. Thus, the rate constant k_s, defined by equation 5.67, is proportional to the overlap of the average solar irradiance $E_{o\lambda}^{av}$ with the molar absorptivity, $\epsilon_{\lambda s}$, summed over all wavelengths where the sensitizers (or sensitizer) absorb light. If $E_{o\lambda}^{av}$ is now defined as the average solar irradiance in a natural water body (e.g., a river or lake), then the rate constant is called k_{sE} and the corresponding half-life is $t_{(1/2)E}$. The subscript E is used to indicate that the chemical is undergoing indirect photoreaction in a natural water body containing dissolved humic acids in sunlight in the environment.

The determination of rates and half-lives in natural humic waters, as described by Zepp et al. (*9, 11*), will be discussed in Chapter 12. The theory of indirect photoreaction will be expanded in Chapters 13–15 and a method will be described for determining environmentally relevant rate constants and half-lives for indirect photoreaction in aqueous humic solution by carrying out experiments outdoors in sunlight with actinometers.

5.B.3. Effect of Attenuation on the Kinetics of Indirect Photoreaction

Attenuation effects must be considered when evaluating indirect photoreaction kinetic data. Consider first the case in which the light is weakly absorbed by the humic acid solution. Under this condition, $A_\lambda = \alpha_\lambda l < 0.05$. The fraction of light absorbed is given by equation 5.45. Conversion of this equation to $F_{s\lambda} = [1 - \exp(-2.303\alpha_\lambda l)]$, expansion of the exponential term in a power series, neglecting all higher power terms (i.e., quadratic and cubic) because $\alpha_\lambda l < 0.05$, and simplification yields

$$F_{s\lambda} = (1 - 10^{-\alpha_\lambda l}) = 2.303\alpha_\lambda l \tag{5.70}$$

Substituting equation 5.70 in equation 5.57 yields

$$E_{o\lambda}^{av} = I_\lambda l \tag{5.71}$$

and substituting equation 5.71 in equation 5.62 yields

$$k_{s\lambda} = 2.303j^{-1}I_\lambda l\epsilon_{\lambda s}Q_\lambda[S] \tag{5.72}$$

where $k_{s\lambda}$, the rate constant for indirect photoreaction at wavelength λ, is predicted to be proportional to the sensitizer concentration $[S]$.

At the other extreme, if the solution absorbs essentially all the light, then $A_\lambda = \alpha_\lambda l > 2$. Under these conditions, $10^{-\alpha_\lambda l} \ll 1$ so that equation 5.45 reduces to

$$F_{s\lambda} = (1 - 10^{-\alpha_\lambda l}) = 1 \tag{5.73}$$

Substituting equation 5.73 in equation 5.57 yields

$$E_{o\lambda}^{av} = I_\lambda(2.303\alpha_\lambda)^{-1} \tag{5.74}$$

and $E_{o\lambda}^{av}$ is inversely proportional to α_λ. Substituting equation 5.74 in equation 5.62 yields

$$k_{s\lambda} = (I_\lambda Q_\lambda j^{-1})(\epsilon_{s\lambda}[S]\alpha_\lambda^{-1}) \tag{5.75}$$

Using equation 5.50 in equation 5.75 yields

$$k_{s\lambda} = I_\lambda Q_\lambda j^{-1}F_{c\lambda} \tag{5.76}$$

If the sensitizers absorb a constant fraction of the total light absorbed by the mixture, then $F_{c\lambda}$, defined by equation 5.49, is constant and independent of $[S]$. Hence, $k_{s\lambda}$, the rate constant for indirect photoreaction at wavelength λ, is independent of $[S]$. If the sensitizers are the only light absorbers in the system, then $\alpha_\lambda' = 0$ and equation 5.47 yields

$$\alpha_\lambda = \epsilon_{\lambda s}[S] \tag{5.77}$$

Substituting equation 5.77 in equation 5.75 yields

$$k_{s\lambda} = I_\lambda Q_\lambda j^{-1} \tag{5.78}$$

and thus $k_{a\lambda}$ is again independent of the sensitizer concentration $[S]$.

At concentrations of humic substances that do not exceed concentrations found in natural waters, the absorbance in a 1-cm cell is sufficiently low so that the indirect photoreaction rate constant $k_{s\lambda}$ should be dependent on the concentration of humic acids (equation 5.72). Attenuation effects in humic natural waters will be discussed in more detail in Chapter 12.

5.C. Theory of the Kinetics of Direct and Indirect Photoreaction in Natural Waters

Consider a well-mixed natural water system composed of dissolved humic acids and a chemical. Assume that the chemical is at a low molar concentration $[C]$ and has an absorbance <0.02. The molar absorptivity of the chemical is ϵ_λ, and the attenuation coefficient of the humic acid solution is α_λ. This solution is placed in a reaction cell of path length l, exposed area A, and volume V. Furthermore, consider that the chemical has a molecular structure such that it absorbs light at $\lambda \geq 290$ nm and can undergo both direct and indirect photoreaction. If the cell is irradiated with a light source emitting radiation from 290 nm and through the visible spectral region (e.g., a xenon lamp), then by using equations 5.23 and 5.66, the direct and indirect photoreactions can be described by the first-order differential equations

$$-(d[C]/dt)_d = k_d[C] \tag{5.79}$$

$$-(d[C]/dt)_s = k_s[C] \tag{5.80}$$

where k_d is the first-order direct photoreaction rate constant defined by equation 5.28 and k_s is the first-order indirect photoreaction rate constant defined by equation 5.67. The overall rate of disappearance of the chemical is then the sum of equations 5.79 and 5.80 and is given by

$$-(d[C]/dt)_p = (k_d + k_s)[C] \tag{5.81}$$

$$-(d[C]/dt)_p = k_p[C] \tag{5.82}$$

where

$$k_p = (k_d + k_s) \tag{5.83}$$

and the subscript p is used to denote the overall (or net) rate of photoreaction and k_p is the corresponding first-order rate constant for the net photoreaction in the units of reciprocal time. Integrating equation 5.82 in the standard manner yields

$$\ln([C_o]/[C_t]) = k_p t \tag{5.84}$$

The corresponding half-life is given by

$$t_{(1/2)} = 0.693/k_p \tag{5.85}$$

If photoreaction takes place in sunlight, then the sunlight photoreaction rate constant and half-life are k_p and $t_{1/2}$, respectively. Equations 5.84 and 5.85 are applicable to sunlight photoreaction in tubes. However, if photoreaction now takes place in sunlight in a natural water body (e.g., a river or lake), then the overall photoreaction rate constant is k_{pE} and the corresponding half-life is $t_{(1/2)E}$. The subscript E denotes that photoreaction is taking place in a natural water body in sunlight in the environment. Experimentally, the rate of photoreaction in tubes is faster than the rate of photoreaction in a natural water body. Rates of photoreaction in tubes and in natural water bodies will be discussed in Section 6.B.4.b.iii. The overall rate of photoreaction in a natural water body can be described by

$$-(d[C]/dt)_{pE} = k_{pE}[C] \tag{5.86}$$

$$\ln([C_o]/[C_t]) = k_{pE}t \tag{5.87}$$

$$k_{pE} = (k_{dE} + k_{sE}) \tag{5.88}$$

$$t_{(1/2)E} = 0.693/k_{pE} \tag{5.89}$$

Because k_{pE} is the sum of k_{dE} and k_{sE}, the theoretical discussions in the first two sections of this chapter apply to the overall, or net, environmental photoreaction rate constant. The transmission and absorption of solar radiation in a natural water body in the environment was described in Chapter 3.

The method of Zepp et al. (*9, 11*) for determining indirect photoreaction rate constants in tubes (k_s) and in a natural water body (k_{sE}) will be discussed in Chapter 12. The theory of photoreaction, specifically indirect photoreaction, will be expanded in Chapters 13–15 and a detailed method will be described for determining k_{dE}, k_{sE}, and k_{pE} by carrying out photoreaction experiments outdoors in sunlight using actinometers.

References

1. Peirce, B. O. *A Short Table of Integrals;* Ginn: New York, 1929; Integral No. 26, p 5; Series Expansion No. 759, p 89.
2. Zepp, R. G. In *The Handbook of Environmental Photochemistry*; Hutzinger, O., Ed.; Springer–Verlag: New York, 1982; Vol. 2, Part B, p 19.
3. Zepp, R. G. *Environ. Sci. Technol.* **1978**, *12*, 327.
4. Zepp, R. G.; Cline, D. M. *Environ. Sci. Technol.* **1977**, *11*, 359.
5. Mill, T.; Mabey, W. R.; Bomberger, D. C.; Chou, T–W.; Hendry, D. G.; Smith, J. H. *Laboratory Protocols for Evaluation of the Fate of Organic Chemicals in Air and Water;* U.S. Environmental Protection Agency. U.S. Government Printing Office: Washington, DC, 1982; EPA 600/3–82–002.

6. Miller, G. C.; Zepp, R. G. *Water Res.* **1979**, *13*, 453.
7. Miller, G. C.; Zepp, R. G. *Environ. Sci. Technol.* **1979**, *13*, 860.
8. Wolfe, N. L.; Zepp, R. G.; Baughman, G. L.; Fincher, R. C.; Gordon, J. A. *Chemical and Photochemical Transformation of Selected Pesticides in Aquatic Systems;* U.S. Environmental Protection Agency. U.S. Government Printing Office: Washington, DC, 1976; EPA 600/3–76–067.
9. Zepp, R. G.; Baughman, G. L.; Schlotzhauer, P. F. *Chemosphere* **1981**, *10*, 119.
10. Zepp, R. G., U.S. Environmental Protection Agency, Office of Research and Development, Athens, GA, personal communication, 1985.
11. Zepp, R. G.; Baughman, G. L.; Schlotzhauer, P. F. *Chemosphere* **1981**, *10*, 109.

Part II.A. Direct Photoreaction Test Methods—Direct Photoreaction Screening Tests: Aqueous Photoreaction Experiments in Sunlight

When assessing a chemical's photoreactivity after it enters aquatic media in the environment, test methods that provide a preliminary idea of its photoreactivity are very desirable. An example is a screening test, as described in Chapters 6–8, that yields environmentally relevant direct photoreaction rate constants and half-lives. To implement the method, solutions of test chemical in pure water in tubes are exposed to sunlight. To correct for variation in solar irradiance during photoreaction, an actinometer solution is simultaneously insolated. From the concentration versus time data, solar irradiance data, molar absorptivity data, and the reaction quantum yield of the actinometer, the reaction quantum yield of the test chemical is obtained. The reaction quantum yield and molar absorptivity data of the test chemical are then combined with solar irradiance data to calculate direct photoreaction rate constants and half-lives as a function of latitude and season of the year in the northern hemisphere.

Chapter 6 describes the theory and development of the screening tests. In developing these tests, primary emphasis has been placed on the theory of the kinetics of photochemical reactions and on the fundamental principles that guide the selection of experimental approaches. Particular emphasis is placed on the photoreaction of chemicals at very low concentrations in aqueous solution. Procedures for obtaining the kinetic data are described in detail in Chapter 7 so

1464–6/88/0097$06.00/1

that a scientist can carry out these experiments easily and precisely. Procedures are described for the extrapolation of these data to a wide variety of environmental conditions. Tables of solar irradiance are listed in Appendix A and examples are given in Chapter 8 that illustrate how to use all these procedures and the experimental data to calculate environmentally relevant rate constants and half-lives.

Chapter 6

Theory and Development of the Screening Test Methods

6.A. Introduction

A screening test method for measuring direct photoreaction rates of chemicals in aqueous solution should meet the following criteria:

1. The test method should be based on the fundamental principles of photochemistry.
2. The test method should yield quantitative data on direct photoreaction rates of chemicals in aqueous media.
3. The light source should be environmentally relevant.
4. The test method should be designed to account for transformation or chemical losses by mechanisms other than photoreaction.

For example, the experiments should be designed to eliminate or minimize biodegradation and volatilization and to account for hydrolysis in estimating chemical losses.

On the basis of these criteria, a screening level test method has been developed to obtain the requisite quantitative data in the form of direct photoreaction rate constants and half-lives. The method was developed through a comprehensive review of the literature on the experimental determination of aqueous photoreaction rate constants, the principles outlined by Zepp and Cline (*1*) and Zepp (*2*), the method developed by the U.S. Environmental Protection Agency, Office of Toxic Substances (*3–5*) and by the research carried out by SRI International (*6–10*). The following sections will describe the development of this screening level test method.

1464–6/88/0099$06.00/1

6.B. Direct Photoreaction Screening Test

6.B.1. Choice of Sunlight as the Irradiating Source

A very critical factor in designing a screening-level aqueous photoreaction test is the choice of the irradiation light source. This factor was analyzed carefully in 1978, and it was decided that sunlight should be used as the light source because of its obvious relevance as well as its low cost in comparison to artificial light sources (*3*). The irradiance of sunlight is variable because it is a function of the time of day, season, latitude, and cloud cover. As a result, many environmental scientists believed that sunlight was not a suitable irradiation source for a screening test. The variability of the sunlight irradiance prompted scientists to use polychromatic artificial light sources (e.g., black lights, sunlamps, and mercury vapor lamps) in screening studies because they are more reproducible than sunlight. The spectral distribution of polychromatic artificial light sources, however, is not comparable to that of sunlight, and large errors are introduced when extrapolating the irradiances of these light sources to the irradiance of sunlight. Only xenon lamps have a spectral distribution similar to sunlight, but they are generally too expensive for use in a screening test. Sunlight photoreaction experiments can be made reproducible if sunlight actinometers are used because actinometers monitor the sunlight irradiance during photoreaction. On the basis of research carried out by SRI International and sponsored by the U.S. Environmental Protection Agency, Office of Toxic Substances, sunlight actinometers were developed and have been incorporated in a screening test which will be described in a later section. Zepp (*11*) analyzed the problem of the choice of the light source and concluded that the best choice for the light source in a screening level test is sunlight. Ref. 11 provides details of this analysis.

6.B.2. Theoretical Discussion

The theory of aqueous photoreaction has been described in Chapters 3–5. These discussions lay the foundation for the aqueous photoreaction screening test, show how the method can be used to obtain direct photoreaction rate constants and half-lives, and indicate which experimental conditions must be standardized in order to obtain environmentally relevant aqueous photoreaction rate data. On the basis of these theoretical discussions in Chapters 3–5, it has been shown that in dilute solution (i.e., absorbance <0.02 at $\lambda \geq 290$ nm), the expressions for the kinetics of direct photoreaction of a chemical at a molar concentration $[C]$ in a pure water body in the environment are given by equations 4.11 and 4.16, which are

$$-(d[C]/dt)_{dE} = k_{dE}[C] \quad (6.1)$$

$$k_{dE} = 2.303j^{-1}\phi_d \sum_{\lambda} \epsilon_\lambda Z_\lambda \quad (6.2)$$

where ϕ_d is the reaction quantum yield of the chemical in dilute solution and is independent of λ, $k_a = \sum_{\lambda} k_{a\lambda}$, the sum of $k_{a\lambda}$ values of all wavelengths of light that are absorbed by the chemical, ϵ_λ is the molar absorptivity of the chemical, Z_λ is the solar irradiance in a water body in the environment in photons cm^{-2} s^{-1}, and j is a constant that makes the units of Z_λ compatible with the units of molar concentration. The term Z_λ corresponds to the midseason solar irradiance at midday under clear sky conditions and is a function of latitude and season. Z_λ is the solar irradiance in a natural water body at shallow depths and for depths up to 0.5 m in a pure water body. The term k_{dE} represents the direct photoreaction rate constant in the units of reciprocal time, and the subscript E is used to indicate that the chemical is undergoing direct photoreaction in a natural water body (e.g., a river or lake) in sunlight in the environment.

Integrating equation 6.1 yields

$$\ln([C_o]/[C_t]) = k_{dE}t \quad (6.3)$$

where $[C_t]$ is the molar concentration at time t during photoreaction and $[C_o]$ is the initial molar concentration. The associated half-life is

$$t_{(1/2)E} = 0.693/k_{dE} \quad (6.4)$$

Because photoreaction of chemicals in sunlight can take longer than a day, equation 6.2 has been modified and takes the form

$$k_{dE} = \phi_d \sum_{\lambda} \epsilon_\lambda L_\lambda \quad (6.5)$$

where L_λ is now the solar irradiance in a natural water body in the units millieinsteins cm^{-2} day^{-1} (*6–8*). The term L_λ is the day-averaged midseason solar irradiance at shallow depths under clear sky conditions and is a function of latitude and season of the year. The solar irradiance term L_λ will be described in more detail in Section 7.B.1.i.

All the equations just described are applicable to a chemical in pure water with an absorbance <0.02 at $\lambda \geq 290$ nm. Because this is a screening-level test, the absorbance has been set at 0.05 for $\lambda \geq 290$ nm; thus k_{dE} is in error by 11%, an acceptable error for this type of test.

In addition, all of the equations that have been described in this chapter are applicable to sunlight incident on a natural water body. The methods used in the screening test, however, measure rate constants in tubes (e.g., 13- × 100-mm) and the rate is faster in tubes. (This observation is discussed in more detail in Section 6.B.4.b.iii.) As a result, equations 6.3 and 6.4 have to be modified slightly to take the effect of measurement of rate constants in tubes into account. For simplicity, the following nomenclature will be used: For natural water bodies (e.g., a lake or river), the rate constant is designated as k_{dE} with the subscript E designating rates in natural water bodies in the environment. For tubes, the rate constant is designated as k_d. The corresponding half-lives for natural water bodies and tubes are $t_{(1/2)E}$ and $t_{(1/2)}$, respectively.

6.B.3. Tier One Test: UV–Visible Absorption Spectra—Estimation of the Maximum Aqueous Photoreaction Rate Constant and Minimum Half-Life

A simple first-tier screening test has been developed using equation 6.5. As an approximation, it is assumed that the reaction quantum yield ϕ_d is at its maximum value, 1.* Therefore, the upper limit for the direct photoreaction sunlight rate constant in aqueous solution is obtained and equation 6.5 becomes

$$(k_{dE})_{\max} = \sum_{\lambda} \epsilon_\lambda L_\lambda \tag{6.6}$$

Substituting equation 6.6 in equation 6.4 yields the lower limit for the half-life. This half-life is then given by

$$[t_{(1/2)}]_{\min} = 0.693/(k_{dE})_{\max} = 0.693/\sum_{\lambda} \epsilon_\lambda L_\lambda \tag{6.7}$$

To estimate $(k_{dE})_{\max}$, it is necessary to obtain data on the molar absorptivity of a chemical and the solar irradiance L_λ (equation 6.6). These two parameters can be obtained in the following way:

*Under certain circumstances, the reaction quantum yield can be greater than 1, for example, when a chain reaction occurs after the absorption of a quantum of light. However, this rarely occurs in dilute aqueous solution.

6.B.3.a. Determination of Molar Absorptivity

The molar absorptivity of a chemical in distilled water can be obtained from the Beer–Lambert law and the UV–vis absorption spectrum. The Beer–Lambert law can be described by equations 1.13 and 1.14

$$A_\lambda = \log_{10}(I_{o\lambda}/I_{t\lambda}) = \log_{10}(1/T_\lambda) = \epsilon_\lambda[C]l \qquad (6.8)$$

where A_λ is the absorbance at wavelength λ, l is the cell path length in cm, $I_{o\lambda}$ is the intensity of light at $l = 0$, $I_{t\lambda}$ is the intensity of light transmitted at l, T_λ is the fraction of light transmitted at l, $[C]$ is the molar concentration, and ϵ_λ is the molar absorptivity in the units M^{-1} cm^{-1}. The absorbance A_λ can be obtained from the UV–vis absorption spectrum of a chemical in aqueous solution by using standard spectroscopic procedures (*12*). These data will be used in equation 6.8 to calculate molar absorptivities for the appropriate wavelength intervals and wavelength centers where the chemical absorbs light. These wavelength intervals and wavelength centers are given in Table A-1 of Appendix A.

Because water is the environmentally relevant medium in which photoreaction takes place, the molar absorptivity should be determined in pure water. However, determination of the molar absorptivity in pure water is difficult for hydrophobic chemicals and is essentially impossible for very hydrophobic chemicals (e.g., DDT, PCBs, and polynuclear aromatic hydrocarbons). As an alternative, Zepp (*2*) recommended determination of molar absorptivities in mixtures of water and acetonitrile or methanol. Because these solvents effectively dissolve many very hydrophobic organic chemicals, have refractive indices very close to that of water and thus cause minimal solvent effects (i.e., minimum shifts and changes in the absorbance of strong absorption bands of nonpolar compounds), and do not absorb radiation at wavelengths greater than 290 nm, they are used in this screening test for determining the molar absorptivity of a chemical. However, a minimum amount of the solvent should be used to measure the molar absorptivity of a chemical.

6.B.3.b. Solar Irradiance

The solar irradiance in water, L_λ, is related to the sunlight intensity in water and is proportional to the average light flux that is available to cause photoreaction in the wavelength interval $\Delta\lambda$ centered at λ over a 24-h day. The term L_λ is the midseason solar irradiance at shallow depths for a water body under clear sky conditions and is a function of latitude and season of the year. The solar irradiance is discussed in more detail in Section 7.B.1.i and L_λ values are given in Tables A-4 to A-11, Appendix A, as a function of latitude and season of the year.

6.B.3.c. Estimation of the Maximum Direct Photoreaction Rate Constant and the Corresponding Minimum Half-Life

The maximum direct photoreaction rate constant is defined by equation 6.6. The L_λ values are chosen from Tables A-4 to A-11, Appendix A, for the appropriate latitude and season. The molar absorptivity can be determined according to the discussions in Sections 6.B.3.a and in 7.A. The product of ϵ_λ and L_λ is calculated at λ(center) for each wavelength interval Δλ where ϵ_λ has a nonzero value. The products of ϵ_λ and L_λ are summed over all wavelength intervals for λ(center) ≥ 297.5 nm. This sum, $\sum_\lambda \epsilon_\lambda L_\lambda$, is equal to $(k_{dE})_{max}$. The maximum rate constant $(k_{dE})_{max}$ is substituted into equation 6.7 to obtain the minimum half-life $[t_{(1/2)E}]_{min}$.

6.B.4. Tier Two Test: Aqueous Photoreaction in Sunlight

6.B.4.a. Phase One Test

The phase one aqueous photoreaction test is based on the method developed by the U.S. Environmental Protection Agency, Office of Toxic Substances (*3*), and uses equations 6.3 and 6.4 in the form applicable to tubes. By measuring the concentration of a chemical as a function of the time *t* in sunlight in tubes, k_{dc} can be determined with equation 6.3. This rate constant is valid for reactions in tubes and is a function of the latitude and season of the year in which the photoreaction experiments were carried out. The corresponding half-life $t_{(1/2)}$ is then calculated by using k_{dc} in equation 6.4.

6.B.4.b. Phase Two Test

The phase two aqueous sunlight photoreaction test is based on the research carried out by SRI International (*6, 8, 10*) and the test method developed by the EPA Office of Toxic Substances (*4, 5*).

6.B.4.b.i. Introduction

The phase one test method is limited because it does not measure the solar irradiance incident on the sample during photoreaction. SRI International studied the use of insolation methods for measuring solar irradiance (e.g., radiometry, photometry, and actinometry). The most suitable method for measuring solar irradiance was found to be actinometry. Radiometry is used to measure total solar flux, but because radiometry involves energy flux rather than photon flux, it is of limited value for studying selective photochemical reactions. Photometers, based on photocells and photomultipliers, have several advantages over radiometers, the most important being that they are sensitive to photon flux rather than energy flux. However, sensitive photometers that use digital pulse-counting methods, are relatively expensive and fragile.

Chemical actinometry offers advantages over instrumental methods. Actinometers conform to the geometry of the reaction vessel and measure the actinic flux under known sensitivity conditions. In general, actinometry is easy to use in aqueous photoreaction experiments that measure solar irradiance. The main complication of actinometry in outdoor experiments is that because the sun is not a monochromatic source, the absorption of light by the actinometer and chemical rarely match. Therefore, a series of actinometers would be needed to match the light absorption of a chemical in the wavelength region $\lambda \geq 290$ nm. Fortunately, the most important wavelength region for sunlight photoreaction is at 290–400 nm. High-energy sunlight is concentrated there and a large number of chemicals absorb light in this region. Therefore, the most useful actinometer is one that absorbs light in the region 290–400 nm.

6.B.4.b.ii. Development of Sunlight Actinometers

Several criteria for chemical and spectral properties should be given high priority in selecting sunlight actinometers. The ideal actinometer should

1. absorb light moderately and uniformly in the spectral region 290–400 nm but be stable in room light
2. have a reaction quantum yield, ϕ_{da}, which is wavelength-independent
3. have a reaction quantum yield that is independent of the dissolved oxygen and concentration of the actinometer

4. have an adjustable half-life ranging from several hours to 90 days
5. have photoreaction products that do not interfere with photoreaction either chemically or spectrally
6. have simple, accurate, methods for its analysis
7. be very stable in water in the dark
8. be moderately soluble in water and relatively nonvolatile
9. have little or no temperature dependence of ϕ_{da}

A review of the literature yielded several hundred examples of photochemical reactions, many of which might fulfill some of the criteria established for the ideal actinometer. To simplify the task, the review was restricted to photoreactions that are bimolecular and are therefore amenable to changes in the rate (i.e., to changes in the value of ϕ_{da}) with changes in concentration of one photoinert reactant. Nucleophilic substitutions constituted one such class. The two most successful actinometers developed to date with a variable reaction quantum yield are *p*-nitroacetophenone and pyridine and *p*-nitroanisole and pyridine (*6, 8, 10*). Both of these actinometers will be described in detail in the following two sections.

6.B.4.b.ii.(1). The *p*-Nitroanisole Actinometer

When *p*-nitroanisole (PNA) and pyridine (PYR) are irradiated in aqueous solution at $\lambda \geq 300$ nm, the following reaction occurs.

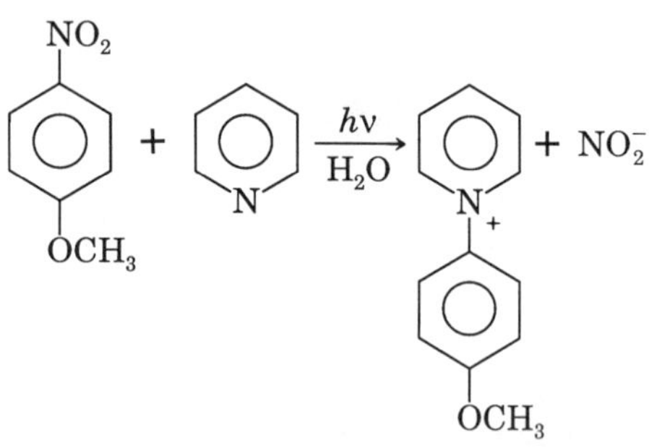

(6.9)

Photoreaction proceeds cleanly over 10 half-lives to give the products. Oxygen does not affect the reaction, nor are secondary reactions of either NO_2^- or *p*-pyridiniumanisole observed. The reaction slows slightly with increasing temperature, indicating a small negative Arrhenius activation energy (–3 kcal/mol). Thus, the rate constant for this actinometer and its reaction quantum yield are not very sensitive to changes in temperature.

The UV–vis absorption spectrum of PNA was obtained in an aqueous solution containing 1% acetonitrile (*8, 13*). The molar absorptivities were calculated by using the Beer–Lambert law as a function of wavelength (i.e., at the wavelength centers corresponding to the interval ranges listed in Table A-1 of Appendix A and at 313.0 and 366.0 nm). The results are summarized in Table 6-1. The compound PNA showed an absorption maximum at 318 nm ($\log_{10}\epsilon = 4.02$). Pyridine in aqueous solution has an absorption maximum at 250 nm ($\log_{10}\epsilon = 3.48$) and essentially no absorption at $\lambda \geq 300$ nm. Consequently, even high concentrations of pyridine in aqueous solution do not absorb significant fractions of sunlight and do not interfere with light absorption in the actinometer.

Table 6-1. Molar Absorptivities $\epsilon_{\lambda a}$ ($M^{-1}cm^{-1}$) of PNA and PNAP

λ *(center), (nm)*	*PNA*	*PNAP*
297.5	8076	3790
300.0	8480	3380
302.5	9030	3070
305.0	9350	2810
307.5	9790	2590
310.0	10000	2380
312.5	10260	2180
315.0	10350	1980
317.5	10370	1790
320.0	10330	1610
323.1	10120	1380
330.0	9210	959
340.0	7165	561
350.0	4906	357
360.0	2968	230
370.0	1618	140
380.0	749	81
390.0	288	45
400.0	87	23
410.0	15	0
420.0	7	
430.0	5	
440.0	0	
313.0	10300	2056
366.0	1990	160

NOTE: The variable $\epsilon_{\lambda a}$ corresponds to the molar absorptivity at λ(center) averaged over the interval range listed in Table A-1 of Appendix A.
PNA is *p*-nitroanisole and PNAP is *p*-nitroacetophenone.

Photoreaction experiments were carried out in the laboratory with the *p*-nitroanisole–pyridine actinometer (PNA–PYR) at 25 °C and 313 nm using a filtered medium-pressure mercury lamp. Concentration ratios of PNA:PYR of 1:10 to 1:100 were used to maintain a large excess of pyridine throughout the reaction to ensure a good approximation to pseudo-first-order kinetics. Nevertheless, water was found to compete with pyridine at the lowest pyridine concentration used. For a concentration of PNA = 7.90×10^{-6} M and [PYR] = 0.012 M, the reaction was well-behaved over two half-lives, with a half-life of 45 min. For these reaction conditions at 313 nm, $k_{da} = 1.54 \times 10^{-2}$ min^{-1} and $\phi_{da} = 5.70 \times 10^{-3}$.

Laboratory experiments at 313 and 366 nm and 25 °C indicated (1) that ϕ_{da} was constant at 313 and 366 nm, (2) the relationship of ϕ_{da} with k_{da} was linear over a 20-fold pyridine concentration, and (3) there was competitive photoreaction of PNA with water at very low pyridine concentrations. Regression analysis of the data yielded

$$\phi_{da} = 0.437[\mathrm{PYR}] + 0.000282 \qquad (6.10)$$

with a correlation coefficient of 0.9998. The PNA–PYR actinometer is extremely fast. The range in reactivity observed is from a half-life of 15 min at [PYR] = 0.01 M to a half-life of 4 h at [PYR] = 0 when exposed to bright summer sunlight.

An important variable for the PNA–PYR actinometer is the specific light absorption rate constant k_a, in days^{-1}, which is defined by

$$k_a = \sum_{\lambda} \epsilon_{\lambda a} L_{\lambda} \qquad (6.11)$$

where $\epsilon_{\lambda a}$ is the molar absorptivity of PNA and L_{λ} is the solar irradiance in water. By using the $\epsilon_{\lambda a}$ data for PNA (Table 6-1) with the L_{λ} values listed in Tables A-4 to A-11, Appendix A, k_a was calculated as a function of latitude and season in the northern hemisphere. These results are summarized in Table A-2, Appendix A. Knowledge of this parameter is essential to the phase two aqueous photoreaction experiments.

6.B.4.b.ii.(2). The *p*-Nitroacetophenone Actinometer

The PNA–PYR actinometer just described is a variable-quantum-yield actinometer that is useful for short-term sunlight experiments because its half-life ranges from approximately 15 min to 4 h in bright summer sunlight. Several aromatic ketones were screened as possible long-term actinometers (i.e., with half-lives greater than 4 h and up to several months). It was found that *p*-nitroacetophenone (PNAP) was very stable to photoreaction in pure water with a half-life of over 5 mo; however, photonucleophilic substitution by pyridine on PNAP was facile. The *p*-nitroacetophenone–pyridine actinometer (PNAP–PYR) formed the basis of a second variable-quantum-yield actinometer with a much larger range of half-lives than the PNA–PYR actinometer.

Laboratory experiments at 313 nm and 25 °C indicated that the reaction quantum yield for transformation of PNAP depended linearly on the molar concentration of pyridine, at a fixed concentration of PNAP. Regression analysis of the reaction quantum yield as a function of the molar concentration of pyridine up to 0.2 M (at a fixed concentration of PNAP equal to 1.00×10^{-5} M) gave the relationship

$$\phi_{da} = 0.0169[\mathrm{PYR}] \qquad (6.12)$$

with a correlation coefficient of 0.9986. In each experiment, PNAP exhibited good first-order kinetics to 75% transformation. At pyridine concentrations greater than 0.2 M, nonlinearity was observed (*14*). Hence, equation 6.12 is only valid to a 0.2 M pyridine concentration.

Laboratory experiments were carried out at 366 nm and 25 °C. The data indicated that ϕ_{da} was essentially the same, within experimental error, at 313 and 366 nm. Further laboratory experiments showed that the reaction slows slightly with increasing temperature indicating that the Arrhenius activation energy is low (–1.3 kcal/mole). The rate constant for this actinomer, and hence its reaction quantum yield, are not very sensitive to changes in temperature. The PNAP–PYR actinometer has a useful half-life range of greater than 4 h up to approximately 3 months when exposed to bright summer sunlight.

Attempts to characterize the photoreaction products have met with limited success. High-performance liquid chromatography (HPLC) has shown that 1 equivalent of NO_3^- is formed on photoreaction, apparently via oxidation of the first-formed NO_2^-. This result points to a process analogous to that found for PNA with pyridine in which the nitro group is displaced to form the pyridinium salt according to the equation

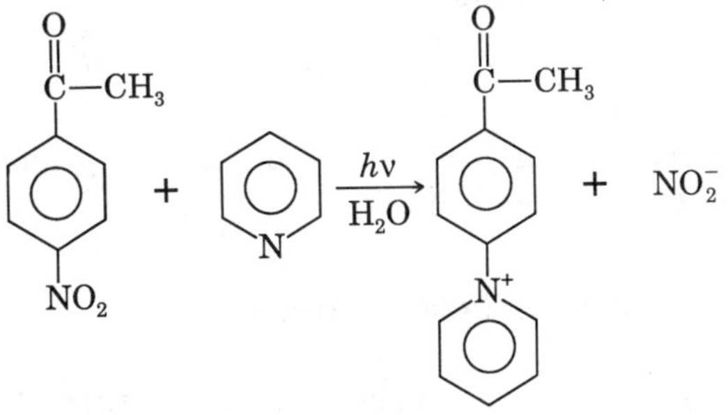

(6.13)

The pyridinium salt has not been isolated. The excellent kinetic behavior exhibited by PNAP indicates that these products do not interfere with the photochemical process of interest.

The UV–vis absorption spectrum of PNAP was obtained in an aqueous solution containing 1% acetonitrile (*8, 13*). The molar absorptivities were calculated using the Beer–Lambert law as a function of wavelength (i.e. at the wavelength centers corresponding to the interval ranges listed in Table A-1 of Appendix A and at 313.0 and 366.0 nm) and the results are summarized in Table 6-1. The chemical PNAP shows absorption of light in the spectral region $\lambda \geq 290$ nm to 410 nm with strong absorption in the region around 300 nm.

An important variable for the PNAP–PYR actinometer is the specific light absorption rate constant k_a, in days^{-1}, which is defined by equation 6.11, where $\epsilon_{\lambda a}$ is the molar absorptivity of PNAP. By using the $\epsilon_{\lambda a}$ data for PNAP from Table 6-1 with the L_λ values listed in Tables A-4 to A-11, Appendix A, k_a has been calculated as a function of latitude and season in the northern hemisphere. The results are summarized in Table A-3 of Appendix A. This parameter is essential to carrying out the phase two aqueous photoreaction experiments.

6.B.4.b.iii. Geometry of the Photoreaction Vessel

The method of Zepp and Cline (*1*) and the method of Mill et al. (*7*) are applicable to sunlight incident on a water surface such as a natural water body. The method developed in this screening test, however, measures rate constants k_d in tubes (e.g., 13- × 100-mm). Such a reaction vessel receives radiation from all directions, because of scattering of the light off the walls, as opposed to a water body (e.g., a lake) that receives sunlight only from above. It was expected, therefore, that photoreactions would be faster in tubes. This was verified experimentally by Mill et al. (*8*) and Dulin and Mill (*10*). Laboratory experiments were performed in tubes and dishes using the actinometers PNA–PYR, PNAP–PYR, and rose bengal. Dishes were used to simulate water bodies. All the data indicated that the ratio of rate constants of tubes to dishes was 2.2 ± 0.3. Thus,

$$k_d = \gamma k_{dE} \tag{6.14}$$

with $\gamma = 2.2$. The factor γ is called the geometry factor. For tubes $\gamma = 2.2$; for natural water bodies $\gamma = 1.0$.

6.B.4.b.iv. Phase Two Test Method

In phase two, a standard actinometer [either PNA–PYR or PNAP–PYR] is used to measure solar irradiance incident on the sample during photoreaction. If the half-life of a chemical is equal to or less than 4 h, as measured in the phase one experiments, then the PNA–PYR actinometer is used. If the half-life of a chemical is greater than 4 h, then the PNAP–PYR actinometer is used.

The rate constant of the actinometer, k_{da}, can be adjusted to match the approximate rate constant of a chemical, k_{dc}, by adjusting the concentration of pyridine. Because the rate constant k_{da} is a function of the reaction quantum yield of the actinometer, the rate constant can be adjusted according to equation 6.10 or 6.12. By using equations 6.5 and 6.14, the rate constant in tubes is given by

$$k_d = \gamma\phi_d \sum_\lambda \epsilon_\lambda L_\lambda \tag{6.15}$$

Therefore, the rate constants for the actinometer, k_{da}, and the chemical, k_{dc}, in tubes are

$$k_{dc} = \gamma\phi_{dc} \sum_\lambda \epsilon_{\lambda c} L_\lambda \tag{6.16}$$

$$k_{da} = \gamma\phi_{da} \sum_\lambda \epsilon_{\lambda a} L_\lambda \tag{6.17}$$

Because the phase two test method involves the simultaneous photoreaction of a chemical and actinometer in identical tubes, equation 6.17 can be divided into equation 6.16. Then solving for ϕ_{dc} yields

$$\phi_{dc} = (k_{dc}/k_{da})\{\sum_\lambda \epsilon_{\lambda a} L_\lambda / \sum_\lambda \epsilon_{\lambda c} L_\lambda\}\phi_{da} \tag{6.18}$$

The reaction quantum yield of the chemical, ϕ_{dc}, can be determined in the following way:

1. Determination of the ratio k_{dc}/k_{da}. The equations for the chemical and actinometer in tubes are found by using equation 6.3.

$$\ln([C_o]/[C_t])_c = k_{dc}t \tag{6.19}$$

$$\ln([C_o]/[C_t])_a = k_{da}t \tag{6.20}$$

Because the phase two test method involves the simultaneous photoreaction of a chemical and actinometer in identical tubes, equation 6.20 can be divided into equation 6.19. Rearranging then yields

$$\ln([C_o]/[C_t])_c = (k_{dc}/k_{da}) \ln([C_o]/[C_t])_a \tag{6.21}$$

The term $\ln([C_o]/[C_t])_c$ is a linear function of $\ln([C_o]/[C_t])_s$ with a slope given by

$$\text{slope} = (k_{dc}/k_{da}) \tag{6.22}$$

The most precise and easiest method of determining the slope is by linear regression analysis of the experimental data in equation 6.21. By simultaneously measuring the concentrations of a chemical and an actinometer as a function of time, the data can be fitted to equation 6.21 by linear regression analysis with $\ln([C_o]/[C_t])_a$ as the independent variable and $\ln([C_o]/[C_t])_c$ as the dependent variable to give the best straight line. The slope, ratio of the rate constants, is given by equation 6.22.

2. The term $\sum_{\lambda} \epsilon_{\lambda a} L_{\lambda}$ for the actinometer used in these experiments has been tabulated as a function of latitude and season of the year in Tables A-2 and A-3, Appendix A.

3. The term $\sum_{\lambda} \epsilon_{\lambda c} L_{\lambda}$ for a chemical can be obtained from the experimentally measured molar absorptivity (Section 7.A). The values of L_{λ} are listed in Tables A-4 to A-11, Appendix A, as a function of latitude and season of the year.

4. The reaction quantum yield of the actinometer ϕ_{da} can be determined from the appropriate equation 6.10 or 6.12, depending upon which actinometer was used in the phase two experiments, and the molar concentration of pyridine.

With the values of ϕ_{dc} and $\epsilon_{\lambda c}$, and the appropriate L_{λ} values, k_{dE} for a chemical can be calculated as a function of latitude and season with equation 6.5. The corresponding half-life can be calculated using k_{dE} in equation 6.4. These rate constants and half-lives are for sunlight photoreaction in natural water bodies under clear sky conditions at shallow depths.

Photoreaction experiments were carried out in the laboratory by using a photochemical "merry-go-round" reactor at 313 nm (Section 10.A.1.a.i) and in sunlight with the chemicals dibenzothiophene, nitrobenzene, and *p*-methoxyacetophenone (*8, 10*). These data were used to calculate the laboratory reaction quantum yield at 313 nm [ϕ_{dc}(313 nm)], the sunlight reaction quantum yield ϕ_{dc}, the half-life calculated from the laboratory data, [$t_{(1/2)E}$(calc)], and the half-life in sunlight, [$t_{(1/2)E}$(obs)]. These data are summarized in Table 6-2. Comparison of the corresponding reaction quantum yield data and half-life data for the laboratory experiments with those in sunlight indicated that the results are in good agreement and confirmed the phase two test method.

6.C. Applicability and Specificity of the Test Method

The phase two test method is applicable to all chemicals that have UV–vis absorptions in the range of 290–800 nm. Solar radiation reaching the earth's surface has a sharp cutoff at a wavelength of approxi-

Table 6-2. Calculated and Observed Reaction Quantum Yields and Environmental Photoreaction Half-Lives for Some Chemicals

Chemical	ϕ_{dc} *(313 nm)* *(Laboratory)*	ϕ_{dc} *(Sunlight)*	$t_{(1/2)E}$ *(calc)* *(Days)*	$t_{(1/2)E}$ *(obs)* *(Days)*
Dibenzothiophene	5.0×10^{-4}	7.3×10^{-4}	9.1	6.4
Nitrobenzene	1.8×10^{-4}	1.9×10^{-4}	19.8	19.3
p-Methoxyacetophenone	9.5×10^{-4}	13.0×10^{-4}	11.9	12.0

mately 290 nm due to its absorption by ozone (*1, 15–17*). The upper wavelength limit is set thermochemically because light of wavelength greater than 800 nm is not of sufficient energy to break the covalent chemical bonds of ground-state molecules (*18, 19*). Photoreaction does not occur without absorption of radiant energy; this is the direct consequence of the Grotthus–Draper law, the first law of photochemistry. If a chemical in aqueous solution absorbs light only at wavelengths below 290 nm it will not undergo direct photoreaction in sunlight. A few examples of chemicals that absorb light only below 290 nm and, therefore, should not be tested by this method are alkanes, alkenes, alkynes, saturated alcohols, and saturated acids. Some chemicals will absorb radiation primarily at wavelengths below 290 nm but also have an absorption tail that extends beyond 290 nm; photoreaction experiments must be carried out for these chemicals.

The phase two test method is only applicable to pure and not to technical-grade chemicals. Overestimates of molar absorptivities often occur when technical-grade substances are tested because the impurities frequently absorb in the same spectral region as the pure chemical.

The phase two test method is a screening test used to obtain the reaction quantum yield of a chemical. Because it does not require sophisticated (and expensive) photochemical equipment, the reaction quantum yield can be obtained easily and at a modest cost. The reaction quantum yield can be obtained more precisely in the laboratory with special photochemical equipment, monochromatic light and procedures described in Chapters 9–11.

The first-tier screening test can be used to estimate $(k_{dE})_{max}$ and $[t_{(1/2)E}]_{min}$. If these data indicate that aqueous photoreaction is an important process relative to other transformation processes (e.g., biodegradation, hydrolysis, and oxidation), then it is recommended that the second-tier photoreaction tests be carried out to determine environmentally relevant rate constants and half-lives in sunlight. The data obtained from the second tier photoreaction test can be used to determine k_{dE} for the test chemical as a function of latitude and season anywhere in the northern hemisphere. These rate constants are then in a form suitable for preliminary mathematical modeling for the environmental fate of a chemical.

The second-tier screening test is applicable to the direct photoreaction of chemicals in a homogeneous dilute aqueous solution with absorbance ≤ 0.05 in reaction cells at all wavelengths greater than 290 nm and at shallow depths (less than 0.5 m). The results are applicable to direct sunlight photoreaction for water bodies and clear sky conditions. In addition, these experiments are limited to the direct photoreaction of chemicals in air-saturated pure water.

This screening test has been designed to determine the molar absorptivity of a chemical, $\epsilon_{\lambda c}$, and its reaction quantum yield, ϕ_{dc}. These parameters can be used to determine environmentally relevant rate constants at low absorbance and shallow depths in pure water as a function of latitude and season. Tables of solar irradiance (Tables A-4 to A-11, Appendix A) have been included in this volume to carry out all the calculations. The method is very general and can be extended to determine the rates of photoreaction over a range of other environmental conditions by using a computer program.

Zepp and Cline (*1*) wrote a computer program (called SOLAR) to calculate the rates of photoreaction as a function of time of the day and day of the year, depth in water, attenuation coefficient of the water for natural water bodies (α_λ), the average ozone layer thickness for the season, location of interest, and latitude and season (Chapters 3 and 4). This program has recently been updated with the best available solar irradiance data and is called the GCSOLAR program (*20*). The GCSOLAR computer program is available on request (*21*).

References

1. Zepp, R. G.; Cline, D. M. *Environ. Sci. Technol.* **1977**, *11*, 359.
2. Zepp, R. G. *Environ. Sci. Technol.* **1978**, *12*, 327.
3. *Fed. Regist.* **1979**, *44*, 16271.
4. U.S. Environmental Protection Agency. *Photolysis in Aqueous Solution in Sunlight*; National Technical Information Service, U.S. Department of Commerce: Springfield, VA, 1984; EPA 560/6–84–003; NTIS PB 84–233287.
5. *Fed. Regist.* **1985**, *50*, 39285, 40 CFR 796.3700.
6. Mill, T.; Davenport, J. E.; Dulin, D. E.; Mabey, W. R.; Bawol, R. *Evaluation and Optimization of Photolysis Screening Protocols;* U.S. Environmental Protection Agency. U.S. Government Printing Office: Washington, DC, 1981; EPA 560/5–81–003.
7. Mill, T.; Mabey, W. R.; Bomberger, D. C.; Chou, T–W.; Hendry, D. G.; Smith, J. H. *Laboratory Protocols for Evaluation of the Fate of Organic Chemicals in Air and Water;* U.S. Environmental Protection Agency. U.S. Government Printing Office: Washington, DC, 1982; EPA 600/3–82–002.

8. Mill, T.; Mabey, W. R.; Winterle, J. S.; Davenport, J. E.; Barich, V. P.; Dulin, D. E.; Tse, D.; Lee, G. "Design and Validation of Screening and Detailed Methods for Environmental Processes"; Final Report; U.S. Environmental Protection Agency, Office of Toxic Substances: Washington, DC, 1982; EPA Contract No. 68–01–6325.
9. Mill, T., SRI International, Menlo Park, CA, personal communication, 1984.
10. Dulin, D.; Mill, T. *Environ. Sci. Technol.* **1982**, *16*, 815.
11. Zepp, R. G. In *The Handbook of Environmental Photochemistry*; Hutzinger, O., Ed.; Springer–Verlag: New York, 1981; Vol. 2, Part B, p 19.
12. *Fed. Regist.* **1985**, *50*, 39472, 40 CFR 796.1050.
13. Mill, T., SRI International, Menlo Park, CA, personal communication, 1985.
14. Winterle, J., SRI International, Menlo Park, CA, personal communication, 1984.
15. Leighton, P. A. *Photochemistry of Air Pollution*; Academic: New York, 1961; Chapter II.
16. Peterson, J. T. *Calculated Actinic Fluxes (290–700 nm) for Air Pollution Photochemistry Applications;* U.S. Environmental Protection Agency. U.S. Government Printing Office: Washington, DC, 1976; EPA 600/4–76–0025.
17. Demerjian, K. L.; Schere, K. L.; Peterson, J. T. *Adv. Sci. Technol.* **1980,** *10*, 389.
18. Calvert, J. G.; Pitts, J. N., Jr. *Photochemistry*; Wiley: New York, 1966; Chapter 7.
19. Benson, S. W. *Thermochemical Kinetics*, 2nd ed.; Wiley–Interscience: New York, 1976.
20. Zepp, R. G., U.S. Environmental Protection Agency, Office of Research and Development, Athens, GA, personal communication, 1982.
21. Zepp, R. G., Environmental Research Laboratory, U.S. Environmental Protection Agency, College Station Road, Athens, GA 30601.

Chapter 7

Experimental Procedures

7.A. Tier One Test: UV–Visible Absorption Spectra—Estimation of the Maximum Aqueous Photoreaction Rate Constant and Minimum Half-Life in Sunlight

Although single-beam spectrophotometers may be used, recording double-beam spectrophotometers are highly recommended. For solution spectra, resolution is not a critical factor because most absorption bands in aqueous solution are very broad. However, the ability to measure weak absorption is extremely important because even weak absorption extended over a large wavelength range from $\lambda \geq 290$ nm can contribute significantly to the rate of absorption of sunlight and therefore can significantly affect the rate of photoreaction. Consequently, the recommended double-beam spectrophotometer should have an absorbance sensitivity of approximately 0.001 at a signal-to-noise ratio of 1. A Cary 219 UV–vis spectrophotometer or an equivalent model is highly recommended.

The UV–vis absorption spectrum of a test chemical in aqueous solution can be determined by standard spectroscopic procedures (e.g., *1*). Long-path-length cells (e.g., 10-cm cells) are recommended when the absorbance is very low. The following additional procedures are also recommended.

1. For test chemicals that ionize or protonate (e.g., carboxylic acids, phenols, or amines), carry out UV–vis absorption studies at pH values at least a factor of 2 above and below the pK_a (i.e., pH $= pK_a \pm 2$). Buffer solutions are prepared at 25 °C with reagent grade chemicals and distilled water as follows:

- pH in the range 3–6* : NaH_2PO_4, HCl
- pH in the range 6–8*: KH_2PO_4, NaOH
- pH > 8: Buffers prepared as described in the CRC *Handbook of Chemistry and Physics* (*2*).

*Use the minimum concentration of buffers to attain the desired pH.

1464–6/88/0117$06.00/1

The pH of all buffer solutions is checked at 25 °C with a pH meter, and adjusted if necessary. These buffer solutions can be added to the test chemical solution until the desired pH is obtained. If these buffers are inadequate, then the pH of the test chemical solution is adjusted with 0.01 M HCl or NaOH at 25 °C.

2. The absorbance, A_λ, is measured as a function of wavelength in the range 290–800 nm in triplicate. If applicable, measurements of A_λ are made at each experimental pH. The baseline is recorded in triplicate when both the sample and reference cells are filled with blank solutions. These data will be used to calculate the molar absorptivities for the appropriate wavelength intervals and wavelength centers (Table A-1, Appendix A) where the test chemical absorbs light. The wavelength center, λ(center), is defined as the midpoint of the interval range.

The molar absorptivities of the test chemical must be carefully determined, especially in the tails of the absorption bands at $\lambda \geq 290$ nm. Large errors will be encountered in calculating photoreaction rate constants and half-lives if these measurements are not carefully carried out.

7.B. Tier Two Test: Aqueous Photoreaction in Sunlight

7.B.1. Experimental Conditions

7.B.1.a. Special Laboratory Equipment

Quartz tubes should be used for the photoreaction of chemicals with appreciable absorption below 340 nm. Chemicals that absorb appreciably at wavelengths greater than 340 nm may be tested in borosilicate tubes. Thin-walled borosilicate or quartz tubes are recommended. Disposable culture tubes (13- × 100-mm) with poly(tetrafluoroethylene) (Teflon)-lined screw caps or quartz tubes with Teflon-lined quartz or borosilicate stoppers may be used as reaction vessels. Tubes of 11-mm i.d. are recommended. For some chemicals, it may be difficult to determine the concentration of the test chemical in reaction tubes of small volume. For these chemicals, larger volume cylindrical reaction vessels are recommended, provided that the cell walls are thin and the path length of radiation through the vessel is less than 0.5 m.

7.B.1.b. Purity of Water

Reagent grade water, (i.e., water meeting the American Society for Testing Materials [ASTM] Type II A standards) or an equivalent grade, is highly recommended to minimize biodegradation. ASTM Type II A water is described in ref. 3. Air-saturated water can be easily prepared by allowing the water to equilibrate in a vessel plugged with sterile cotton.

7.B.1.c. Sterilization

To eliminate or minimize biodegradation, all glassware must be sterilized and aseptic techniques must be used in the preparation of all solutions and in carrying out all photoreaction experiments. Glassware can be sterilized in an autoclave or by any other suitable nonchemical method.

7.B.1.d. pH Effects

All photoreaction experiments should be carried out at pH levels at least 2 units above and below the pK_a (i.e., pH $= pK_a \pm 2$) for any test chemical that ionizes or protonates (e.g., carboxylic acids, phenols, and amines). Buffers described in Section 7.A. should be used.

7.B.1.e. Volatile Chemical Substances

Special care should be taken when testing a volatile chemical to ensure that the chemical is not lost as a result of volatilization during the course of the photoreaction experiment. Therefore, the reaction vessels must be effectively sealed. Disposable culture tubes, as described in Section 7.B.1.a, are recommended. Volatile compounds can be conveniently studied in culture tubes equipped with gas-tight caps (Mininert, Alltech Associates, Inc., Arlington Heights, IL). Samples can be introduced into or removed from the tubes through the septum in the cap with no loss of substrate. Alternatively, the tubes can be sealed with a torch. In addition, the reaction vessels should be filled as completely as possible to prevent volatilization to any air space.

7.B.1.f. Control Solution

Precautions must be taken to prevent loss of chemical from the reaction vessels by processes other than photoreaction. For example, biodegradation and volatilization can be eliminated or minimized by using sterile techniques and by allowing minimal air space in sealed vessels.

Hydrolysis, however, is a process that cannot be minimized by such techniques. Thus, control vessels containing test chemical that are not exposed to sunlight are required. In this way, the loss of test chemical for processes other than photoreaction may be determined and corrections may be made.

For simplicity, if the loss of the chemical in the control is small (i.e., approximately 10% or less), the following corrections should be made. For the phase one experiments, on the basis of the loss of the chemical in the controls, a first-order rate constant k_{control} can be calculated by using equation 6.3. On the basis of the disappearance of the test chemical in the photoreaction experiments, a first-order rate constant $(k_{dc})_{\mathrm{obs}}$ can be calculated. The corrected term is then given by

$$(k_{dc})_{\mathrm{corr}} = (k_{dc})_{\mathrm{obs}} - k_{\mathrm{control}} \tag{7.1}$$

The same procedure can be used to calculate a corrected rate constant for the PNAP–PYR or PNA–PYR actinometer.

For the phase two experiments, ln $([C_o]/[C_t])_{\mathrm{control}}$ is calculated at each time point on the basis of loss of the test chemical in the controls. On the basis of the disappearance of the test chemical in the photoreaction experiments, ln $([C_o]/[C_t])_{\mathrm{obs}}$ can be calculated. The corrected term at each time point is then given by

$$\ln([C_o]/[C_t])_{\mathrm{corr}} = \ln([C_o]/[C_t])_{\mathrm{obs}} - \ln([C_o]/[C_t])_{\mathrm{control}} \tag{7.2}$$

The same procedure can be applied to obtain the corrected term for the PNAP–PYR or PNA–PYR actinometer. These results are then used to calculate the ratio of the rate constants $(k_{dc}/k_{da})_{\mathrm{corr}}$ in equation 6.21.

If a loss of test chemical due to hydrolysis is 10%, then hydrolysis studies should be carried out first (*4–6*).

7.B.1.g. Absorption Spectrum as a Criterion for Performing the Aqueous Photoreaction Test

This aqueous photoreaction screening test applies to all chemicals that have UV–vis absorption maxima in the range 290–800 nm. Some test chemicals have absorption maxima significantly below 290 nm but have measureable absorption tails above the baseline in their absorption spectra at wavelengths greater than 290 nm. Photoreaction experiments must be carried out for these chemicals. The absorption spectrum of the test chemical in aqueous solution can be measured by standard spectroscopic procedures (e.g., *1*).

7.B.1.h. Sunlight Actinometers

To quantify the rate of photoreaction more precisely, the solar irradiance incident on the sample during photoreaction must be measured. Two standard variable-quantum-yield actinometers have been developed to measure the solar irradiance incident on the sample during photoreaction (*see* Section 6.B.4.b.ii.), and these actinometers have been incorporated in this test method.

According to equation 6.5, the rate constant is a function of the reaction quantum yield. Furthermore, the reaction quantum yield of the actinometers can be adjusted by varying the molar concentration of pyridine according to equations 6.10 and 6.12. Hence, by varying the pyridine concentration, the actinometer rate constant can be adjusted so that the corresponding half-life can be varied over a large range of times. With the standard PNA–PYR actinometer, the pyridine concentration can be adjusted so that the half-life can range from 15 min to 4 h (equation 6.10). With the standard PNAP–PYR actinometer, the pyridine concentration can be adjusted so that the half-life can range from 4 h to greater than 3 months (equation 6.12). The concentration of PNA or PNAP is set to be $\leq 1 \times 10^{-5}$ M. If, on the basis of the tier two, phase one experiments, the half-life of the test chemical is ≤ 4 h, the PNA–PYR actinometer is used. If the half-life of the test chemical is >4 h, the PNAP–PYR actinometer is appropriate.

The concentration required to adjust the rate of the actinometer to equal the rate of the test chemical can be derived as follows. Because the rate of photoreaction of the test chemical must be adjusted to equal that of the actinometer,

$$k_{dc} = k_{da} \tag{7.3}$$

According to equation 6.14, the rate constant for the actinometer in tubes is

$$k_{da} = 2.2k_{dE} \tag{7.4}$$

However, k_{dE} for the actinometer can be obtained from equation 6.5 and is given by

$$k_{dE} = \phi_{da} \sum_{\lambda} \epsilon_{\lambda a} L_{\lambda} \tag{7.5}$$

Because k_a for the actinometer is $\sum_{\lambda} \epsilon_{\lambda a} L_{\lambda}$ (equation 6.11), equation 7.5

becomes

$$k_{dE} = \phi_{da} k_a \tag{7.6}$$

Substituting equation 7.6 in 7.4 yields

$$k_{da} = 2.2\phi_{da} k_a \tag{7.7}$$

According to equation 6.12, for the PNAP–PYR actinometer, $\phi_{da} = 0.0169[\text{PYR}]$. Substituting this result in equation 7.7 yields

$$k_{da} = 0.0372[\text{PYR}]k_a \tag{7.8}$$

Finally, substituting equation 7.8 in 7.3 and solving for [PYR] for the PNAP–PYR actinometer yield

$$[\text{PYR}] = 26.9(k_{dc}/k_a) \tag{7.9}$$

According to equation 6.10, for the PNA–PYR actinometer, $\phi_{da} = 0.437[\text{PYR}] + 2.82 \times 10^{-4}$. Substituting this result in equation 7.7 yields

$$k_{da} = 0.961[\text{PYR}]k_a + 6.20 \times 10^{-4} k_a \tag{7.10}$$

and substituting equation 7.10 in 7.3 and solving for [PYR] for the PNA–PYR actinometer yield

$$[\text{PYR}] = 1.04(k_{dc}/k_a) - 6.45 \times 10^{-4} \tag{7.11}$$

For 1 L of solution, the mass of pyridine required to adjust the rate of the actinometer is

$$\text{mass(grams)} = \text{fw}[\text{PYR}] \tag{7.12}$$

where fw = 79.10 g/mol, the formula weight of pyridine. Because the density of pyridine is 0.9819 g/mL at 20 °C (*2*), the volume of pyridine (at 20 °C) required to equal this mass is

$$V(\text{mL}) = 80.6[\text{PYR}] \tag{7.13}$$

7.B.1.i. Solar Irradiance Data

To calculate the reaction quantum yield of a test chemical, ϕ_d, and then to calculate k_{dE} and $t_{(1/2)E}$, solar irradiance data, L_λ, must be used.

The generalized theory of solar irradiance in natural water bodies was described in Chapter 3. A discussion of the theoretical basis for L_λ data follows.

The direct photoreaction rate constant k_{dE} is a function of the solar irradiance in water (equation 6.2). The term Z_λ is the midday solar irradiance in water at shallow depths in the units photons cm^{-2} s^{-1}. The Z_λ values can be obtained from the GCSOLAR computer program of Zepp (*7*).

Because many chemicals photoreact over a period of days, the Z_λ solar irradiance values are not applicable. Therefore, to handle chemicals that photoreact over longer periods of time, L_λ solar irradiance values (day-averaged values) have been introduced (*8*). To obtain the L_λ data, Mill used the Zepp and Cline computer program (SOLAR) with $\epsilon_\lambda = 1$, $\phi_d = 1$, and the absorption coefficient α_λ for pure water. The L_λ values were obtained for the midseasons of winter, spring, summer, and fall for the latitudes 20° to 50° N. The annual midseason dates for the seasons of the year can be obtained from the *Astronomical Almanac* (*9*), which is published annually. The values of L_λ are in the units of millieinsteins cm^{-2} day^{-1}. The L_λ values are for clear sky conditions, for water bodies with shallow depths, and for chemicals with very weak absorbance in pure water.

The Zepp and Cline computer program has been updated recently and is called GCSOLAR (*10*). Mill et al. (*11*) have used this program with $\epsilon_\lambda = 1$, $\phi_d = 1$, and the absorption coefficient α_λ for pure water to obtain updated L_λ values as a function of latitude and season. These L_λ values are summarized in Tables A-4 to A-11 of Appendix A for latitudes 0° to 70° N. in 10° increments, and will be used extensively to calculate environmentally relevant direct photoreaction rate constants and half-lives.

The solar irradiance data, Z_λ, are midday, midseason values in the units of photons cm^{-2} s^{-1}. The solar irradiance data, L_λ, are day-averaged, midseason values in the units of millieinsteins cm^{-2} day^{-1}. The L_λ data are used in equation 6.5 to calculate day-averaged, midseason, direct photoreaction rate constants. The Z_λ data are used in equation 6.2 to calculate midday, midseason, direct photoreaction rate constants.

7.B.1.j. Geometry of the Photoreaction Vessel

The methods of Zepp and Cline (*12*) and Mill et al. (*8, 13*) are applicable to sunlight incident on a water surface such as a natural water body. However, the screening-level test method measures rate constants in tubes (e.g., 13- × 100-mm). A tube is exposed to radiation from all directions, because light is scattered off the walls, in contrast to a water body, such as a lake, which receives sunlight only from above. It is expected, therefore, that photoreactions would be faster in tubes. This expectation was verified experimentally (*13, 14*) and has been discussed in detail in Section 6.B.4.iii. For these experiments, tubes are the simplest reaction vessels to use because they can be sealed easily to minimize or eliminate volatilization and can be handled easily. Hence, the use of tubes as reaction vessels is recommended and the method has been modified to take into account the increased rate in tubes (equation 6.14).

7.B.1.k. Chemical Analysis of Solutions

To determine the concentrations of a test chemical in solution, an analytical method should be selected that is most applicable to the analysis of that specific substance. Chromatographic methods are generally recommended because of their specificity in the analysis of the parent chemical without interference from impurities. Whenever practicable, the chosen analytical method should have a precision of at least ±5%.

The *p*-nitroacetophenone in the chemical actinometer solution is conveniently analyzed by HPLC with a 30-cm C_{18} reverse-phase column and a UV detector set at 280 nm. The mobile phase is 2.5 vol % acetic acid, 50 vol % acetonitrile, and 47.5 vol % water, which is passed through the column at a flow rate of 2 mL/min (*14*).

The *p*-nitroanisole in the chemical actinometer solution is conveniently analyzed by HPLC using a 30-cm C_{18} reverse-phase column and a UV detector set at 280 nm. The mobile phase is 50 vol % acetonitrile and 50 vol % water which is passed through the column at a flow rate of 2 mL/min (*14*).

7.B.2. Preparations

7.B.2.a. Preparation of Test Chemical Solution

Homogeneous solutions of the test chemical should be prepared at less than one-half its solubility in water at a concentration such that the absorbance is ≤0.05 in the photoreaction vessel at wavelengths greater than 290 nm. For very hydrophobic chemicals, it is difficult and time-consuming to prepare aqueous solutions. To facilitate the preparation of aqueous solutions containing very hydrophobic chemicals and to allow for easier analytical procedures, the following procedure may be used to aid in the dissolution of the chemical. The pure chemical is dissolved in reagent grade acetonitrile. Pure water is added to an aliquot of the acetonitrile solution, as described in Section 7.B.1, or buffer solution, as described in Section 7.A., step 1, for chemicals that ionize or protonate. An amount of 1 vol % acetonitrile should not be exceeded in the final solution. The reaction solution is placed in the appropriate photoreaction reaction tubes as described in Section 7.B.1.a.

7.B.2.b. Preparation of Buffer Solutions

Buffer solutions should be prepared according to the procedures outlined in Section 7.A., step 1, with reagent grade chemicals and pure water.

7.B.2.c. Preparation of Actinometer Solutions

The appropriate actinometer solution to be prepared for the tier two, phase two test is a function of the half-life of the test chemical as determined in the tier two, phase one test. If the half-life of the test chemical is greater than 4 h, then the PNAP–PYR actinometer solution should be prepared. If the half-life of the test chemical is ≤4 h, then the PNA–PYR actinometer solution should be prepared.

7.B.2.c.i. The PNAP–PYR Actinometer Solution

By using the test chemical photoreaction rate constant, k_{dc}, determined in tier two, phase one, and the variable k_a listed in Table A-3, Appendix A, the molar concentration of pyridine needed to adjust the rate of disappearance of PNAP to match the rate of disappearance of the test chemical can be obtained from equation 7.9. The variable $k_a(= \sum_{\lambda} \epsilon_{a\lambda} L_{\lambda})$ is equal to the day-averaged rate constant for sunlight absorption by PNAP, which changes with season and latitude. The value of k_a is selected from Table A-3 for the season nearest the midexperiment date of the tier two, phase one, studies and the decadic latitude nearest that of the experimental site.

After the molar concentration of pyridine has been determined, an actinometer solution can be prepared by using the following procedure:

1. 0.0165 g of PNAP is dissolved in 10 mL of 0.01 M acetonitrile
2. 1 mL of this solution plus the the mass in grams or the volume (V) of pyridine at 20 °C, obtained from equation 7.12 or 7.13, is added to a 1-L volumetric flask.

3. The volumetric flask is filled with pure water to give 1 L of solution and is shaken vigorously to ensure homogeneity

The resultant concentration of PNAP is 1.00 × 10^{-5} M. The PNAP–PYR solution should be wrapped with aluminum foil and kept from bright light.

7.B.2.C.ii. The PNA–PYR Actinometer Solution

By using the test-chemical photoreaction rate constant, k_{dc}, determined in tier two, phase one, and the variable k_a listed in Table A-2, Appendix A, the molar concentration of pyridine needed to adjust the rate of disappearance of PNA to match the rate of disappearance of the test chemical can be obtained from equation 7.11. The value of k_a is selected from Table A-2 for the season nearest the midexperiment date of the tier two, phase one, studies and the decadic latitude nearest that of the experimental site.

After the molar concentration of pyridine has been determined, an actinometer solution can be prepared as follows:

1. 0.0153 g of PNA is dissolved in 10 mL of 0.01 M acetonitrile
2. 1 mL of this solution is added to a 1-L volumetric flask
3. The mass in grams or the volume (V) of pyridine at 20 °C obtained from equation 7.12 or 7.13 is added to the volumetric flask
4. The volumetric flask is filled with pure water to give 1 L of solution and is shaken vigorously to ensure homogeneity

The resulting concentration of PNA is 1.00 × 10^{-5} M. The PNA–PYR solution should be wrapped with aluminum foil and kept from bright light.

7.B.3. Performance of the Tests

7.B.3.a. Phase One Experiments

For all experiments, an aqueous solution of the test chemical is prepared, as described in Section 7.B.2.a, along with a sufficient number of samples in quartz or borosilicate glass tubes to perform all the required tests. The tubes are filled as completely as possible and sealed. A sufficient number of control samples are prepared in the absence of ultraviolet light and light is totally excluded by wrapping the tubes with aluminum foil or by other suitable methods. The samples are then analyzed after each time point t to measure the loss of chemical in the absence of light. The samples, including the controls, are placed outdoors in an area free of shade and reflections of sunlight from windows and buildings. The samples are placed on a black, nonreflective background inclined at approximately 30° from the horizontal with the upper end pointing due north (in the northern hemisphere). The photoreaction experiments must be conducted during a frost-free time of year (i.e., May–September in the northern hemisphere—weather permitting), and all the experiments are started initially at noon (1200). The date and time the experiment was begun, the date and time completed, the time of sunrise and sunset on all days when photoreaction experiments were performed, the times exposure was stopped and restarted for intermittent exposure, the weather conditions during the period, and the latitude of the site, are all recorded. For chemicals that ionize or protonate, photoreaction experiments are carried out at the required pH levels as described in Section 7.B.1.

If a significant loss of test chemical has occurred in the control samples, the cause should be determined and eliminated or minimized. If hydrolysis is found to be significant, hydrolysis studies should be carried out first (*4–6*).

One of the following procedures is used, depending on how fast the chemical substance photoreacts.

7.B.3.a.i. Procedure 1

If the test chemical transforms 50–80% within 28 days, the concentration of the chemical should be measured, in triplicate, at time t = 0 and periodically at noon (1200) until at least 50% of the chemical has been consumed. To save time, the sampling times can be estimated while the photoreaction experiments are in progress. The concentration of test chemical is determined from three freshly opened reaction tubes for each time point. The concentration in each of the three control solutions is determined for each time point and as soon as the photoreaction experiments are completed.

7.B.3.a.ii. Procedure 2

If the test chemical transforms in the range of 20–50% in 28 days, the concentration of the chemical is determined, in triplicate, at time t =

0. The concentration of the chemical in the three separate reaction tubes and the three control tubes is determined after 28 days of photoreaction.

7.B.3.a.iii. Procedure 3

For test chemicals that transform in sunlight 50–80% within 1 or 2 days, the samples are placed outside at noon (1200) and three samples are analyzed for concentration of the chemical at $t = 0$, three freshly opened reaction tubes at are analyzed at noon (1200) the next day, and again, three freshly opened reaction tubes at are analyzed at noon (1200) the second day. The concentration of the test chemical in each of the three control solutions is determined after the first day of photoreaction and as soon as the photoreaction experiments have been completed on the second day.

7.B.3.a.iv. Analytical Methodology

An analytical method is selected that is most applicable to the analysis of the specific chemical being tested (see Section 7.B.1.k).

7.B.3.b. Phase Two Experiments

By using the test chemical photoreaction rate constant, k_{dc}, determined in tier two, phase one, the appropriate actinometer solution is prepared, as described on Section 7.B.2, as well as a sufficient number of samples in quartz tubes to perform all the required tests. The tubes are filled as completely as possible, sealed, and covered with aluminum foil immediately after preparation. An aqueous solution of test chemical is prepared, as described in Section 7.B.2.a, as well as a sufficient number of samples in quartz or borosilicate tubes to perform all the required tests. These tubes are filled as completely as possible, sealed, and covered with aluminum foil as soon as possible after preparation. All the samples are placed outdoors in an area free of shade and reflections of sunlight from windows and buildings. The samples are placed on a black, nonreflective, background inclined at approximately 30° from the horizontal with the upper end pointing due north (in the northern hemisphere). The foil is removed from all samples except the test chemical control solutions and the actinometer control solutions at noon (1200). On the basis of the results of the phase one experiments, the concentration of test chemical and actinometer (PNAP or PNA), are determined, in triplicate, at time $t = 0$ and periodically thereafter (at least five data points at approximately equal time intervals). The concentration of PNAP or PNA in the three actinometer control solutions is determined as well as concentration of test chemical in the three control solutions for each time point.

An analytical method should be selected that is most applicable to the analysis of the specific chemical tested (see Section 7.B.1.k), and follow the procedure given in 7.B.1.k for the analysis of PNAP or PNA.

References

1. *Fed. Regist.* **1985**, *50*, 39472, 40 CFR 796.1050.
2. *Handbook of Chemistry and Physics*; Weast, R. C., Ed.; 57th ed.; Chemical Rubber Company: Cleveland, OH, 1976–1977.
3. American Society for Testing and Materials. "Standard Specification for Reagent Water"; *Annual Book of ASTM Standards*; American Society for Testing Materials: Philadelphia, PA, 1983.
4. U.S. Environmental Protection Agency. "Hydrolysis as a Function of pH", CG, CS–5000; National Technical Information Service, U.S. Department of Commerce: Springfield, VA, 1982; EPA 560/6–82–003; NTIS PB82–233008.
5. *Fed. Regist.* **1985**, *50*, 39283, 40 CFR 796.3500.
6. *Fed. Regist.* **1987**, 36334, 40 CFR 796.3510.
7. Zepp, R. G. Environmental Research Laboratory, U.S. Environmental Protection Agency, College Station Road, Athens, GA 30601.
8. Mill, T.; Mabey, W. R.; Bomberger, D. C.; Chou, T–W.; Hendry, D. G.; Smith, J. H. *Laboratory Protocols for Evaluation of the Fate of Organic Chemicals in Air and Water;* U.S. Environmental Protection Agency. U.S. Government Printing Office: Washington, DC, 1982; EPA 600/3–82–002.
9. *Astronomical Almanac*; Nautical Almanac Office. U.S. Naval Observatory: Washington, DC, 1987.
10. Zepp, R. G., U.S. Environmental Protection Agency, Office of Research and Development, Athens, GA, personal communication, 1982.
11. Mill, T., SRI International, Menlo Park, CA, personal communication, 1984.
12. Zepp, R. G.; Cline, D. M. *Environ. Sci. Technol.* **1977**, *12*, 327.
13. Mill, T.; Mabey, W. R.; Winterle, J. S.; Davenport, J. E.; Barich, V. P.; Dulin, D. E.; Tse, D.; Lee, G. "Design and Validation of Screening and Detailed Methods for Environmental Processes"; Final Report; U.S. Environmental Protection Agency, Office of Toxic Substances: Washington, DC, 1982; EPA Contract No. 68–01–6325.
14. Dulin, D.; Mill, T. *Environ. Sci. Technol.* **1982**, *16*, 815.

Chapter 8

Illustrative Examples

8.A. Tier One Test: UV–Visible Absorption Spectra—Estimation of Maximum Photoreaction Rate Constant and Minimum Half-Life in Sunlight

8.A.1. Illustrative Example 1

A neutral organic chemical A was dissolved in water at a concentration of 1.00×10^{-3} M. The UV–vis absorption spectrum was obtained in a 10.0-cm quartz absorption cell and no absorbance was detected above the baseline in the region 290 nm and higher (i.e., $A_{\lambda c} = 0$ for $\lambda \geq 290$ nm). Because $A_{\lambda c} = 0$, $\epsilon_{\lambda c} = 0$ (equation 6.8). Using this result in equation 6.6, yields $(k_{dE})_{max} = 0$. This result indicates that no direct photoreaction can take place in sunlight at any latitude or during any season of the year. This example illustrates the principle of the Grotthus–Draper law, the first law of photochemistry: for direct photoreaction to take place in sunlight, the chemical must absorb sunlight in the wavelength region $\lambda \geq 290$ nm.

8.A.2. Illustrative Example 2

Consider a plant located in Columbus, GA, on the Chattahoochee River that produces an organic chemical B that is neither an acid nor a base. The waste effluent passes through primary and secondary treatment plants and is then discharged directly into the river. The plant produces chemical B continuously every day of the year. The plant is located at 32.5° N. latitude. Estimate the maximum direct sunlight photoreaction rate constant and the corresponding minimum half-life for this chemical in the river for the winter and summer seasons under clear skies.

1464–6/88/0131$06.00/1

8.A.2.a. Laboratory Experiments and Data

The water solubility of chemical B is 1.00×10^{-3} M at 25 °C. Chemical B was dissolved directly in water and a 1.00×10^{-4} M solution was prepared at 25 °C. The UV–vis absorption spectrum was obtained in triplicate according to the tier one procedure in a 10.0-cm quartz absorption cell. By using the wavelength interval range (from Table A-1, Appendix A), the average absorbance of the triplicate runs at λ(center) was obtained and the results are summarized in Table 8-1.

8.A.2.b. Calculations

Substituting the data $l = 10.0$ cm and $c = 10^{-4}$M in equation 6.8, the average molar absorptivity, A_λ (in $M^{-1}cm^{-1}$), is

$$\epsilon_{\lambda c} = 10^3 A_{\lambda c} \tag{8.1}$$

From the average $A_{\lambda c}$ value at λ(center) in Table 8-1, the average molar absorptivity can be obtained from equation 8.1; the results are summarized in Table 8-1. The plant is located at 32.5° N. latitude; therefore, the closest L_λ values are at 30° N. latitude. These values are obtained from Table A-7, Appendix A, and are summarized in Table 8-2 for the summer and winter seasons. By using the data from Table 8-2 and equations 6.6 and 6.7, the following results are obtained.

For summer,

$$(k_{dE})_{\max} = \sum_{\lambda} \epsilon_{\lambda c} L_\lambda = 13.0 \text{ day}^{-1} \tag{8.2}$$

$$[t_{(1/2)E}]_{\min} = 0.053 \text{ day} \tag{8.3}$$

Table 8-1. Spectral Data for Chemical B in Illustrative Example 2

λ *(center)(nm)*	$A_{\lambda c}$	$\epsilon_{\lambda c}(M^{-1}cm^{-1})$
297.5	1.684	1684
300.0	1.434	1434
302.5	1.221	1221
305.0	0.919	919
307.5	0.742	742
310.0	0.208	208
312.5	0.138	138
315.0	0.094	94
317.5	0.057	57
320.0	0.009	9
323.1	0.002	2
330.0	0.000	0

Table 8-2. Light Absorption Data for Chemical B in Illustrative Example 2

	Summer		*Winter*	
λ*(center)(nm)*	L_λ^a	$\epsilon_{\lambda c}L_\lambda(day^{-1})$	L_λ^a	$\epsilon_{\lambda c}L_\lambda(day^{-1})$
297.5	1.09 (-4)	0.18	6.78 (-6)	0.01
300.0	4.11 (-4)	0.59	4.23 (-5)	0.06
302.5	1.14 (-3)	1.39	1.71 (-4)	0.21
305.0	2.46 (-3)	2.26	4.95 (-4)	0.45
307.5	4.45 (-3)	3.30	1.11 (-3)	0.82
310.0	7.02 (-3)	1.46	2.04 (-3)	0.42
312.5	1.00 (-2)	1.38	3.26 (-3)	0.45
315.0	1.32 (-2)	1.24	4.69 (-3)	0.44
317.5	1.64 (-2)	0.94	6.21 (-3)	0.35
320.0	1.95 (-2)	0.18	7.76 (-3)	0.07
323.1	3.46 (-2)	0.07	1.43 (-2)	0.03
330.0	1.18 (-1)	0.00	5.17 (-2)	0.00
	$\Sigma_\lambda\ \epsilon_{\lambda c}L_\lambda$ = 12.99		$\Sigma_\lambda\ \epsilon_{\lambda c}L_\lambda$ = 3.31	

[a]The units of L_λ are in millieinsteins cm^{-1}day^{-1}. The number in parentheses is the power of 10 by which the first number is multiplied.

For winter,

$$(k_{dE})_{\max} = \sum_{\lambda} \epsilon_{\lambda c}L_\lambda = 3.31 \text{ day}^{-1} \tag{8.4}$$

$$[t_{(\frac{1}{2})E}]_{\min} = 0.21 \text{ day} \tag{8.5}$$

Because the chemical may transform rapidly in the summer and winter seasons, tier two experiments must be performed to more accurately define direct photoreaction rates in the river as a function of the season.

8.B. Tier Two, Phase One: Aqueous Photoreaction in Sunlight—Illustrative Example 3

Consider the same scenario as described in Illustrative Example 2. Using the tier two, phase one procedure, experiments are performed to estimate the rate of direct photoreaction and half-life in the river in the spring at the latitude of the site of the plant.

8.B.1. Photoreaction Experiments—Conditions

Because chemical B absorbs appreciably below 340 nm, 11-mm i.d. quartz tubes were used (this size tube has an approximate path length of 1 cm). Chemical B was dissolved directly in pure water and a 1.0×10^{-5} M solution was prepare at 25 °C. Because the water solubility is 1.00×10^{-3} M at 25 °C, this sample solution was well below one-half its water solubility. The UV–vis spectrum of this solution in a 1-cm absorption cell indicated that $A_{\lambda c}$ was less than 0.05 at $\lambda \geq 290$ nm. Hence, first-order kinetics are applicable.

A series of quartz tubes were filled with this aqueous solution and sealed, and photoreaction experiments were carried out in sunlight according to the appropriate procedure described in Section 7.B.3. The experiments were started at noon (1200) on May 8, 1982. The weather conditions are summarized as follows for this period, and the concentration data given are the means of triplicate determinations.

1. May 2, 1982: at $t = 0$ (noon, 1200), $[C_o] = 1.0 \times 10^{-5}$ M
2. May 2, 1982: noon to sunset—clear and sunny
3. May 3, 1982: noon (1200), $[C_t] = 0.84 \times 10^{-5}$ M; control, $[C_t] = 1.0 \times 10^{-5}$ M
4. May 3, 1982: at 1400 the weather conditions were cloudy with rain. The rain and cloudy weather continued until 2200
5. From sunrise, May 4, 1982, through 1200 May 8, 1982, the weather was clear and sunny. At 1200, May 8, 1982, analysis of the samples gave an average concentration of $[C_t] = 0.40 \times 10^{-5}$ M. Because 60% of chemical B transformed, the photoreaction experiments were terminated and the control samples were analyzed. The average concentration of the control samples was 0.99×10^{-5} M, essentially the same as $[C_o]$. Hence, no adventitious processes occurred and the loss of chemical was due only to sunlight photoreaction.

8.B.2. Data and Calculations

Listed in Table 8-3 are the times of sunrise and sunset along with the total number of hours of sunlight for the dates sunlight photoreaction experiments were carried out.

Table 8-4 summarizes the dates photoreacted, the times exposed to sunlight, the total sunlight photoreaction time for each date in days, and the total number of days of sunlight photoreaction. k_{dc} and $t_{(1/2)}$ can be calculated using equations 6.3 and 6.4 as follows: Substituting $t = 5.6$ days; $[C_o] = 1.0 \times 10^{-5}$ and $[C_t] = 0.40 \times 10^{-5}$ into

$$k_{dc} = (1/t) \ln ([C_o]/[C_t]) = (1/5.6) \ln (1.0 \times 10^{-5}/0.40 \times 10^{-5}) \quad (8.6)$$

Table 8-3. Summary of Times for Sunrise and Sunset for May 2–8, 1982, in Illustrative Example 3

Date (1982)	*Sunrise*	*Sunset*	*Total Sunlight Hours*
5/2	0600	2010	14.2
5/3	0559	2011	14.2
5/4	0558	2012	14.2
5/5	0557	2013	14.3
5/6	0556	2014	14.3
5/7	0555	2015	14.3
5/8	0554	2016	14.4

yields

$$k_{dc} = 0.16 \text{ days}^{-1} \tag{8.7}$$

and

$$t_{½} = 0.693/0.16 \text{ days}^{-1} = 4.3 \text{ days} \tag{8.8}$$

The rate constant for direct photoreaction of chemical B in tubes in pure water is 0.16 days^{-1} and the corresponding half-life is 4.3 days for the period of photoreaction May 2–8, 1982, at 32.5° N. latitude. By using equation 6.14, the direct photoreaction rate constant (k_{dE}) in the river is calculated to be 0.073 days^{-1} and the corresponding half-life $[t_{(½)E}]_{min}$ is 9.5 days.

8.C. Tier Two, Phase Two: Aqueous Photoreaction in Sunlight—Illustrative Example 4

Consider the same scenario as described in Illustrative Examples 2 and 3. Using the tier two, phase two procedure, experiments are per-

Table 8-4. Experimental Results in Illustrative Example 3

Date	*Times Photoreacted*	*Hours Photoreacted/ Total Hours of Sunlight*	*Sunlight Photoreaction Time for Each Date (days)*
5/2	1200 to 2010	(8.2/14.2)	0.58
5/3	0559 to 1200	(6.0/14.2)	0.42
5/3	1200 to 1400	(2.0/14.2)	0.14
5/4	0558 to 2012		1.00
5/5	0557 to 2013		1.00
5/6	0556 to 2014		1.00
5/7	0555 to 2015		1.00
5/8	0554 to 1200	(6.1/14.4)	0.42
			Total 5.56 days

formed to determine the sunlight reaction quantum yield of the chemical and to estimate the rate constants for direct photoreaction in aqueous solution and the half-lives in the river at Columbus, GA, for clear sky conditions in the summer and winter seasons.

8.C.1. Photoreaction Experiments—Conditions

The sunlight photoreaction experiments were carried out in the beginning of May, 1982 at 32.5° N. latitude. The results from the tier two, phase one experiments indicated that k_{dc} for the test chemical was 0.16 days^{-1} and the half-life was 4.3 days in tubes. Thus, the PNAP–PYR actinometer was chosen to carry out the tier two, phase two experiments. Because the experiments were carried out in early May at 32.5° N. latitude, the value of k_a for PNAP was chosen from Table A-3, Appendix A, for the spring season and at 30° N. latitude. The value chosen was 483 days^{-1}. Using equation 7.9, the molar concentration of pyridine required to adjust the actinometer rate to match the disappearance of the test chemical is [PYR] = 26.9(0.16/483) = 8.91×10^{-3} M. With this concentration of pyridine, a PNAP–PYR actinometer solution was prepared according to the procedure described in Section 7.B.2.c.i. The quantum yield for this actinometer is calculated using equation 6.12.

$$\phi_{da} = 0.0169[\text{PYR}] = 0.0169(8.91 \times 10^{-3}) = 1.51 \times 10^{-4} \quad (8.9)$$

Procedures for tier two, phase two experiments were followed, and sunlight experiments were initiated at 1200 on May 9, 1982. The mean initial concentration of test chemical was 1.0×10^{-5} M, and the mean initial concentration of PNAP was 1.0×10^{-5} M. Samples of the chemical, actinometer, and the controls were analyzed in triplicate periodically at 1200 on May 10, 11, 13, 15, and 16. On May 16, the photoreaction experiments were terminated. The mean of all samples are summarized in Table 8-5.

Table 8-5. Concentrations of All Samples in Illustrative Example 4

Date	*Conc. of Chem. (M)*	*Conc. of Act. (M)*	*Conc. of Chem. Control (M)*	*Conc. of Act. Control (M)*
May 9	1.0×10^{-5}	1.0×10^{-5}	1.0×10^{-5}	1.0×10^{-5}
May 10	0.82×10^{-5}	0.86×10^{-5}	1.0×10^{-5}	1.0×10^{-5}
May 11	0.65×10^{-5}	0.71×10^{-5}	1.0×10^{-5}	0.99×10^{-5}
May 13	0.44×10^{-5}	0.52×10^{-5}	0.99×10^{-5}	0.99×10^{-5}
May 15	0.30×10^{-5}	0.38×10^{-5}	0.99×10^{-5}	1.0×10^{-5}
May 16	0.23×10^{-5}	0.30×10^{-5}	1.0×10^{-5}	0.99×10^{-5}

Because no significant loss of PNAP or test chemical was observed in the control samples, no adventitious processes occurred and the loss of test chemical and PNAP was due only to sunlight photoreaction.

8.C.2. Calculations

By using the data in Table 8-5, ln $([C_o]/[C_t])$ for the test chemical and actinometer can be calculated. The results are summarized in Table 8-6.

By using all the data in equation 6.21 (including the time point $t = 0$) and linear regression analysis with ln $([C_o]/[C_t])_a$ as the independent variable and ln $([C_o]/[C_t])_c$ as the dependent variable, the slope is 1.23 with a correlation coefficient of 1.00. Therefore $(k_{dc}/k_{da}) = 1.23$.

By using the molar absorptivities obtained in Illustrative Example 2, and the L_λ values for spring at 30° N. latitude, Table A-7, Appendix A, and the data in Table 8-7, the value of $\sum_\lambda \epsilon_{\lambda c} L_\lambda$ can be calculated and is

$$\sum_\lambda \epsilon_{\lambda c} L_\lambda = 9.96 \text{ days} \tag{8.10}$$

For this experiment, $k_a (= \sum_\lambda \epsilon_{\lambda a} L_\lambda)$ is 483 days^{-1} (Appendix A, Table A-3). All the pertinent data are summarized as follows:

$$k_{dc}/k_{da} = 1.23$$

$$\sum_\lambda \epsilon_{\lambda c} L_\lambda = 9.96 \text{ days}^{-1}$$

$$\sum_\lambda \epsilon_{\lambda a} L_\lambda = 483 \text{ days}^{-1}$$

$$\phi_{da} = 1.51 \times 10^{-4}$$

Table 8-6. Molar Concentrations and Direct Photoreaction Rate Constants for the Test Chemical and Actinometer in Illustrative Example 4

	Test Chemical		*Actinometer*	
t (*days*)	$[C_t] \times 10^5 (M)$	*ln* $([C_o]/[C_t])_c$	$[C_t] \times 10^5 (M)$	*ln* $([C_o]/[C_t])$
0	1.0	0.000	1.0	0.000
1	0.82	0.198	0.86	0.151
2	0.65	0.431	0.71	0.342
4	0.44	0.821	0.52	0.654
6	0.30	1.20	0.38	0.968
7	0.23	1.47	0.30	1.20

Table 8-7. Data for Calculating $\epsilon_{\lambda c}L_\lambda$ Values for Illustrative Example 4

λ(*center*), *nm*	$\epsilon_{\lambda c}$ $(M^{-1}cm^{-1})$	L_λ^a	$\epsilon_{\lambda c}L_\lambda$ $(days^{-1})$
297.5	1684	5.73 (-5)	0.10
300.0	1434	2.50 (-4)	0.36
302.5	1221	7.65 (-4)	0.93
305.0	919	1.79 (-3)	1.65
307.5	742	3.43 (-3)	2.55
310.0	208	5.64 (-3)	1.17
312.5	138	8.27 (-3)	1.14
315.0	94	1.12 (-2)	1.05
317.5	57	1.41 (-2)	0.80
320.0	9	1.70 (-2)	0.15
323.1	2	3.04 (-2)	0.06
330.0	0	1.05 (-1)	0.00

[a]The values of L_λ correspond to spring at 30° N in the units millieinsteins cm^{-2} day^{-1}. The number in parentheses is the power of 10 by which the first number is multiplied.

Substituting these results into equation 6.18 yields

$$\phi_{dc} = (1.23)(483/9.96)(1.51 \times 10^{-4}) \tag{8.15}$$

$$\phi_{dc} = 9.01 \times 10^{-3} \tag{8.16}$$

Therefore, the reaction quantum yield of the test chemical is 9.0×10^{-3}.

The rate constants and half-lives for direct photoreaction of the test chemical in the river under clear sky conditions for the winter and summer seasons can be calculated as follows. The values of $\sum_\lambda \epsilon_{\lambda c}L_\lambda$ have been calculated from Illustrative Example 2, equations 8.2 and 8.4. For summer $\sum_\lambda \epsilon_{\lambda c}L_\lambda = 13.0$ days^{-1}; for winter $\sum_\lambda \epsilon_{\lambda c}L_\lambda = 3.31$ days^{-1}. The reaction quantum yield for the test chemical is 9.0×10^{-3}. Using these data in equation 6.5 yields

$$\text{for summer: } k_{dE} = 9.0 \times 10^{-3}\,(13.0) = 0.117 \text{ days}^{-1}$$

$$\text{for winter: } k_{dE} = 9.0 \times 10^{-3}\,(3.31) = 0.0298 \text{ days}^{-1}$$

thus, k_{dE} for the summer and winter seasons under clear sky conditions in Columbus, GA, is 0.12 and 0.030 days^{-1}, respectively.

These values can be substituted into equation 6.4 to obtain the corresponding half-lives for these two seasons:

$$\text{for summer: } t_{(1/2)E} = (0.693/0.12) = 5.8 \text{ days}$$

$$\text{for winter: } t_{(1/2)E} = (0.693/0.030) = 23 \text{ days}$$

Part II.B. Direct Photoreaction Test Methods—Laboratory Determination of the Direct Photoreaction Quantum Yield in Aqueous Solution and Sunlight Photoreaction Rates

Chapters 9–11 continue the assessment of the photoreactivity of a chemical when it enters aquatic media in the environment by describing carefully controlled laboratory photoreaction procedures that yield a precise value of the reaction quantum yield of a chemical. Chapter 9 describes the theory and development of these test procedures. In developing these laboratory experiments, emphasis has been placed on the theory of the kinetics of direct photoreaction and the fundamental principles that guide the selection of experimental approaches. Detailed discussions are presented in Chapters 9 and 10 on the theory and practice of chemical actinometry. Procedures for obtaining kinetic data are described so that the environmental scientist can carry out these experiments easily and precisely. An example is given in Chapter 11 to illustrate how the experimental data are used to obtain the reaction quantum yield of a test chemical and environmentally relevant direct photoreaction rate constants and half-lives as a function of latitude and season in the northern hemisphere.

1464–6/88/0139$06.00/1

Chapter 9

Theory and Development of the Tests

9.A. Introduction

The test method described in Chapters 6–8, the first in a series of aqueous photoreaction tests, was designed to determine the molar absorptivity of a test chemical by using standard spectroscopic techniques and to determine the reaction quantum yield of a test chemical in aqueous solution in sunlight. These parameters were then combined with solar irradiance data to determine environmentally relevant direct photoreaction rate constants and half-lives in aqueous solution as a function of latitude and season anywhere in the northern hemisphere.

The test method described in Chapters 6–8 was developed as a screening test to obtain the direct photoreaction quantum yield of a chemical in aqueous solution by carrying out photoreaction experiments in sunlight. This method does not require sophisticated and expensive equipment and thus the reaction quantum yield can be determined easily and at modest cost. However, the method does have some limitations. For example, the screening test can be carried out only during the warm seasons of the year in the northern hemisphere. Unfortunately, in many parts of the northern hemisphere, the temperature falls below 0 °C during the winter months. Therefore, if photoreaction experiments were carried out in the winter in certain parts of the northern hemisphere, the dilute aqueous solutions would freeze, the tubes would break, and the samples would be destroyed. Furthermore, under certain circumstances, a more precise value of the reaction quantum yield may be required than given by this test method. For example, a more precise value of the reaction quantum yield is necessary for photochemical research. If the photoreaction test data indicate that photoreaction is a relatively important transformation process, and an initial risk assessment indicates that the health of humans or the environment is threatened, then an upper-tier test may be required to obtain a more precise value of the reaction quantum

1464–6/88/0141$08.25/1

yield for a specific test chemical. Finally, many laboratories have photochemical equipment that uses monochromatic light and may prefer to use it. Thus, more comprehensive test methods are needed to determine the reaction quantum yield in the laboratory using specialized photochemical equipment, and these methods will be described in detail in this chapter.

9.B. Environmentally Relevant Direct Aqueous Photoreaction Rate Constants and Half-Lives

Chapters 6–8 have described in detail the theory and practice of determining environmentally relevant direct photoreaction rate constants and half-lives in sunlight. A brief summary will be presented here because these principles will be used extensively in this chapter.

The direct sunlight photoreaction of a chemical in an optically thin aqueous solution (i.e., the absorbance is less than 0.02 at $\lambda \geq 290$ nm) can be described by the following equations:

$$\ln([C_o]/[C_t]) = k_{dE}t \tag{9.1}$$

$$t_{(1/2)E} = 0.693/k_{dE} \tag{9.2}$$

$$k_{dE} = \phi_d k_a \tag{9.3}$$

where ϕ_d is the reaction quantum yield in sunlight and is an average value over all wavelengths absorbed by the chemical, $k_a = \Sigma_\lambda k_{a\lambda}$ the sum of $k_{a\lambda}$ values for all wavelengths of sunlight that are absorbed by the chemical (i.e., the light absorption rate constant), t is the time, $[C_o]$ and $[C_t]$ are the molar concentrations of chemical at $t = 0$ and t, and $t_{(1/2)E}$ is the half-life. The term k_{dE} is the first-order direct photoreaction rate constant for a chemical in a natural water body in sunlight in units of reciprocal time. In general, the reaction quantum yield in sunlight, ϕ_d, is equivalent to the reaction quantum yield, $\phi_{d\lambda}$, measured in the laboratory at wavelength λ, because the reaction quantum yield of most organic chemicals in dilute aqueous solution is independent of λ (*1*).

Furthermore, under the conditions just cited, the first-order direct photoreaction rate constant in aquatic media in the environment, k_{dE}, is given by the equation

$$k_{dE} = \phi_d \sum_\lambda \epsilon_{\lambda c} L_\lambda \tag{9.4}$$

where ϕ_d is the reaction quantum yield in sunlight and is dimensionless, $\epsilon_{\lambda c}$ is the molar absorptivity of the chemical in the units M^{-1} cm^{-1},

L_λ is the solar irradiance in aquatic media in the units millieinsteins cm^{-2} day^{-1}. The sum is taken over the wavelength range 290–800 nm. The term L_λ is the day-averaged, midseason, solar irradiance at shallow depths for a water body under clear sky conditions and is a function of latitude and season.

A simple screening test was developed in Chapters 6–8 that uses equation 9.4. As an approximation, it has been assumed that the reaction quantum yield ϕ_d is equal to 1, the maximum value. The upper limit for the direct photoreaction sunlight rate constant in aquatic media is then obtained and equation 9.4 becomes

$$(k_{dE})_{\max} = \sum_\lambda \epsilon_{\lambda c} L_\lambda \tag{9.5}$$

Using equation 9.5 in equation 9.2, the lower limit for the half-life is given by

$$[t_{(\frac{1}{2})E}]_{\min} = 0.693/(k_{dE})_{\max} = 0.693/\sum_\lambda \epsilon_{\lambda c} L_\lambda \tag{9.6}$$

The molar absorptivity of a chemical can be determined experimentally by the method outlined in Section 10.B.1. The solar irradiance data are tabulated in Tables A-4 to A-11, Appendix A. These data can then be substituted into equation 9.5 to calculate $(k_{dE})_{\max}$. Finally, $(k_{dE})_{\max}$ can be substituted into equation 9.6 to calculate $[t_{(\frac{1}{2})E}]_{\min}$. On the basis of these data, a decision can be made whether to determine the reaction quantum yield of the chemical in the laboratory.

Once the reaction quantum yield has been determined in the laboratory by procedures to be described later, it can be combined with $\epsilon_{\lambda c}$ and the appropriate L_λ values to calculate k_{dE} as a function of latitude and season anywhere in the northern hemisphere by using equation 9.4. The corresponding half-life can be calculated by using k_{dE} in equation 9.2.

9.C. Theoretical Aspects for Determining the Reaction Quantum Yield in the Laboratory

9.C.1. Basic Equations

On the basis of the first and second laws of photochemistry (equations 1.27 and 1.28), the general rate equation for direct aqueous photoreaction of a chemical in a cell containing volume V (in cm^3) of solution with exposed area A (in cm^2) is given by the relationship

$$\text{average rate} = -(d[C]/dt)_{d\lambda} = \phi_{d\lambda} I_{o\lambda} (A/V) F_{s\lambda} F_{c\lambda} \tag{9.7}$$

where the subscript d denotes that the chemical is undergoing direct photoreaction, λ is the wavelength of photoreaction in nm, $[C]$ is the molar concentration of the chemical, t is the time, $\phi_{d\lambda}$ is the reaction quantum yield of the chemical at wavelength λ, $I_{o\lambda}$ is the incident light intensity in photons (or einsteins) $cm^{-2}\ s^{-1}$ at wavelength λ, $F_{s\lambda}$ is the fraction of light absorbed by the system at λ, and $F_{c\lambda}$ is the fraction of light absorbed by the chemical at λ. The terms $F_{s\lambda}$ and $F_{c\lambda}$ are defined by equations 1.23 and 1.29 and are

$$F_{s\lambda} = 1 - 10^{-(\alpha_\lambda + \epsilon_\lambda[C])l} \tag{9.8}$$

$$F_{c\lambda} = \epsilon_\lambda[C]/(\alpha_\lambda + \epsilon_\lambda[C]) \tag{9.9}$$

where α_λ is the attenuation coefficient of water in cm^{-1}, ϵ_λ is the molar absorptivity of the chemical in $M^{-1}\ cm^{-1}$, $[C]$ is the molar concentration, and l is the cell path length in cm.

If $I_\lambda = I_{o\lambda}(A/V)$, equation 9.7 becomes

$$-(d[C]/dt)_{d\lambda} = \phi_{d\lambda} I_\lambda F_{s\lambda} F_{c\lambda} \tag{9.10}$$

where I_λ is now the incident light intensity in photons (or einsteins) $cm^{-3}\ s^{-1}$.

9.C.2. Kinetics When Incident Light Is Weakly Absorbed

For an aqueous solution of a chemical with low optical density or absorbance only a small fraction of the light in the solution is absorbed by the system. The fraction of light absorbed by the system is given by equation 9.8. Conversion of this equation to $\{1 - \exp[-2.303(\alpha_\lambda + \epsilon_\lambda[C])l]\}$ expansion of the exponental term in a power series, neglecting all higher power terms (i.e., quadratic, cubic, etc.) because the absorbance is equal to $(\alpha_\lambda + \epsilon_\lambda[C])l$ and is less than 0.02, and simplification yield

$$F_{s\lambda} = 2.303(\alpha_\lambda + \epsilon_\lambda[C])l \tag{9.11}$$

Substituting equations 9.11 and 9.9 in equation 9.10 yields

$$-(d[C]/dt)_{d\lambda} = 2.303\phi_{d\lambda} I_\lambda \epsilon_\lambda [C] l \tag{9.12}$$

$$-(d[C]/dt)_{d\lambda} = k_{d\lambda}[C] \tag{9.13}$$

where $k_{d\lambda}$ is the first-order direct photoreaction rate constant at wavelength λ and is given by the expression

$$k_{d\lambda} = 2.303\phi_{d\lambda}I_{\lambda}\epsilon_{\lambda}l \tag{9.14}$$

Integrating equation 9.13 between ($t = 0$, $[C_o]$) and (t, $[C_t]$) yields

$$\ln([C_o]/[C_t]) = k_{d\lambda}t \tag{9.15}$$

and the associated half-life is

$$t_{(1/2)} = 0.693/k_{d\lambda} \tag{9.16}$$

9.C.3. Kinetics When Incident Light Is Strongly Absorbed

For an aqueous solution of a chemical (e.g., an actinometer or test chemical) with high optical density or absorbance, essentially all the light is absorbed. Under this condition the absorbance is >2. Because the absorbance is equal to $(\alpha_{\lambda} + \epsilon_{\lambda}[C])l$ and is greater than 2, the term $10^{-(\alpha_{\lambda} + \epsilon_{\lambda}[C])l} \ll 1$, and $F_{s\lambda}$, defined by equation 9.8, is then equal to 1. Because essentially all the light is absorbed by the chemical, $\epsilon_{\lambda}[C] \gg \alpha_{\lambda}$ and $F_{c\lambda}$, defined by equation 9.9, is then equal to 1. Substituting these results in equation 9.10 yields the zero-order differential equation

$$-(d[C]/dt)_{d\lambda} = \phi_{d\lambda}I_{\lambda} \tag{9.17}$$

Integrating equation 9.17 between ($t = 0$, $[C_o]$) and (t, $[C_t]$) yields

$$([C_o] - [C_t]) = k_{d\lambda}t \tag{9.18}$$

where $k_{d\lambda}$ is the zero-order direct photoreaction rate constant, in the units of molar time^{-1}, and is given by the expression

$$k_{d\lambda} = \phi_{d\lambda}I_{\lambda} \tag{9.19}$$

Equations 9.17, 9.18, and 9.19 are applicable to the disappearance of a chemical (i.e., test chemical or actinometer) and the reaction kinetics are studied by measuring the disappearance of a chemical as a function of time. However, there are instances in which the reaction kinetics are studied by following the formation of one of the products (e.g., the ferrous ion in the ferrioxalate actinometer discussed in Section 9.D.2.a.). If the rate of disappearance of chemical equals the rate of formation of one of the products, then

$$-(d[C]_r/dt)_{d\lambda} = (d[C]_p/dt)_{d\lambda} \tag{9.20}$$

where $[C]_r$ is the molar concentration of reactant and $[C]_p$ is the molar concentration of one of the products. By using equation 9.20 in equation 9.10 and proceeding in the manner just described, the following equations are obtained:

$$(d[C]_p/dt)_{d\lambda} = \phi_{d\lambda} I_\lambda \tag{9.21}$$

$$[C_t]_p = k_{d\lambda} t \tag{9.22}$$

$$k_{d\lambda} = \phi_{d\lambda} I_\lambda \tag{9.23}$$

where $[C_t]_p$ is the molar concentration of the product (p) formed at time t.

9.D. Chemical Actinometry

Chemical actinometers are used in reaction quantum yield experiments to measure the integrated light intensity incident on the sample during irradiation. Chemical actinometers are photochemical reactions that have been calibrated directly or indirectly with light sources of known light flux and have well-defined reaction quantum yields $\phi^a_{d\lambda}$ at specific wavelengths λ. For simplicity the superscript a and the subscript λ will be dropped from the symbol ϕ and the term ϕ_{da} will be used to denote the direct photoreaction quantum yield of the actinometer. Thus, the term ϕ_{da} is the reaction quantum yield for direct photoreaction of the actinometer at wavelength λ.

In the environment, photoreaction takes place in aqueous media. Complications are encountered when reaction quantum yield experiments are carried out in the cylindrical cells recommended in Section 9.E. with water as the solvent for the chemical and an organic solvent for the actinometer. These complications arise because the light flux measured is a function of the refractive index of the water and the organic solvent relative to the type of glass used for the cylindrical cells. As a result, the reaction quantum yield measured can be in error. This effect will be discussed in more detail in Section 9.E. Therefore, to measure precise reaction quantum yields of chemicals in cylindrical cells, actinometers in aqueous solution are highly desirable. In the following sections, four aqueous actinometers will be described that can be used to obtain environmentally relevant reaction quantum yields in the spectral region 290–750 nm. In cases where aqueous solutions are not practical, the recommended actinometer solvent is acetonitrile. Acetonitrile has an index of refraction $n^{20°} = 1.344$ that is close to the refractive index of water ($n^{20°} = 1.332$). Thus, the *o*-nitrobenzaldehyde actinometer dissolved in acetonitrile is also described and can be used in these reaction quantum yield experiments. Other actinometers can be used in these experiments. Detailed information on these actinometers is given in references 2–4.

9.D.1. Low Optical Density Actinometers

Two low optical density actinometers that can be used to determine the reaction quantum yield of a chemical at low optical density (i.e., absorbance <0.02) will be described. These actinometers are ideal for environmental actinometry experiments because they are in aqueous solution and have variable reaction quantum yields that are well-defined. These two low-optical-density actinometers, composed of the systems *p*-nitroacetophenone–pyridine and *p*-nitroanisole–pyridine, have been recently developed for the Office of Toxic Substances, U.S. Environmental Protection Agency, by SRI International (*5–7*). These actinometers have been described thoroughly in Section 6.B.4.b.ii. Because they will be used extensively in this chapter, some pertinent information follows.

9.D.1.a. *p*-Nitroacetophenone–Pyridine Actinometer

Laboratory experiments with the *p*-nitroacetophenone–pyridine actinometer (PNAP–PYR) indicated that the reaction quantum yield for the photoreaction of PNAP was a linear function of the molar concentration of pyridine, [PYR], at a fixed molar concentration of PNAP (equation 6.12), and is

$$\phi_{da} = 0.0169[\text{PYR}] \tag{9.24}$$

The reaction quantum yield of this actinometer is independent of the wavelength.

The UV–vis absorption spectrum of PNAP was obtained in an aqueous solution containing 1% acetonitrile and the molar absorptivities at 313 and 366 nm (ϵ_{313} and ϵ_{366a}) were 2056 and 160 M^{-1} cm^{-1}, respectively (*see* Table 6-1). With a 450-W medium-pressure mercury lamp in an Ace-type photochemical "merry-go-round" reactor (PMGRR) (described in Section 10.A.1.a.i.), this actinometer can be adjusted with pyridine to have half-lives that range from greater than 12 h to several weeks (described in Section 10.A.2.a.).

9.D.1.b. *p*-Nitroanisole–Pyridine Actinometer

Laboratory experiments with the *p*-nitroanisole–pyridine actinometer (PNA–PYR) indicated that the reaction quantum yield for the photoreaction of PNA was a linear function of the molar concentration of pyridine [PYR], at a fixed molar concentration of PNA (equation 6.10), and is

$$\phi_{da} = 0.437[\text{PYR}] + 2.82 \times 10^{-4} \tag{9.25}$$

This actinometer also has a reaction quantum yield that is independent of the wavelength.

The UV–vis absorption spectrum of PNA was obtained in an aqueous solution containing 1% acetonitrile, and the molar absorptivities at 313 and 366 nm (ϵ_{313a} and ϵ_{366a}) were 10,300 and 1990 M^{-1} cm^{-1}, respectively (Table 6-1). With a 450-W medium-pressure mercury lamp in an Ace-type PMGRR, this actinometer can be adjusted with pyridine to have half-lives that range from approximately 15 min to 12 h (Section 10.A.2.b.).

9.D.2. High Optical Density Actinometers

Three excellent high optical density actinometers (absorbance >2) can be used to determine the reaction quantum yield of a chemical: the ferrioxalate actinometer, the Reinecke's salt actinometer, and the *o*-nitrobenzaldehyde actinometer. The ferrioxalate actinometer is ideal for environmental actinometry because it is in aqueous solution and is most suitable for use in the spectral range 290–500 nm. This actinometer is highly recommended and has been adopted for use in this test method. The one disadvantage of this actinometer is that actinometry experiments must be carried out in a dark room with photographic "safe lights" because the actinometer is sensitive to visible light.

The *o*-nitrobenzaldehyde actinometer is only sensitive in the UV and blue spectral regions, so actionmetry experiments can be carried out in normal laboratory light. This actinometer follows zero-order kinetics, and the kinetics are applicable up to 25% transformation (*8*).

Reinecke's salt is recommended for use in the spectral region 500–750 nm. Because this actinometer is sensitive to visible light, actinometry experiments must be carried out in a dark room with photographic "safe lights".

All the recommended actinometers are high-efficiency actinometers that have well-defined quantum yields at fixed wavelengths.

9.D.2.a. Potassium Ferrioxalate Actinometer

The potassium ferrioxalate actinometer developed by Parker (*9*) and Hatchard and Parker (*10*) is a solution-phase actinometer that is widely used by many photochemists. It is used at high optical density (absorbance >2) and is applicable over a wide range of wavelengths (254–577 nm). However, it is most useful in the range 254–500 nm. Because it is an excellent actinometer and has a large wavelength range of sensitivity in the environmentally relevant region 290–500 nm, it has been adopted for use in these test methods.

Quantum yields for the ferrioxalate actinometer are given in Table 9-1 as a function of wavelength (*3, 10–13*). At 0.15 M ferrioxalate, the quantum yield at 313 nm is 1.20, and the average quantum yield at 366 nm is 1.18. The net photochemical reaction is

$$2Fe^{3+} + (C_2O_4)^{2-} \xrightarrow{h\nu} 2Fe^{2+} + 2CO_2 \tag{9.26}$$

From the stoichiometry of equation 9.26

$$-d[Fe^{3+}]/dt = d[Fe^{2+}]/dt \tag{9.27}$$

Table 9-1. Quantum Yields (ϕ_λ) of Fe^{2+} at 22 °C

Wavelength (nm)	$[K_3Fe(C_2O_4)_3]$	ϕ_λ	F_λ
577/9	0.15	0.013	0.118
546	0.15	0.15	0.061
509	0.15	0.86	0.132
480	0.15	0.93	0.578
468	0.15	0.92	0.850
436	0.15	1.01	0.997
416	0.006	1.12[a]	
	0.006	1.11	
405	0.006	1.14	0.962
392	0.006	1.13[b]	
365/6	0.006	1.21, 1.26[b]	1.00
	1.15	1.15, 1.20[b]	1.00
358	0.006	1.25[a]	
334	0.006	1.23	
313	0.006	1.24	1.00
	0.15	1.20[c]	1.00
297/302	0.006	1.24	
254	0.006	1.25	

[a]Wegner and Adamson (*12*).
[b]Lee and Selinger (*11*).
[c]Zepp (*13*).
SOURCE: Adapted from refs. 3 and 10–13.

At high optical density, equations 9.21, 9.22, and 9.23 are applicable and the reaction kinetics are zero-order.

The kinetics of this actinometer are followed by measuring the molar concentration of Fe^{2+} formed as a function of time. The molar concentration of Fe^{2+} is measured spectrophotometrically via the formation of the red phenanthroline complex and the determination of the absorbance of the complex at 510 nm. Ferric ion forms only a weak complex with phenanthroline which is transparent at 510 nm. Murov (*4*) gives a detailed procedure for using the ferrioxalate actinometer, and this procedure is highly recommended.

Zepp (*13*) has found that it is most useful to use the ferrioxalate actinometer at a concentration of 0.15 M. At 0.15 M, the ferrioxalate actinometer can accept a much greater dose of light than at 0.006 M. The irradiated solution is diluted 100-fold prior to analysis of Fe^{2+}, which must not be allowed to exceed 0.005 M. Thus, it is recommended that the Murov procedure be modified as described by Zepp.

The fraction of the light absorbed (F_λ) is listed in Table 9-1 as given by Hatchard and Parker (*10*). At wavelengths equal to or less than 436 nm, F_λ is essentially 1. Only at wavelengths greater than 436 nm does F_λ decrease significantly below one, and all actinometry experiments at $\lambda > 436$ nm must incorporate these values of F_λ.

With a 450-W medium-pressure mercury lamp filtered at 313 or 366 nm in an Ace-type PMGRR, ferrioxalate decomposes rapidly and samples should be taken for analysis in the range of 1 to 15 min.

9.D.2.b. *o*-Nitrobenzaldehyde Actinometer

The *o*-nitrobenzaldehyde actinometer was originally developed by Pitts et al. (*14*) for use in a KBr matrix. This actinometer was then used in polymethacrylate film, and procedures for use in this matrix are fully described by de Mayo and Shizuka (*3*).

Smith et al. (*15, 16*) described the use of the *o*-nitrobenzaldehyde actinometer with acetonitrile as the solvent. Dulin and Mill (*7*) reported the use of this actinometer in acetonitrile as a high-optical-density actinometer. The photochemical reaction is

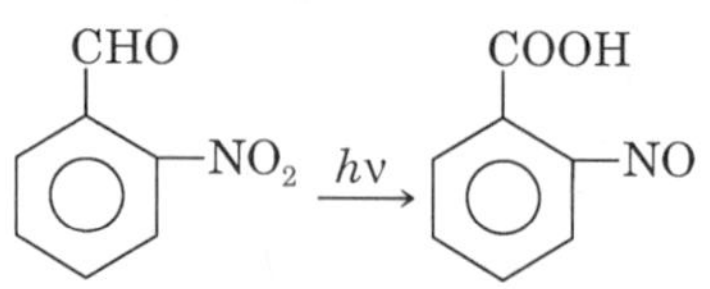

(9.28)

and the reaction quantum yield was reported to be 0.505 and is independent of the wavelength. It is sensitive in the spectral region 280–400 nm.

The photochemical reaction is studied by measuring the concentration of *o*-nitrobenzaldehyde (*o*-NB) remaining at time *t* using reverse-phase high-pressure liquid chromatography (HPLC) with 50% acetonitrile in water as the eluent and the UV detector set at 280 nm. In the calibration experiments by Dulin and Mill (*7*), a solution of approximately 1×10^{-2} M *o*-NB in acetonitrile was irradiated in quartz or borosilicate tubes for periods up to 16 min at 313 and 366 nm in an Ace-type PMGRR with a 450-W medium-pressure mercury lamp. They reported that the light intensity I_λ at 313 nm varied from 2.5×10^{-6} to 3.5×10^{-6} einsteins s^{-1} L^{-1} and at 366 nm varied from 6.7×10^{-6} to 8.2×10^{-6} einsteins s^{-1} L^{-1}.

Mill (*8*) reported that the zero-order kinetics were obeyed up to 25% transformation and the molar absorptivities at 313 and 366 nm (ϵ_{313a} and ϵ_{366a}) were 1500 and 181 M^{-1} cm^{-1}, respectively. At 313 nm all the light is absorbed by *o*-NB in a 1-cm cell. At 366 nm, essentially all (98.5%) the light is absorbed by *o*-NB in a 1-cm cell. If this actinometer is used at $\lambda > 366$ to 400 nm, the concentration of *o*-NB must be adjusted in the reaction cell so that the optical density is >2 in the reaction cell throughout irradiation (*13*).

The *o*-nitrobenzaldehyde actinometer in acetonitrile solution is an excellent actinometer for use in laboratory photoreaction experiments. It is easy to use but is only sensitive in the spectral region 280–400 nm. Therefore, it is highly recommended for use with chemicals that absorb in the region 290–400 nm. Because the photochemical reaction is followed by measuring the concentration of *o*-NB as a function of time, equations 9.17, 9.18, and 9.19 are applicable.

9.D.2.c. Reinecke's Salt Actinometer

The ferrioxalate actinometer is most useful in the environmentally relevant spectral range 290–500 nm. On the other hand, the Reinecke's salt actinometer [$KCr(NH_3)_2(NCS)_4$], developed by Wegner and Adamson (*12*), can be used in the spectral range 500–750 nm. It is recommended for use to determine the reaction quantum yield of chemicals that absorb in that range.

Upon irradiation of this actinometer, the Reinecke's salt gives the hydrate [$Cr(NH_3)_2(NCS)_3(H_2O)$] and NCS^-. The formation of thiocyanate ion can be determined spectrophotometrically with ferric ion. The actinometer is used in aqueous solution with an optical density ≥2.2. De Mayo and Shizuka (*3*) give a detailed and highly recommended procedure for using this actinometer.

9.E. Determination of the Cell Path Length

The theoretical discussion given in Section 9.C. indicates that the path length of the light is an important variable, especially when low-optical-density actinometers are used (equation 9.14). For a 1-cm square cell, the path length is easily defined. However, for a cylindrical cell where the light falls on rounded surfaces, the path length is not easily defined because of refraction and reflection in the cell. Zepp (*1*) developed an experimental method for determining the effective path length for any cell, in particular for the cylindrical cells that are used in environmental studies.

If the molar concentration $[C]$ of the light-absorbing substance is high, then $\epsilon_\lambda[C] \gg \alpha_\lambda$ and, to a good approximation, $(\alpha_\lambda + \epsilon_\lambda[C]) = \epsilon_\lambda[C]$. Using this result in equations 9.8 and 9.9 yields

$$F_{s\lambda} = 1 - 10^{-\epsilon_\lambda[C]l} \tag{9.29a}$$

$$F_{c\lambda} = 1 \tag{9.29b}$$

Under these conditions, equation 9.10 becomes

$$(\text{rate})_C = -(d[C]/dt)_{d\lambda} = \phi_{d\lambda} I_\lambda (1 - 10^{-\epsilon_\lambda[C]l}) \tag{9.30}$$

For the special case when essentially all the light is absorbed by the chemical, $\epsilon_\lambda[C]l > 2$ and $10^{-\epsilon_\lambda[C]l}$ is essentially 0, and thus equation 9.30 becomes

$$(\text{rate})_{\max} = \phi_{d\lambda} I_\lambda \tag{9.31}$$

Dividing equation 9.31 into 9.30 yields

$$(\text{rate})_C/(\text{rate})_{\max} = 1 - 10^{-\epsilon_\lambda[C]l} = X \tag{9.32}$$

Rearranging equation 9.32 yields

$$1 - X = 10^{-\epsilon_\lambda[C]l} \tag{9.33a}$$

and

$$-\log_{10}(1 - X) = \epsilon_\lambda[C]l \tag{9.33b}$$

A plot of $-\log_{10}(1 - X)$ versus $\epsilon_{\lambda c}$ yields a straight line with a slope equal to the path length of the cell, l.

Zepp (*1*) measured the effective cell path length of small cylindrical cells used in his environmental photochemical studies by employing the benzophenone-sensitized isomerization of *cis*-1,3-pentadiene. This procedure has been used to measure the effective path length of 13- × 100-mm (Corning) glass culture tubes, and it was found that l was 1.12 cm (*17*). In a similar manner, reaction tubes made from borosilicate stock of 12-mm o.d. had an effective path length of 1.00 cm (*17*).

The following procedure can be used to determine the path length of cylindrical cells using the *o*-nitrobenzaldehyde actinometer (*o*-NB) which is described in Section 9.D.2.b.

1. *o*-NB solution is prepared with optical density > 2 and with known molar concentration, C_m.
2. *o*-NB solutions with optical densities (ODs) 1 to 0.1 (e.g., ODs approximately 1, 0.8, 0.6, 0.4, 0.2, and 0.1) with known molar concentrations, C_1, C_2, C_3, C_4, C_5, and C_6, are prepared by sequential dilutions of the solution prepared in step 1.
3. The seven solutions (C_m, C_1, C_2, C_3, C_4, C_5, and C_6) are irradiated for equal lengths of time in a PMGRR and analyzed for *o*-NB by the procedure outlined in Section 10.A.3.b. There should be less than 25% transformation of *o*-NB.
4. The ratios $\Delta C_1/\Delta C_m$, $\Delta C_2/\Delta C_m$, $\Delta C_3/\Delta C_m$, ... , $\Delta C_6/\Delta C_m$ are determined and designated X_1, X_2, X_3, ... , X_6.
5. The quantity $[-\log_{10}(1 - X_1)]$ versus $\epsilon_\lambda C_1$, $[-\log_{10}(1 - X_2)]$ versus $\epsilon_\lambda C_2$, etc., is plotted for each set of data points; ϵ_λ is the absorption coefficient for *o*-NB at the irradiation wavelength λ.
6. The slope of the best straight line is calculated. This slope is equal to the effective cell path length l (equation 9.33b). A more accurate and precise value of l may be obtained by a fit of the data to equation 9.33b by using linear regression analysis.

These photoreaction experiments can be carried out at wavelengths 313 or 366 nm. The molar absorptivities of *o*-NB at these wavelengths are $\epsilon_{313a} = 1500\ M^{-1}\ cm^{-1}$ and $\epsilon_{366a} = 181\ M^{-1}\ cm^{-1}$ (*8*).

When reaction quantum yield experiments are carried out in cylindrical cells (i.e., tubes) and different solvents are used for actinometer and chemical, the light flux measured in an Ace-type PMGRR will be solvent-dependent because of the different refractive indices of the solvents used. Vesley (*18*) measured the reaction quantum yield for the photoisomerization of *cis*-1,3-pentadiene in benzene at 365 nm in an Ace-type PMGRR by using the standard actinometers ferrioxalate in water and benzophenone–benzhydrol in benzene. The reaction quantum yield of the photoisomerization reaction was 0.64 when the ferrioxalate actinometer (in water) was used, while the reaction quantum yield was found to be 0.55 when the benzophenone–benzhydrol actinometer (in benzene) was used.

Vesley attributed the differences in the values of the reaction quantum yield to the differences in the refractive indexes of water and benzene relative to the refractive index of glass. Benzene has a refractive index $n^{20°} = 1.501$ that is similar to the refractive index of Pyrex $n^{20°} = 1.52$. Thus, all the incident light passed into the solution; whereas with water, with a refractive index $n^{20°} = 1.332$, some of the light is refracted and reflected at the inner cell walls. As a result, the samples in which the benzene was used as the solvent "see" more photons than the aqueous actinometer solutions. Thus, the ferrioxalate actinometer underestimated the light that actually entered the solution in the cell when benzene was the solvent for the photoisomerization reaction and, therefore, overestimated the true reaction quantum yield for the *cis*-1,3-pentadiene by 15%. This discussion is valid only for the Ace-type PMGRR. The Ace-type PMGRR is designed so that the entire reaction cell is irradiated. This type of PMGRR is the only one commercially available. There is a Moses-types PMGRR (*19*) that differs slightly from the Ace-type PMGRR in that it contains windows (i.e., slits) so that only a narrow portion of the solution in the reaction cell is irradiated. For a description of these photochemical reactors, see Section 10.A.1.a.i.

Because water is the medium in which photoreaction takes place in environmental photochemistry, actinometry experiments should be performed in aqueous solution whenever practical. In those cases where aqueous actinometers are not practical, the actinometer solvent acetonitrile ($n^{20°} = 1.344$) may be used because its refractive index is close to that of pure water.

9.F. Development of the Test Methods

In general, the reaction quantum yield is determined under experimental conditions in which all the incident light is absorbed by the chemical (absorbance >2). The concentrations of chemicals used in these experiments range from 10^{-3} M to greater than 0.1 M. However, most organic pollutants are hydrophobic, and the water solubility is generally below 10^{-3} M. For these chemicals in aqueous solution, only a small fraction of the incident light is absorbed (absorbance <0.02). Zepp (*1*) described two procedures to determine the reaction quantum yields of organic pollutants in very dilute aqueous solution with absorbances <0.02. One procedure measures the light intensity incident on the sample using a high-optical-density actinometer (e.g., the ferrioxalate actinometer). This procedure is used extensively by photochemists and has been incorporated in this test method.

In addition, Zepp (*1*) described a second procedure for the determination of the reaction quantum yield by comparing the photoreaction rate of the chemical and actinometer under conditions in which the light is only weakly absorbed by both compounds. Such experiments are conveniently performed in a PMGRR. This elegant method has not been used by environmental photochemists because the standard chemical actinometers cannot be analyzed at the low concentrations required for this technique. However, two excellent low-optical-density variable reaction quantum yield actinometers have been developed recently for the Office of Toxic Substances, U.S. Environmental Protection Agency, by SRI International (*5–7*). These actinometers have been discussed in Chapters 6–8 and in Section 9.D.1. These actinometers have already been incorporated in the screening test method, (*see* Chapters 6–8).

Another procedure that can be used to determine the reaction quantum yield of a test chemical at high optical density is by measuring the light incident on the test chemical with a high-optical-density actinometer (e.g., the ferrioxalate actinometer). This procedure is the conventional procedure used by most photochemists for measuring the reaction quantum yield in the laboratory. This excellent procedure is very limited and is appplicable only to those test chemicals that are very soluble in water and have high optical density (i.e., absorbance >2). For this reason, this procedure has not been incorporated in this test method. However, it is discussed in this section and may be used for certain organic chemicals.

9.F.1. Procedure One: Determination of the Reaction Quantum Yield by the Low Optical Density Test Chemical and Actinometer Method

9.F.1.a. Theoretical Discussion

Because aqueous solutions of the test chemical and actinometer will be irradiated at low optical density (absorbance <0.02), equation 9.14 is applicable. By using this equation, the equations for test chemical and actinometer can be written as

$$k_{dc} = 2.303\phi_{dc} I_{\lambda} \epsilon_{\lambda c} l \qquad (9.34)$$

$$k_{da} = 2.303\phi_{da} I_{\lambda} \epsilon_{\lambda a} l \qquad (9.35)$$

For simplicity the subscript λ has been deleted from the rate constants and reaction quantum yields. For the remainder of this chapter it will be assumed that the rate constants are measured at a fixed wavelength λ.

If the aqueous solutions of test chemical and actinometer are irradiated simultaneously with the same lamp in identical cells with the same path length l and at a fixed wavelength λ (or in a narrow wavelength band) in a PMGRR, then equation 9.35 can be divided into 9.34. [The light source must be monochromatic (Section 10.A.1.c. presents details)]. This operation yields

$$\phi_{dc} = (k_{dc}/k_{da})(\epsilon_{\lambda a}/\epsilon_{\lambda c})\phi_{da} \tag{9.36}$$

By using equation 9.15, the equations for test chemical and actinometer can be written as

$$\ln([C_o]/[C_t])_c = k_{dc}t \tag{9.37}$$

$$\ln([C_o]/[C_t])_a = k_{da}t \tag{9.38}$$

The half-life of the test chemical and actinometer can be written as $[t_{(1/2)}]_c$ and $[t_{(1/2)}]_a$ and can be obtained from equation 9.16. Again, because both the actinometer and the test chemical solutions are irradiated simultaneously with the same lamp in identical cells in a PMGRR, equation 9.38 can be divided into equation 9.37. Carrying out this operation and rearranging the result yield

$$\ln([C_o]/[C_t])_c = (k_{dc}/k_{da}) \ln([C_o]/[C_t])_a \tag{9.39}$$

During irradiation at wavelength λ, the concentrations of test chemical and actinometer are measured periodically as a function of the time t. These data are then used to determine the ratio of the rate constants k_{dc}/k_{da} by using linear regression analysis on equation 9.39. The slope of the best straight line is the ratio of the rate constants.

By using standard spectroscopic procedures, the molar absorptivities of test chemical and actinometer ($\epsilon_{\lambda c}$ and $\epsilon_{\lambda a}$) can be determined at the irradiating wavelength λ. By combining these data with the ratio of the rate constants k_{dc}/k_{da} and the reaction quantum yield of the actinometer (ϕ_{da}), the reaction quantum yield of the test chemical (ϕ_{dc}) can be determined using equation 9.36.

9.F.1.b. Procedure One

Procedure one involves the simultaneous irradiation of aqueous solutions of test chemical and actinometer at low optical density (absorbance <0.02) in an Ace-type PMGRR to determine the reaction quantum yield of the test chemical. The excitation source is a 450-W medium-pressure mercury lamp with appropriate filters to isolate the 313- or 366-nm bands. The procedure has been divided into three phases. In phase one, the molar absorptivity of the test chemical is determined at 313 and 366 nm (ϵ_{313c} and ϵ_{366c}) by using UV–vis absorption spectroscopy. On the basis of these results, photoreaction experiments are carried out at the wavelength corresponding to the higher value of ϵ_{313c} or ϵ_{366c}. The phase two procedure is composed of trial experiments in a PMGRR at 313 or 366 nm to determine the approximate rate constant and half-life of the test chemical and to choose the appropriate actinometer (which has a rate constant approximately the same as the rate constant of the test chemical). In the phase three procedure, aqueous low-optical-density solutions of test chemical and actinometer (at a fixed molar concentration of pyridine [PYR]) are irradiated simultaneously in a PMGRR at 313 or 366 nm. The data obtained from the phase one, two, and three experiments are then used to determine the reaction quantum yield of the test chemical.

9.F.1.b.i. Phase One Experiments

In the phase one experiments, the molar absorptivities of the test chemical are determined at 313 and 366 nm (ϵ_{313c} or ϵ_{366c}) with standard spectroscopic procedures outlined in Section 10.B.1. Photoreaction experiments are then carried out in the phase two and three experiments at 313 or 366 nm corresponding to the higher value of ϵ_{313c} or ϵ_{366c}. By irradiating the test chemical at the wavelength corresponding to the higher value of molar absorptivity, the rate of photoreaction is maximized.

9.F.1.b.ii. Phase Two Experiments

First, a dilute aqueous solution of test chemical at a low optical density (absorbance <0.02) and at an initial molar concentration $[C_o]_c$ is irradiated in an Ace-type PMGRR at wavelength 313 or 366 nm with a 450-W medium-pressure mercury lamp and appropriate filters to isolate the monochromatic wavelength λ. At time t, after at least 50% transformation, the molar concentration of test chemical $[C_t]_c$ is measured. These data are used in equation 9.37 to determine an approximate rate constant k_{dc}. This result is used in equation 9.16 to obtain the half-life of the test chemical $[t_{(1/2)}]_c$. If $[t_{(1/2)}]_c$ is equal to or less than 12 h, use the PNA–PYR actinometer (Sections 9.D.1.b. and 10.A.2.b.) should be used. If $[t_{(1/2)E}]_c$ is greater than 12 h, the PNAP–PYR actinometer (Sections 9.D.1.a. and 10.A.2.a.) should be used. If a Moses-type PMGRR was used, new criteria have to be defined to determine which of these two actinometers should now be used. Details are given in Section 10.A.2.c.

Trial photoreaction experiments are then carried out to determine the concentration of pyridine needed to adjust the rate constant of the actinometer to approximately equal the rate constant of the test chemical (within ±50%). On the basis of the discussions given in Section 10.A.2.a. for the PNAP–PYR actinometer in a PMGRR with a 450-W medium-pressure mercury lamp, the concentration of pyridine needed to adjust the rate constant of the actinometer to approximately equal the rate constant of the test chemical at 313 and 366 nm is given by equations 9.40 and 9.41, respectively (*see* Section 10.A.2.a. for the derivation of these two equations.)

$$\text{at 313 nm } [\text{PYR}] = 1.16k_{dc} \tag{9.40}$$

$$\text{at 366 nm } [\text{PYR}] = 5.95k_{dc} \tag{9.41}$$

where k_{dc} is in the units of h^{-1} and [PYR] is the molar concentration of pyridine.

A trial experiment is then carried out in the phase two procedure by preparing the actinometer with the molar concentration of pyridine [PYR] estimated from equation 9.40 or 9.41 and irradiating the actinometer solution at a molar concentration of PNAP, $[C_o]_a$ in the PMGRR at 313 or 366 nm. After at least 50% transformation, the concentration of PNAP, $[C_t]_a$, is recorded at time t and the approximate rate constant k_{da} is calculated by using these data in equation 9.38. This trial experiment is repeated by adjusting the molar concentration of pyridine [PYR] until the rate constant of the actinometer and test chemical are approximately equal (within ±50%).

The same trial experiments can be used for the PNA–PYR actinometer by starting with equation 10.10 or 10.12 (Section 10.A.2.b.) to determine the concentration of pyridine needed to adjust the rate constant of this actinometer to approximately equal the rate constant of the test chemical.

9.F.1.b.iii. Phase Three Experiments

An aqueous solution of test chemical at low optical density and at a molar concentration $[C_o]_c$ is prepared and placed in small cylindrical tubes. On the basis of the results of the phase two experiments, the appropriate actinometer is chosen and the concentration of pyridine is adjusted so that the rate constant of the actinometer is approximately equal to the rate constant of the test chemical (within ±50%). An aqueous actinometer solution is prepared at molar concentration $[C_o]_a$ with a molar concentration of pyridine [PYR]. The actinometer solution is placed in identical tubes used for the test chemical solution. The volumes of actinometer and test chemical solution must be the same.

Aqueous solutions of test chemical and actinometer are simultaneously irradiated in the PMGRR at the chosen wavelength (i.e., 313 or 366 nm) using the 450-W medium-pressure mercury lamp and appropriate filters. The molar concentrations of test chemical $[C_t]_c$ and actinometer $[C_t]_a$ are measured as a function of the time t. By using these data in equation 9.39 and performing linear regression analysis with ln $([C_o]/[C_t])_a$ as the independent variable and ln $([C_o]/[C_t])_c$ as the dependent variable, the ratio of the rate constants (k_{dc}/k_{da}) is obtained.

The reaction quantum yield of the actinometer (ϕ_{da}) can be determined at the molar concentration of pyridine employed in this experiment (i.e., substituting [PYR] in equation 9.24 or 9.25). Using these data along with the molar absorptivities of test chemical (ϵ_{313c} or ϵ_{366c}) and the actinometer (ϵ_{313a} or ϵ_{366a}), given in Sections 9.D.1.a. and 9.D.1.b., in equation 9.36, the reaction quantum yield of the test chemical can be determined.

9.F.2. Procedure Two: Determination of the Reaction Quantum Yield by the Low Optical Density Test Chemical and High Optical Density Actinometer Method

9.F.2.a. Theoretical Discussion

Because the test chemical is at low optical density, equations 9.14 and 9.15 are applicable in the form:

$$k_{dc} = 2.303\phi_{dc}I_\lambda\epsilon_{\lambda c}l \qquad (9.42)$$

$$\ln([C_o]/[C_t])_c = k_{dc}t \qquad (9.43)$$

Furthermore, because the actinometer is at high optical density and the ferrioxalate actinometer (Section 9.D.2.a.) is widely used by photochemists, equations 9.22 and 9.23 are applicable in the form:

$$[C_t]_a = k_{da}t \qquad (9.44)$$

$$k_{da} = \phi_{da}I_\lambda \qquad (9.45)$$

where $[C_t]_a$ now represents the molar concentration of the product formed, which is $[Fe^{2+}]$. Solving for ϕ_{dc} in equation 9.42 and I_λ in equation 9.45 yields the following two equations

$$\phi_{dc} = k_{dc}/[2.303 I_\lambda \epsilon_{\lambda c} l] \tag{9.46}$$

$$I_\lambda = k_{da}/\phi_{da} \tag{9.47}$$

and substituting equation 9.47 in equation 9.46 yields

$$\phi_{dc} = \phi_{da}(k_{dc}/k_{da})(2.303\epsilon_{\lambda c} l)^{-1} \tag{9.48}$$

The irradiating source must be monochromatic. (*See* Section 10.A.1.c. for details.)

9.F.2.b. Procedure Two

Procedure two involves the photoreaction of an aqueous solution of test chemical at low optical density and an aqueous solution of an actinometer at high optical density in a photochemical optical bench (POB, described in detail in Section 10.A.1.a.ii.) or in a PMGRR to determine the reaction quantum yield of a test chemical. The excitation source is a 450-W medium-pressure mercury lamp with appropriate filters to isolate the 313- or 366-nm bands. The procedure has been divided into two phases. In the phase one experiments, the molar absorptivities of test chemical (ϵ_{313c} and ϵ_{366c}) are determined at 313 and 366 nm using UV–vis absorption spectroscopy. On the basis of these results, photoreaction experiments are carried out at either 313 or 366 nm, corresponding to the higher value of ϵ_{313c} or ϵ_{366c}. In the phase two experiments, an aqueous low-optical-density solution of test chemical and an aqueous high-optical-density solution of the ferrioxalate actinometer are irradiated sequentially in a POB or in a PMGRR at 313 or 366 nm. These data, along with the phase one data, are used to determine the quantum yield of the test chemical.

9.F.2.b.i. Phase One Experiments

The phase one experiments are identical to those described in Section 9.F.1.b.i.

9.F.2.b.ii. Phase Two Experiments

In the first series of experiments, a dilute aqueous solution of the ferrioxalate actinometer at high optical density (i.e., at absorbance >2) is placed in a cell (or cells) and irradiated in a POB (or a PMGRR) at wavelength 313 or 366 nm (on the basis of the results of the phase one experiments). A 450-W medium-pressure mercury lamp and appropriate filters are used. The molar concentration of ferrous ion formed $[C_t]_a$ is measured as a function of the time t. These data are fitted to equation 9.44 by using linear regression analysis to obtain k_{da}. No data should be used after the absorbance of the actinometer falls below 2. In a third set of experiments at the end of this procedure, k_{da} is determined again in the same manner as just described. The results of these two experiments are averaged to obtain $(k_{da})_{\text{av}}$.

In a second set of experiments, a dilute aqueous solution of test chemical at low optical density (absorbance <0.02) and at an initial molar concentration $[C_o]_c$ is then irradiated in the same POB or PMGRR as used in the actinometry experiments. The reaction cell or cells must be identical to those used in the actinometry experiments and contain the same volume of solution. The molar concentration of test chemical $[C_t]_c$ is measured as a function of time and these data are fitted to equation 9.43. Linear regression analysis is used to obtain k_{dc}. By using k_{dc} and $(k_{da})_{\text{av}}$, the molar absorptivity of the test chemical (ϵ_{313c} or ϵ_{366c}), and the path length l of the reaction cell or cells (which must be known or determined by the procedure described in Section 9.E.) in equation 9.48, the reaction quantum yield of the test chemical ϕ_{dc} can be determined.

9.F.3. Procedure Three: Determination of the Reaction Quantum Yield by the High Optical Density Test Chemical and Actinometer Method

9.F.3.a. Simultaneous Photoreaction of Test Chemical and Actinometer

9.F.3.a.i. Procedure Involving the Disappearance of Test Chemical and Actinometer

Because the test chemical and actinometer solutions are irradiated at high optical density (i.e., absorbance >2), equation 9.19 is applicable. By using this equation, the equations for test chemical and actinometer can be written

$$k_{dc} = \phi_{dc} I_\lambda \tag{9.49}$$

$$k_{da} = \phi_{da} I_\lambda \tag{9.50}$$

If solutions of the test chemical and actinometer are irradiated simultaneously with the same lamp in identical cells at a fixed wavelength λ in a PMGRR, then equation 9.50 can be divided into equation 9.49. The light source must be monochromatic. (Section 10.A.1.c. gives the details.) Carrying out this operation and rearranging the result yield

$$\phi_{dc} = \phi_{da}(k_{dc}/k_{da}) \tag{9.51}$$

If the concentrations of test chemical and actinometer are measured as a function of time during irradiation, then equation 9.18 is applicable. Thus, the equations for test chemical and actinometer can be written as follows:

$$([C_o] - [C_t])_c = k_{dc}t \tag{9.52}$$

$$([C_o] - [C_t])_a = k_{da}t \tag{9.53}$$

Because both the test chemical and actinometer solutions are irradiated at wavelength λ in a PMGRR with the same lamp in identical cells, then equation 9.53 can be divided into equation 9.52. Carrying out this operation and rearranging the result yield

$$([C_o] - [C_t])_c = (k_{dc}/k_{da})([C_o] - [C_t])_a \tag{9.54}$$

During irradiation, the molar concentrations of test chemical and actinometer are measured periodically as a function of time t. These data are then used to determine the ratio of the rate constants (k_{dc}/k_{da}) by performing linear regression analysis on equation 9.54. The slope of the best straight line is the ratio of the rate constants.

Finally, by using the ratio of the rate constants (k_{dc}/k_{da}) and the reaction quantum yield of the actinometer ϕ_{da} at wavelength λ in equation 9.51, the reaction quantum yield of the test chemical ϕ_{dc} can be obtained.

9.F.3.a.ii. Procedure Involving the Disappearance of Test Chemical and the Formation of Ferrous Ion in the Ferrioxalate Actinometer

If it is now assumed that the high optical density actinometer is the ferrioxalate actinometer, then equation 9.22 is applicable. The equations for test chemical and actinometer at high optical density can then be written as follows:

$$([C_o] - [C_t])_c = k_{dc}t \qquad (9.55)$$

$$([C_t])_a = k_{da}t \qquad (9.56)$$

Dividing equation 9.56 into equation 9.55 and rearranging the result yields

$$([C_o] - [C_t])_c = (k_{dc}/k_{da})[C_t]_a \qquad (9.57)$$

If the molar concentration of ferrous ion, $[C_t]_a$, is measured simultaneously with the molar concentration of test chemical, $[C_t]_c$, as a function of the time t, these data can be substituted into equation 9.51, and the ratio of the rate constants (k_{dc}/k_{da}) can be obtained with linear regression analysis.

By using the ratio of the rate constants (k_{dc}/k_{da}) as just described and the reaction quantum yield of the actinometer ϕ_{da} at wavelength λ in equation 9.51, the reaction quantum yield of the test chemical ϕ_{dc} can be obtained.

9.F.3.b. Sequential Photoreaction of Test Chemical and Actinometer

Another procedure, which is extensively used by photochemists, can be used to determine the reaction quantum yield of a test chemical (ϕ_{dc}) by sequentially irradiating actinometer and test chemical solutions. In this method, solutions of test chemical and actinometer at high optical density (absorbance >2) are irradiated sequentially at a wavelength λ in a POB (or PMGRR). In the first series of experiments, the actinometer solution is irradiated in a cell (or cells) in a POB (or PMGRR) and the molar concentration of the actinometer is measured as a function of the time t. By using these data in equation 9.53 and linear regression analysis, the actinometer rate constant is obtained because the slope equals k_{da}. If the ferrioxalate actinometer is used, then equation 9.56 is applicable. By measuring the molar concentration of ferrous ion formed, $[C_t]_a$, as a function of the time t, k_{da} can then be determined by using these data in equation 9.56 and linear regression analysis. In the third series of experiments, the preceding procedure is repeated and another actinometer rate constant is obtained. The results of these two experiments are averaged to give $(k_{da})_{av}$. Because the reaction quantum yield of the actinometer (ϕ_{da}) is known at wavelength λ and the actinometer rate constant $(k_{da})_{av}$ has been obtained by the procedure described, I_λ can be determined with equation 9.50.

In the second series of experiments, an aqueous solution of test chemical at high optical density is irradiated at wavelength λ in a cell (or cells) in a POB (or PMGRR). The cell (or cells) must be identical to those used in the actinometer experiments and must contain the same volume of solution. The molar concentration of test chemical is measured as a function of the time t and these data are used in equation 9.52 to obtain k_{dc} with linear regression analysis. By substituting the values of k_{dc} and I_λ in equation 9.49, the reaction quantum yield of test chemical (ϕ_{dc}) can be determined.

9.G. Quantum Yield as a Function of Wavelength

In Section 9.B., it was shown that the rate constant k_{dE} in the environment is a function of the reaction quantum yield. It has been assumed that the reaction quantum yield is independent of the wavelength (i.e., the reaction quantum yield $\phi_{d\lambda} = \phi_d$. (Section 5.A.1.b presents the theoretical details.) This assumption is valid for most complex molecules in solution because the photoreaction from second or higher excited states usually cannot compete with the rapid radiationless decay of excited molecules to their first excited state (*1*). Thus, this test method is recommended only for the measurement of the reaction quantum yield of a test chemical at one wavelength. However, exceptions to this rule (e.g., organometallics and α-diketones) must be considered when performing these experiments. Determination of the reaction quantum yield at more than one wavelength may be necessary for specific chemicals or classes of chemicals.

If the reaction quantum yield is independent of the wavelength, environmentally relevant direct photoreaction rate constants and half-lives can be calculated by the method described in Section 9.I. If the reaction quantum yield is a function of the wavelength, environmentally relevant rate constants and half-lives can be calculated by the method described in Section 9.J.

9.H. Photoreaction and Molar Absorptivity in Water

Because water is the environmentally relevant medium in which photoreaction takes place, these reaction quantum yield experiments should be carried out in pure water. Up to 1% acetonitrile, however, is acceptable. Furthermore, whenever possible, the molar absorptivity of the test chemical should be determined in pure water. However, determination of the molar absorptivity in pure water is difficult for hydrophobic chemicals and is essentially impossible for very hydrophobic chemicals (e.g., DDT, PCBs, and polynuclear aromatic hydrocarbons). As an alternative, Zepp (*1*) recommended the determination of the molar absorptivity in mixtures of water and acetonitrile or methanol. Because these solvents effectively dissolve many very hydrophobic organic chemicals, have refractive indexes very close to that of water and thus cause minimal solvent effects (i.e., minimum shifts and changes in the absorbance of strong absorption bands of nonpolar compounds), and do not absorb radiation at wavelengths greater than 290 nm, they have been recommended for use in this test method for determining the molar absorptivity of test chemicals. However, a minimum amount of the solvent should be used.

9.I. Solar Irradiance Data and the Calculation of Environmentally Relevant Direct Photoreaction Rates

The solar irradiance parameter L_λ must be used to calculate the sunlight direct photoreaction rate constant k_{dE} and the corresponding half-life $t_{(1/2)E}$ for a test chemical as a function of latitude and season of the year anywhere in the northern hemisphere. The L_λ values are proportional to the average light flux that is available to cause photoreaction in the wavelength interval $\Delta\lambda$ over a 24-h day at a specific latitude and season date. The L_λ values are defined by the angle of declination of the sun at $-20°$ for winter, $-10°$ for fall, $+10°$ for spring, and $+20°$ for summer. The annual midseason dates that correspond to these angles of declination can be obtained from the *Astronomical Almanac* (*20*), which is published annually. The L_λ values are listed in Tables A-4 to A-11 of Appendix A as a function of latitude and season and are applicable to clear sky conditions, shallow depths in water bodies, and for chemicals whose absorbance is less than 0.02 in pure water (*17, 21*). The theoretical basis for the L_λ data has been discussed in detail in Section 7.B.1.i., and in Chapters 3 and 4.

If the reaction quantum yield is independent of the wavelength, as is generally the case for most chemicals, the measured reaction quantum yield can be combined with the measured $\epsilon_{\lambda c}$ data and the appropriate L_λ values to calculate k_{dE} as a function of latitude and season anywhere in the northern hemisphere with equation 9.4. The corresponding half-life can be calculated using k_{dE} in equation 9.2.

9.J. Calculation of Environmentally Relevant Direct Photoreaction Rates When the Reaction Quantum Yield Is a Function of the Wavelength

When the reaction quantum yield is a function of the wavelength, the determination of direct photoreaction rates must be modified. Under these circumstances, equation 5.28 must take the form

$$k_d = 2.303 j^{-1} \sum_{\lambda} \phi_{d\lambda} \epsilon_{\lambda} E_{o\lambda}^{av} \tag{9.58}$$

When $E_{o\lambda}^{av}$ is the solar irradiance in the environment (i.e., $E_{o\lambda}^{av} = L_{\lambda}$), equation 9.58 takes the form

$$k_{dE} = \sum_{\lambda} \phi_{d\lambda} \epsilon_{\lambda} L_{\lambda} \tag{9.59}$$

The reaction quantum yield, $\phi_{d\lambda}$, is determined as a function of the wavelength λ by the methods described in this chapter.

These data are then combined with the measured molar absorptivity data ϵ_{λ} and the appropriate L_{λ} data to calculate k_{dE} as a funtion of latitude and season anywhere in the northern hemisphere. These rate constants are substituted in equation 9.2 to calculate the associated half-lives $[t_{(1/2)E}]$.

9.K. Applicability and Specificity of the Test Methods

For environmental photochemistry, the general procedures outlined in this test method are applicable to all chemicals that have UV–vis absorptions in the region 290–800 nm. Some chemicals have absorption maxima significantly below 290 nm and consequently cannot undergo direct photoreaction in sunlight (e.g., alkanes, alkenes, alkynes, saturated alcohols, and saturated acids). This condition is a direct consequence of the Grotthus–Draper law of photochemistry. Some chemicals have absorption maxima significantly below 290 nm but have measurable absorption tails above the baseline in their absorption spectra at wavelengths greater than 290 nm. Photoreaction experiments must be carried out for these chemicals.

These test methods are applicable to pure chemicals and not to technical grade chemicals.

The molar absorptivity data ($\epsilon_{\lambda c}$) can be combined with the appropriate L_{λ} data to estimate $(k_{dE})_{max}$ and $[t_{(1/2)E}]_{min}$ at shallow depths in water bodies as a function of latitude and season in the northern hemisphere. If the data indicate that direct aqueous photoreaction is an important process relative to other transformation processes (e.g., biodegradation, hydrolysis, and oxidation), then it is recommended that the reaction quantum yield ϕ_{dc} be determined in the laboratory by procedures one or two outlined in this chapter. Once ϕ_{dc} has been determined, it can be combined with the molar absorptivity data $\epsilon_{\lambda c}$ and the appropriate L_λ data to calculate k_{dE} at shallow depths in water bodies as a function of latitude and season in the northern hemisphere. The corresponding half-life can then be calculated from k_{dE}.

Procedure one is applicable only to solutions of test chemicals and actinometers that have low optical densities (i.e., absorbance <0.02). Procedure two is applicable only to solutions of test chemicals with low optical densities and actinometers with high optical densities (i.e., absorbance >2).

Procedure one is limited to the spectral region 290–400 nm because the recommended low-optical-density actinometers are only sensitive to light in this region. However, procedure two using the ferrioxalate actinometer, is useful in the spectral region 290–500 nm. This procedure can be extended up to approximately 750 nm using the Reinicke's salt actinometer (*3*).

A third procedure could be used to determine ϕ_{dc} in the laboratory if both the test chemical and actinometer solutions have high optical densities. In general, this procedure is limited to very soluble and strongly light-absorbing organic compounds. This method is briefly described in Section 9.F.3.

Complications can be encountered in procedures two and three if a test chemical photoreacts slowly and the intensity of the light source (measured twice, once before and once after irradiation of the test chemical) decreases dramatically. Under these circumstances, $(I_\lambda)_{ave}$ will not be a valid measure of the light intensity of the lamp. The error introduced under these circumstances can be minimized either by using a more intense light source or measuring rates of photoreaction at wavelengths where the rate of light absorption of the test chemical is a maximum (i.e., irradiate the test chemical at λ where ϵ_λ is a maximum), or both.

This test method has been designed to determine the molar absorptivity of a test chemical, $\epsilon_{\lambda c}$, and its quantum yield, ϕ_{dc}. These parameters can be used to determine environmental direct photoreaction rate constants at low absorbance and shallow depths in a pure water body as a function of latitude and season. Tables of solar irradiance (Tables A-4 to A-11) have been included in Appendix A to carry out all the calculations. However, the method is very general and can be extended to determine the rates of direct photoreaction over a range of other environmental conditions using a computer program. Zepp has written a computer program called GCSOLAR to calculate the rates of direct photoreaction as a function of time of the day and day of the year, depth in water, as a function of the attenuation (or absorption) coefficient of the water (α_λ) for natural water bodies, the average ozone layer thickness that pertains to the seasons and location of interest, and as a function of latitude and season of the year. The computer program is available on request (*22*).

References

1. Zepp, R. G. *Environ. Sci. Technol.* **1978**, *12*, 327.
2. Calvert, J. G.; Pitts, J. N., Jr. *Photochemistry*; Wiley: New York, 1966; Chapter 7.
3. de Mayo, P.; Shizuka, P. In *Creation and Detection of Excited States*; Ware, W. R., Ed.; Marcel Dekker: New York, 1976; Vol. 4, Chapter 4.
4. Murov, S. L. *Handbook of Photochemistry*; Marcel Dekker: New York, 1973; Sections 9–14.
5. Mill, T.; Davenport, J. E.; Dulin, D. E.; Mabey, W. R.; Bawol, R. *Evaluation and Optimization of Photolysis Screening Protocols;* U.S. Environmental Protection Agency. U.S. Government Printing Office: Washington, DC, 1981; EPA 560/5–81–003.
6. Mill, T.; Mabey, W. R.; Winterle, J. S.; Davenport, J. E.; Barich, V. P.; Dulin, D. E.; Tse, D.; Lee, G. "Design and Validation of Screening and Detailed Methods for Environmental Processes"; Final Report; U.S. Environmental Protection Agency, Office of Toxic Substances: Washington, DC, 1982; EPA Contract No. 68–01–6325.
7. Dulin, D.; Mill, T. *Environ. Sci. Technol.* **1982**, *16*, 815.
8. Mill, T., SRI International, Menlo Park, CA, personal communication, 1984.
9. Parker, C. A. *Proc. R Soc. London* **1953**, *A–220*, 104.
10. Hatchard, C. G.; Parker, C. A. *Proc. R Soc. London* **1956**, *235*, 518.
11. Lee, J.; Selinger, H. H. *J. Chem. Phys.* **1964**, *40*, 519.
12. Wegner, E. E.; Adamson, A. W. *J. Am. Chem. Soc.* **1966**, *88*, 394.
13. Zepp, R. G., U.S. Environmental Protection Agency, Office of Research and Development, Athens, GA, personal communication, 1984.
14. Pitts, J. N., Jr.; Wan, J. K. S.; Schuck, E. A. *J. Am. Chem. Soc.* **1964**, *86*, 3606.
15. Smith, J. H.; Mabey, W. R.; Bohonos, N.; Holt, B. R.; Lee, S. S.; Chou, T.-W.; Bomberger, D. C.; Mill, T. *Environmental Pathways of Selected Chemicals in Fresh Water Systems. Part I: Background and Experimental Procedures;* U.S. Environmental Protection Agency. U.S. Government Printing Office: Washington, DC, 1977; EPA 600/7–77–113.
16. Smith, J. H.; Mabey, W. R.; Bohonos, N.; Holt, B. R.; Lee, S. S.; Chou, T.-W.; Bomberger, D. C.; Mill, T. *Environmental Pathways of Selected Chemicals in Fresh Water Systems. Part II: Laboratory Studies;* U.S. Environmental Protection Agency. U.S. Government Printing Office: Washington, DC, 1978; EPA 600/7–78–074.
17. Mill, T.; Mabey, W. R.; Bomberger, D. C.; Chou, T–W.; Hendry, D. G.; Smith, J. H. *Laboratory Protocols for Evaluation of the Fate of Organic*

Chemicals in Air and Water; U.S. Environmental Protection Agency. U.S. Government Printing Office: Washington, DC, 1982; EPA 600/3–82–002.
18. Vesley, C. F. *Mol. Photochem.* **1971**, *3*, 193.
19. Moses, F. G.; Liu, R. S. H.; Monroe, B. M. *Mol. Photochem.* **1969**, *1*, 245.
20. *Astronomical Almanac*; Nautical Almanac Office. U.S. Naval Observatory: Washington, DC, 1987.
21. Mill, T. SRI International, Menlo Park, CA, personal communication, 1984.
22. Zepp, R. G., Environmental Research Laboratory, U.S. Environmental Protection Agency, College Station Road, Athens, GA 30601; personal communication, 1982.

Chapter 10

Experimental Procedures

10.A. Experimental Conditions

Most of the experimental conditions described in Section 7.B.1.a. to 7.B.1.k. are applicable to the experiments described in this chapter. In addition, the following experimental conditions are applicable to the laboratory determination of the reaction quantum yield of a chemical.

10.A.1. Special Photochemical Laboratory Equipment

A number of different designs of photochemical equipment can be used to measure the reaction quantum yield in the laboratory. The one that is chosen will depend on the equipment available and, to a certain extent, on the light source and filter system chosen for the measurement. The apparatus chosen must contain a light source, appropriate filters, sample holders, and cells that allow solutions of test chemical and actinometer to be irradiated reproducibly with a uniform and constant amount of light at a discrete or narrow-band wavelength λ. The temperature of the reaction cells must be reasonably constant ($t_i \pm 2$ °C) in the range 20–30 °C. The apparatus should be housed in a separate part of the laboratory and properly shielded so that extraneous light that could contribute to the photoreaction of chemicals is excluded and laboratory personnel are not exposed to UV light. If the ferrioxalate actinometer is used, then actinometry experiments must be carried out in a dark room with photographic "safe lights". Excellent reviews describing a wide variety of photochemical equipment used to measure the reaction quantum yield in the laboratory are given in references 1–4.

1464–6/88/0171$06.00/1

10.A.1.a. Design of the Apparatus

10.A.1.a.i. Photochemical "Merry-Go-Round" Reactor

The design of the photochemical "merry-go-round" reactor (PMGRR) has been described in the literature (*2, 3, 5*). In a PMGRR, the light source is in the center of the apparatus with reaction tubes arranged in a ring around it. The ring rotates around the light source to give a uniform irradiation of the reaction tubes, as shown in Figure 10.1. Glass filters may be inserted between the light source and the reaction tubes. Filter solutions may be contained in the immersion well holding the light source or in a glass donut that surrounds the light source. The PMGRR may be immersed in a water bath or a stream of air may be passed through the space between the light source and the filters to dissipate the heat generated by the light source. The filter solutions can be circulated to an external cooling source.

Two types of PMGRR equipment are available and are very similar; they are the Ace-type and the Moses-type. The Ace-type PMGRR is designed so that the entire reaction cell is irradiated. The Moses-type PMGRR (*5*) is designed with windows that act as slits with fixed apertures, so that only a narrow portion of the reaction cell is irradiated. Because the Ace-type PMGRR is the only one commercially available, this test method has been specifically designed to use this PMGRR. If the Moses-type PMGRR is available, this test method must be slightly modified as described in Section 10.A.2.c.

Small cylindrical tubes are used in the PMGRR to hold small volumes of test chemical and actinometer solutions. One tube is used for each data point measured.

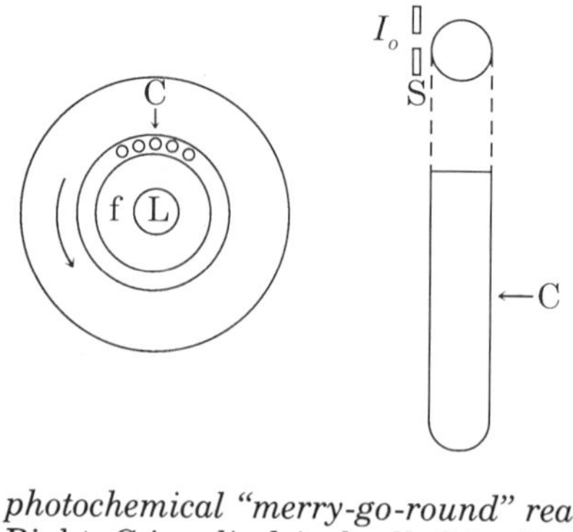

Figure 10.1. Left: *photochemical "merry-go-round" reactor (PMGRR). L is lamp; f is filter.* Right: *C is cylindrical cell; S is slit;* I_o *is incident light.*

10.A.1.a.ii. Photochemical Optical Bench

The photochemical optical bench (POB) is composed of a light source, lenses, glass and/or chemical solution filters, and a reaction vessel. The component parts are mounted on a rail on the optical bench. A typical POB apparatus is shown in Figure 10.2. The light source is at the end of the bench and is surrounded by a housing. A lens is used to collimate the beam. The light passes through lenses and glass filters or through cells containing filter solutions, and transmits wavelength λ. Filter solutions in the cell may be circulated through a cooling system, and the glass filters should be cooled with a stream of air to prevent heat buildup. The photochemical reaction vessel containing the test chemical or actinometer solution is mounted coaxially with the lamp so that the filtered light enters the window of the reaction vessel. The temperature of the reaction vessel should be controlled by circulating constant-temperature water through side walls. The light flux is usually measured before and after the irradiation of the test chemical solution with an actinometer solution. Commercially available optical benches are highly recommended but simple "homemade" benches can be used (*6*). The POB equipment is commercially available from a number of sources. One type of POB that can be used is a Schoeffel Reaction Chemistry System (Schoeffel Industries, Inc.) that contains a 1000-W xenon lamp, a high intensity monochromator, and a sample compartment.

10.A.1.b. Light Sources

A number of light sources are available for use in photochemical studies, for example, low-, medium-, and high-pressure mercury lamps; xenon lamps; and lasers. The characteristics and applications of these

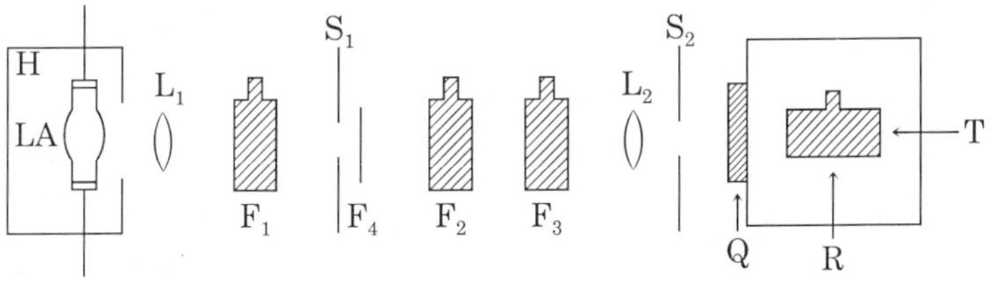

Figure 10.2. A typical optical bench: H is a housing; LA is the point source lamp; L_1 and L_2 are lenses; F_1, F_2, and F_3 are filter solutions; F_4 is a glass filter; S_1 and S_2 are slits; Q is a quartz window; R is reaction cell and T is a temperature-controlled vessel.

lamps are described in detail in references 1–3. The use of a POB or PMGRR will often narrow the choice of lamp with regard to use of a point source or a tube-type lamp because the POB can be focused to give a collimated beam while the PMGRR cannot. The light source should emit light at a constant and high intensity. The 450-W medium-pressure mercury lamp is highly recommended for use in this test method.

10.A.1.c. Light Filtering Systems

Monochromatic or narrow-band wavelength light is essential for the accurate determination of the reaction quantum yield of a test chemical. A serious error can arise if an incorrectly filtered light source that allows light (i.e., a light leak) at other than the irradiation wavelength λ. If $\epsilon_{\lambda c}$ is dramatically larger at the leak wavelength, a small fractional leak can produce a large error because the fraction of light absorbed is a linear function of $\epsilon_{\lambda c}$. Various systems for the isolation of monochromatic or narrow-band wavelengths are described in references 1–3. Two filter systems commonly used to isolate the 313- and 366-nm bands from a 450-W medium-pressure mercury lamp are highly recommended and have been incorporated in this test method. These are

1. 313-nm filtering system: Corning Glass CS–754 filter with 0.005 M potassium chromate solution containing 3% sodium carbonate. The 754 filter is designed to block out visible light. If the test chemical does not absorb light at wavelengths greater than 400 nm, this filter is not needed.
2. 366-nm filtering system: Corning Glass CS 0–52 and CS 7–60 filters.

Reagent grade chemicals should be used to prepare the chemical filter solution. Because this filter solution (and in general any filter solution) degrades slowly over prolonged periods of irradiation, the solution should be carefully monitored. Even when tap water is used to cool the lamp, it must be checked repeatedly because the buildup of solid material or algae may reduce the light intensity.

10.A.1.d. Reaction Cells

In general, large reaction cells are appropriate for POB equipment while small reaction cells are used for PMGRR equipment. These cells should be constructed of borosilicate glass or quartz. Actinometer and test chemical solutions must be irradiated in identical cells containing the same volumes of actinometer and test chemical solution.

For the PMGRR equipment, disposable culture tubes (13- × 100-mm) with Teflon-lined screw caps or quartz tubes (13- × 100-mm) with quartz ground-glass stoppers or Teflon-lined borosilicate screw caps are recommended for use as reaction tubes. Volatile compounds can be conveniently studied in culture tubes equipped with Mininert caps (Section 7.B.1.e.). The path length, l, of these cells has been measured as described in Section 9.E. and was found to be 1.12 cm. Grease should be avoided because hydrophobic chemicals might adsorb to it. One tube is used for each data point measured.

For the POB equipment, the most common and functional design for reaction cells is a cylindrical shape with optically flat circular windows fused to each end of the cylinder and at right angles to its axis. The windows should be made of material that will transmit 100% of the light at the desired wavelengths. Optically flat quartz windows are recommended. The size of the reaction cell and windows should be consistent with the dimensions of the light beam used in the equipment. The reaction cell should be of sufficient volume to permit removal of samples for analysis without significantly altering the volume of solution in the cell. The cell path length can be measured directly or it can be determined by the procedure discussed in Section 9.E. Details for the construction of these reaction cells may be found in references 1 and 2.

10.A.1.e. Temperature Control

The effect of temperature on photochemical processes in solution is usually small because photochemical reactions occur from electronic excited states and usually have small or essentially zero activation energies. However, a number of photochemical reactions have activation energies that are in the 3–7-kcal/mol range (7). Thus, the reaction rate may increase or decrease slightly with temperature and the reaction cell temperature should be controlled to $t_i \pm 2$ °C in the range 20–30 °C.

10.A.2. Chemical Actinometry

A detailed discussion of chemical actinometry was given in Section 9.D. However, some further details must be considered for the two low-optical-density actinometers PNAP–PYR and PNA–PYR discussed in

Sections 9.D.1.a and 9.D.1.b, respectively. To carry out procedure one, it is necessary to choose the appropriate actinometer and to adjust the concentration of pyridine so that the rate constant of the actinometer approximately equals the rate constant of the test chemical. The rate constant of the actinometer is a function of the molar concentration of pyridine [PYR], the intensity of light incident on the actinometer (I_λ), and the molar absorptivity of the actinometer $\epsilon_{\lambda a}$. The pertinent equations for the low optical density actinometer are equations 9.35 and 9.16.

For procedure one, it is necessary to derive equations by letting the rate constant of the actinometer equal the rate constant of the test chemical; that is,

$$k_{da} = k_{dc} \tag{10.1}$$

Substituting equation 9.35 for the actinometer in equation 10.1 yields

$$k_{dc} = 2.303\phi_{da} I_\lambda l \epsilon_{\lambda a} \tag{10.2}$$

Substituting equation 9.35 for the actinometer in equation 9.16 yields

$$(t_{(1/2)})_a = 3.01 \times 10^{-1}[\phi_d I_\lambda l \epsilon_{\lambda a}]^{-1} \tag{10.3}$$

Equations 10.2 and 10.3 are perfectly general and are applicable for use in a POB or PMGRR using any light source filtered to yield a monochromatic wavelength band λ. Equations will be derived for a typical 450-W medium-pressure mercury lamp in an Ace-type or a Moses-type PMGRR. The pertinent information has been taken from the paper by Dulin and Mill (*8*); the data taken from Sections 9.D.1.a. and 9.D.1.b. and from Mill (*9*).

10.A.2.a. *p*-Nitroacetophenone–Pyridine Actinometer (PNAP–PYR) in an Ace-Type PMGRR

For the PNAP–PYR actinometer, the reaction quantum yield, ϕ_{da}, is a function of the molar concentration of pyridine [PYR] and is given by equation 9.24. For the 450-W medium-pressure mercury lamp in the Ace-type PMGRR used by Dulin and Mill (*8*), the average light intensities at 313 and 366 nm were $I_{(313)} = 3.0 \times 10^{-6}$ einsteins s^{-1} L^{-1} and $I_{(366)} = 7.5 \times 10^{-6}$ einsteins s^{-1} L^{-1}. Conversion of these intensities to hours yields $I_{(313)} = 1.08 \times 10^{-2}$ einsteins h^{-1} L^{-1} and

$I_{(366)} = 2.70 \times 10^{-2}$ einsteins h^{-1} L^{-1}. For the cells used by Dulin and Mill (*8*), the path length $l = 1.0$ cm (*9*). The molar absorptivities of PNAP at 313 and 366 nm are: $\epsilon_{313a} = 2056$ $M^{-1}cm^{-1}$ and $\epsilon_{366a} = 160$ M^{-1} cm^{-1}.

At 313 nm, the following relationships can be derived. Substituting equation 9.24 in equation 10.2 and using the pertinent data yields

$$k_{dc} = 0.864[\text{PYR}] \tag{10.4}$$

$$[\text{PYR}] = 1.16k_{dc} \tag{10.5}$$

where k_{dc} is in the units h^{-1} and [PYR] is in the units of molar concentration. Furthermore, because $k_{da} = k_{dc}$ (equation 10.1), equation 10.4 becomes

$$k_{da} = 0.864[\text{PYR}] \tag{10.6}$$

Substituting this result into equation 9.16 yields

$$[t_{(1/2)}]_a = 0.802/[\text{PYR}] \tag{10.7}$$

where $[t_{(1/2)}]_a$ is the half-life of the actinometer in hours. The half-life of the actinometer can range from 16 h at [PYR] = 0.05 M to 33 days at [PYR] = 10^{-3} M. Thus, this actinometer is suitable for use with test chemicals having half-lives from greater than 12 h to several weeks.

This derivation can be repeated at 366 nm to give:

$$[\text{PYR}] = 5.95k_{dc} \tag{10.8}$$

where k_{dc} is in the units of h^{-1} and [PYR] is in mol L^{-1}. The half-life of the actinometer in hours is given by

$$[t_{(1/2)}]_a = 4.12/[PYR] \tag{10.9}$$

By using the molar absorptivity data for PNAP at 313 and 366 nm, the absorbances A_{313a} and A_{366a} are found to be approximately equal to or less than 0.02 in a 1.00-cm absorption cell containing a PNAP solution of concentration $\leq 1 \times 10^{-5}$ M. The recommended tubes used for these photoreaction experiments have path lengths of approximately 1 cm. PNAP solutions of concentration $\leq 1 \times 10^{-5}$ M have absorbances of approximately ≤ 0.02 in the recommended tubes and thus correspond to low optical density. PNAP solutions of concentration $\leq 1 \times 10^{-5}$ M are recommended for use in this test method.

10.A.2.b. *p*-Nitroanisole–Pyridine Actinometer (PNA–PYR) in an Ace-Type PMGRR

The calculations just described can be repeated by using equation 9.25 for the PNA–PYR actinometer, with $I_{(313)} = 1.08 \times 10^{-2}$ einsteins h^{-1} L^{-1} and $I_{(366)} = 2.70 \times 10^{-2}$ einsteins h^{-1} L^{-1}, $l = 1.0$ cm, $\epsilon_{313a} = 1.03 \times 10^4$ M^{-1} cm^{-1}, and $\epsilon_{366a} = 1.99 \times 10^3$ M^{-1} cm^{-1}. Thus, at 313 nm:

$$[\text{PYR}] = 8.93 \times 10^{-3}(k_{dc} - 0.0722) \qquad (10.10)$$

where k_{dc} is in the units h^{-1} and [PYR] is in mol L^{-1}. The half-life of the actinometer in hours is given by

$$[t_{(1/2)}]_a = (162[\text{PYR}] + 0.104)^{-1} \qquad (10.11)$$

The half-life of the actinometer can range from 15 min at [PYR] = 0.024 M to 8.3 h at [PYR] = 10^{-4} M. Thus, this actinometer is suitable for use with test chemicals having half-lives of greater than 15 min to approximately 12 h. At 366 nm

$$[\text{PYR}] = 1.85 \times 10^{-2}(k_{dc} - 0.0349) \qquad (10.12)$$

where k_{dc} is in the units of h^{-1} and [PYR] is in mol L^{-1}. The half-life of the actinometer is given by the equation

$$[t_{(1/2)}]_a = (78.1[\text{PYR}] + 0.0504)^{-1} \qquad (10.13)$$

By using the molar absorptivity of PNA at 366 nm, the absorbance A_{366a} is less than 0.02 in a 1.00-cm absorption cell containing a PNA solution at a concentration of of $\leq 1 \times 10^{-5}$ M. The recommended tubes used for these photoreaction experiments have path lengths of approximately 1 cm. Hence, PNA solutions of concentration $\leq 1 \times 10^{-5}$ M have absorbances less than 0.02 in the recommended tubes; this corresponds to low optical density at 366 nm. However, at 313 nm, the molar absorptivity is large and A_{313a} is less than 0.02 in a 1.00-cm absorption cell containing a PNA solution of concentration $\leq 0.2 \times 10^{-5}$ M; however, PNA is difficult to analyze at concentrations significantly below 0.2×10^{-5} M. As a compromise, the allowable absorbance has been raised to 0.04 and the concentration of PNA has been set at 0.4×10^{-5} M for photoreaction experiments at 313 nm in the recommended tubes. PNA can be analyzed in this concentration range and only a small error (<10%) has been introduced in the method (*9*).

10.A.2.c. The PNAP–PYR and PNA–PYR Actinometers in a Moses-Type PMGRR

This test method has been designed for use with a commercially available Ace-type PMGRR. If a Moses-type PMGRR is available, the discussions given in Section 9.F.1. have to be modified. As indicated in Section 10.A.1.a.i., the Ace-type PMGRR is designed so that the reaction tubes are irradiated completely, while the Moses-type PMGRR is designed with windows, which act as slits; as a result, irradiation occurs only in a narrow portion of the reaction cells. Thus, the light intensity is reduced in a Moses-type PMGRR, as is the reaction rate of a chemical. Because the reaction rate of the chemical is reduced, the corresponding half-life is increased.

Mill (*9*) indicated that his Ace-type PMGRR had an effective aperture of 5.35 cm^2, while his Moses-type PMGRR (which he had specially built) had an aperture of 1.21 cm^2. Because $t_{(1/2)}$ is proportional to the reciprocal of the rate constant k_d, and k_d is proportional to the light intensity, $t_{(1/2)}$ is proportional to the reciprocal of the light intensity. Furthermore, because the light intensity is proportional to the aperture (A_p in cm^2), $t_{(1/2)}$ is proportional to the reciprocal of the aperture.

With the subscript notation "Ace" to represent the Ace-type PMGRR and the subscript notation "Test" to represent the Moses-type PMGRR, one can write

$$\frac{[t_{(1/2)}]_{c,\text{Ace}}}{[t_{(1/2)}]_{c,\text{Test}}} = \frac{[1/(A_p)_{\text{Ace}}]}{[1/(A_p)_{\text{Test}}]} \quad (10.14a)$$

with the assumption that the same reaction cells are placed in both types of PMGRR (and thus have the same path length). Rearranging equation 10.14 yields

$$\{[t_{(1/2)}]_{c,\text{Ace}}/[t]_{(1/2)c,\text{Test}}\} = \{(A_p)_{\text{Test}}/(A_p)_{\text{Ace}}\} \quad (10.14b)$$

$$[t_{(1/2)}]_{c,\text{Test}} = [t_{(1/2)}]_{c,\text{Ace}}\{(A_p)_{\text{Ace}}/(A_p)_{\text{Test}}\} \quad (10.14c)$$

From the previous discussions, the criteria for the choice of actinometer were: if $[t_{(1/2)}]_c$ was equal to or less than 12 h, then the PNA–PYR actinometer was used; if $[t_{(1/2)}]_c$ was greater than 12 h, then the PNAP–PYR actinometer was used. Therefore, $[t_{(1/2)}]_{c,\text{Ace}} = 12$ h, and because $(A_p)_{\text{Ace}} = 5.35$ cm^2 (*9*), equation 10.14c becomes

$$[t_{(1/2)}]_{c,\text{Test}} = 64(A_p)_{\text{Test}} \quad (10.15)$$

where $(t_{(1/2)})_{c,\mathrm{Test}}$ is in hours and A_p is the aperture in cm^2. Therefore, if a Moses-type PMGRR is used in this test method, the aperture A_p must be measured and substituted in equation 10.15. In this case, choice of actinometer will be based on the following criteria: if $(t_{(1/2)})_{c,\mathrm{Test}} \leq 64/(A_p)_{\mathrm{Test}}$, then the PNA–PYR actinometer is used; if $(t_{(1/2)})_{c,\mathrm{Test}} > 64/(A_p)_{\mathrm{Test}}$, then the PNAP–PYR actinometer is used.

By using the approach just described, equations 10.5 and 10.8 (for the PNAP–PYR actinometer) and equations 10.10 and 10.12 (for the PNA–PYR actinometer) can be modified as follows:

$$\frac{(k_{dc})_{\mathrm{Test}}}{(k_{dc})_{\mathrm{Ace}}} = \frac{(A_p)_{\mathrm{Test}}}{(A_p)_{\mathrm{Ace}}} \tag{10.16a}$$

$$(k_{dc})_{\mathrm{Ace}} = (k_{dc})_{\mathrm{Test}}\{(A_p)_{\mathrm{Ace}}/(A_p)_{\mathrm{Test}}\} \tag{10.16b}$$

For the Ace-type PMGRR, the aperture is 5.35 cm^2 (*9*). Thus, equation 10.16b becomes

$$(k_{dc})_{\mathrm{Ace}} = 5.35(k_{dc})_{\mathrm{Test}}/(A_p)_{\mathrm{Test}} \tag{10.17}$$

where the rate constants are in the units of h^{-1} and $(A_p)_{\mathrm{Test}}$ is the aperture of the Moses-type PMGRR in cm^2.

For the PNAP–PYR actinometer, equations 10.5 and 10.8 define the pyridine concentration needed to adjust the rate constant of the actinometer to equal the rate constant of the test chemical in the Ace-type PMGRR at 313 and 366 nm, respectively. The corresponding equations for the Moses-type PMGRR can be obtained by substituting equation 10.17 in equations 10.5 and 10.8 yielding

$$313\ \mathrm{nm:}\ [\mathrm{PYR}] = 6.21(k_{dc})_{\mathrm{Test}}/(A_p)_{\mathrm{Test}} \tag{10.18}$$

$$366\ \mathrm{nm:}\ [\mathrm{PYR}] = 31.8(k_{dc})_{\mathrm{Test}}/(A_p)_{\mathrm{Test}} \tag{10.19}$$

where $(k_{dc})_{\mathrm{Test}}$ is in the units of h^{-1}, $(A_p)_{\mathrm{Test}}$ is the aperture of the Moses-type PMGRR in the units of cm^2, and [PYR] is in mol L^{-1}. Following the procedure just described for the PNA–PYR actinometer, equations 10.10 and 10.12 can be modified by using equation 10.17.

$$313\ \mathrm{nm:}\ [\mathrm{PYR}] = 0.0478[(k_{dc})_{\mathrm{Test}}/(A_p)_{\mathrm{Test}}] - 6.46 \times 10^{-4} \tag{10.20}$$

$$366\ \mathrm{nm:}\ [\mathrm{PYR}] = 0.0990[(k_{dc})_{\mathrm{Test}}/(A_p)_{\mathrm{Test}}] - 6.46 \times 10^{-4} \tag{10.21}$$

where $(k_{dc})_{\mathrm{Test}}$ is in the units of h^{-1}, $(A_p)_{\mathrm{Test}}$ is the aperture of the Moses-type PMGRR in cm^2, and [PYR] is in the units of molar concentration.

10.A.3. Chemical Analysis of Solutions

The chemical analyses of test chemical solutions, PNAP, and PNA have been discussed in Section 7.B.1.k. and the reader is referred to this section for the details.

10.A.3.a. Chemical Analysis of Ferrous Ion in the Ferrioxalate Actinometer

The molar concentration of Fe^{2+} formed during the irradiation of the ferrioxalate actinometer is usually measured spectrophotometrically via the formation of a red phenanthroline complex and determination of the absorbance at 510 nm. The procedure described by Murov (*3*) is excellent and is recommended. The procedure has been modified slightly and now uses a 0.15 M concentration of ferrioxalate. As a result, the irradiated solution has to be diluted 100-fold prior to analysis for Fe^{2+}, which must not be allowed to exceed 0.005 M (*10*).

10.A.3.b. Chemical Analysis of *o*-Nitrobenzaldehyde

o-Nitrobenzaldehyde is analyzed by using reverse-phase high-pressure liquid chromatography (HPLC) with 50% acetonitrile in water as the eluent and the UV detector set at 280 nm (*9*).

10.B. Procedure One: Determination of the Reaction Quantum Yield by the Low Optical Density Test Chemical and Actinometer Method

10.B.1. Phase One Experiments: Ultraviolet–Visible Absorption Spectra

The UV–vis absorption spectra in aqueous solution can be determined by standard spectroscopic procedures (*11*). It is recommended that the following additional procedures be followed:

For chemicals that ionize or protonate (e.g., carboxylic acids, phenols, amines), UV–vis absorption studies must be carried out at pH's at least a factor of 2 above and below the pK_a (i.e., pH = $pK_a \pm 2$). Buffer solutions are prepared at a temperature t_i (where t_i is the temperature in °C at which the photoreaction experiments are carried out) by using reagent grade chemicals and air-saturated distilled water as follows:

- pH in the range 3–6* : NaH_2PO_4, HCl
- pH in the range 6–8*: KH_2PO_4, NaOH
- pH > 8: 20 Buffers prepared as described in the CRC *Handbook of Chemistry and Physics* (*12*).

The pH of all the buffer solutions at t_i °C should be checked with a pH meter and adjusted to the proper pH, if necessary. These buffer solutions can then be added to the test chemical solution until the desired pH is obtained. If these buffers are inadequate, the pH of the test chemical solution should be adjusted with 0.01 M HCl or NaOH at t_i °C.

The absorbance, $A_{\lambda c}$, should be measured in triplicate as a function of wavelength in the range 290–800 nm. If applicable, $A_{\lambda c}$ should be measured at each experimental pH. Triplicate measures of the baseline are needed when both the sample and reference cells are filled with blank solutions. These data will be used to calculate the molar absorptivities for the appropriate intervals and wavelength centers where the test chemical absorbs light (see Table A-1, Appendix A). The wavelength center, λ(center), is the midpoint of the interval range listed in Table A-1.

$A_{\lambda c}$ at 313.0 and 366.0 nm should be measured in triplicate. These data will be used to calculate the molar absorptivities of the test chemical, ϵ_{313c} and ϵ_{366c}. Photoreaction experiments should be carried out at the wavelength corresponding to the higher value of the molar absorptivity.

The molar absorptivities of the test chemical $\epsilon_{\lambda c}$ must be carefully determined, especially in the tails of the absorption bands at $\lambda \geq 290$ nm. Large errors will occur when calculating photoreaction rate constants and half-lives if these measurements are not carefully carried out.

10.B.2. Phase Two Experiments: Trial Photoreaction Experiments

10.B.2.a. Determination of the Approximate Rate Constant of the Test Chemical

* Use the minimum concentration of buffers to attain the desired pH.

10.B.2.a.i. Preparation of Buffer Solutions

Buffer solutions are prepared according to the procedures outlined in Section 10.B.1. with reagent grade chemicals and air saturated pure water for chemicals that ionize or protonate reversibly.

10.B.2.a.ii. Preparation of Test Chemical Solution

A homogeneous solution of test chemical is prepared at a concentration less than one-half its water solubility with an absorbance less than 0.02 at 313 or 366 nm in the photoreaction vessel. For very hydrophobic chemicals, it is difficult and time-consuming to prepare aqueous solutions. To facilitate the preparation of aqueous solutions containing very hydrophobic chemicals and to allow for easier analytical procedures, the following procedure may be used as an aid in the dissolution of the test chemical. The pure test chemical is dissolved in reagent grade acetonitrile. Air-saturated pure water (as described in Section 7.B.1.b.), or buffer solution (as described in Section 10.B.1., for chemicals that ionize or protonate) is added to an aliquot of the acetonitrile solution. In the final solution, 1 vol % of acetonitrile must not be exceeded.

10.B.2.a.iii. Performance of the Test

An aqueous solution of test chemical, as well as a sufficient number of samples are prepared in quartz or borosilicate glass tubes (Section 10.A.1.d.) to perform all the required tests. The initial molar concentration of test chemical $[C_o]$ is measured in duplicate (Section 7.B.1.k.). The tubes are filled as completely as possible and sealed. Grease must not be used. Two control samples are prepared in the absence of UV light and light is totally excluded by wrapping the tubes with aluminum foil or by any other suitable method. The samples, including the controls, are placed in an Ace-type PMGRR with a 450-W medium-pressure mercury lamp with appropriate filters as described in Sections 10.A.1.b. and 10.A.1.c. The reaction tubes should be controlled to a temperature of $t_i \pm 2$ °C within the range 20–30 °C. The samples are irradiated at 313 or 366 nm corresponding to the higher value of the molar absorptivity of the test chemical (ϵ_{313c} or ϵ_{366c} determined in the phase one experiments) until at least 50% of the test chemical has transformed. The molar concentration of the test chemical $[C_t]$ is measured in duplicate at time t (Section 7.B.1.k.). For test chemicals that ionize or protonate, the photoreaction experiments are carried out at the required pH's as described in Section 10.B.1.

After the photoreaction experiments are completed, the concentration of test chemical in the controls are determined in duplicate. If a significant loss of test chemical has occurred in the controls, the cause must be determined and the loss minimized or eliminated. If hydrolysis is significant, hydrolysis experiments should be carried out first (*13–15*). More details on the use of control samples in these experiments are given in Section 7.A.1.f.

The data obtained in this section will be used to determine an approximate k_{dc} and $[t_{(1/2)}]_c$ by using equations 9.37 and 9.16, respectively.

10.B.2.b. Chemical Actinometry Experiments

These experiments are designed to allow the determination of the appropriate low-optical-density actinometer [i.e., PNAP–PYR, (Section 9.D.1.a.) or PNA–PYR (Section 9.D.1.b.) and to adjust the rate of the chosen actinometer so that the rate constant of the actinometer k_{da} is approximately equal to the rate constant of the test chemical k_{dc}. On the basis of the trial photoreaction experiments just described, if $[t_{(1/2)}]_c$ is equal or less than 12 h, the PNA–PYR actinometer is used; if $[t_{(1/2)}]_c$ is greater than 12 h, the PNAP–PYR actinometer is used.

10.B.2.b.i. Preparation of Actinometer Solutions

PNAP–PYR Actinometer: By using the test chemical photoreaction rate constant k_{dc} determined as just described and equations 10.5 or 10.8, the molar concentration of pyridine needed to adjust the rate constant of the actinometer to approximately equal the rate constant of the test chemical at the appropriate wavelength 313 or 366 nm (chosen based on the results of the phase one experiments) can be determined. Once the molar concentration of pyridine [PYR] has been estimated, the actinometer solution can be prepared as follows: 0.0165 g of PNAP is dissolved in 10 mL of acetonitrile to yield a 0.01 M solution. 1 mL of this solution is added to a 1-L volumetric flask. To the volumetric flask is added the mass in grams, or the volume of pyridine at 20 °C, obtained from the equations

$$\text{mass (grams)} = 79.1[\text{PYR}] \qquad (10.22)$$

$$\text{volume (mL)} = 80.6[\text{PYR}] \qquad (10.34)$$

The volumetric flask is filled with pure water to give 1 L of solution and shaken vigorously to ensure homogeneity. The resultant concentration of PNAP is 1.00×10^{-5} M. The PNAP–PYR solution should be wrapped with aluminum foil and kept from bright light.

PNA–PYR Actinometer: By using the test chemical photoreaction rate constant k_{dc} determined as described in Section 10.B.2.a.iii. and equations 10.10 or 10.12, the concentration of pyridine needed to adjust the rate constant of the actinometer to approximately equal the rate constant of the test chemical at the appropriate wavelength 313 or 366 nm (chosen based on the results of the phase one experiments) can be determined. Once the molar concentration of pyridine has been determined, an actinometer solution can be prepared as follows. If photoreaction experiments are carried out at 366 nm , 0.0153 g PNA is dissolved in 10 mL of acetonitrile to yield a 0.01 M solution and 1 mL of this solution is added to a 1-L flask. The mass in grams, or the volume of pyridine at 20 °C, as defined by equations 10.22 and 10.23, is added to the volumetric flask. The volumetric flask is filled with pure water to give 1 L of solution and shaken vigorously to make ensure homogeneity. The resultant concentration is 1.00×10^{-5} M. The PNA–PYR solution should be wrapped with aluminum foil and kept from bright light.

If photoreaction experiments are carried out at 313 nm, 6.12×10^{-3} g of PNA is dissolved in 10 mL of acetonitrile to yield a 4×10^{-3} M solution, and the procedure just described is followed. The resultant PNA concentration is 0.400×10^{-5} M. The PNA–PYR solution should be wrapped with aluminum foil and kept from bright light.

10.B.2.b.ii. Performance of the Test

A solution of the appropriate actinometer is prepared. The initial molar concentration of the PNAP or PNA actinometer $[C_o]$ is measured (Section 7.B.1.k.). A sufficient number of samples are prepared in borosilicate glass or quartz tubes to perform all the required tests. The tubes should be identical to the tubes used to determine the photoreaction rate constant of that test chemical, k_{dc}. The tubes are filled as completely as possible and sealed. The actinometer tubes must contain the same volume as that used in the test chemical tubes. Two control samples are prepared in the absence of UV light and light is totally excluded by wrapping the tubes with aluminum foil or by using any other suitable method. The samples, including the controls, are placed in the Ace-type PMGRR with the 450-W medium-pressure mercury lamp and the appropriate filters. The tubes should be controlled to a temperature $t_i \pm 2$ °C within the range 20–30 °C. The samples are irradiated at the chosen wavelength λ (313 or 366 nm) until at least 50% of the actinometer has transformed. The molar concentration of the actinometer $[C_t]$ is determined in duplicate at time t. The molar concentration of the actinometer controls at the end of the experiment is determined. If a significant loss of actinometer has occurred, the cause is determined and the loss must be eliminated or minimized. A rate constant is calculated by using equation 9.38. If the rate constant k_{da} is not approximately the same as the rate constant of the test chemical k_{dc}, then the experiment must be repeated with an appropriate concentration of pyridine until k_{da} is approximately equal to k_{dc} (±50%).

10.B.3. Phase Three Experiments: Determination of the Reaction Quantum Yield of the Test Chemical

On the basis of the results of the phase one and two experiments, the appropriate actinometer is used with the required concentration of pyridine to make the rate constant of the actinometer solution approximately equal the rate constant of test chemical solution. The procedure to be followed is outlined in Section 10.B.2.b.i. The actinometer solution and a sufficient number of samples are prepared in borosilicate glass or quartz tubes to perform all the required tests. The initial molar concentration of the actinometer $(C_o)_a$ is determined in duplicate. All the tubes are filled as completely as possible, sealed, and half are covered with aluminum foil as soon as possible after preparation. Aqueous solutions of test chemical are prepared as described in Section 10.B.2.a.ii., and the initial molar concentration of test chemical, $[C_o]_c$ is determined in duplicate. A sufficient number of samples in borosilicate or quartz tubes to perform the required tests are prepared. All the tubes are filled as completely as possible, sealed, and half are covered with aluminum foil as soon as possible after preparation. The reaction tubes for actinometer and test chemical must be identical and contain the same volume of solution.

The tubes to be irradiated are placed in the Ace-type PMGRR containing a 450-W medium-pressure mercury lamp and the appropriate filters, and all the control tubes are placed close to the PMGRR. The tubes should be temperature controlled to t_i ± 2 °C within the range 20–30 °C. On the basis of the results of the phase two experiments, the concentration of test chemical and actinometer are determined, in duplicate, periodically between 10 and 80% transformation (There should be at least six data points at approximately equal intervals.) The concentration of actinometer and test chemical in the controls is determined in duplicate at each time point.

10.C. Procedure Two: Determination of the Reaction Quantum Yield by the Low Optical Density Test Chemical and High Optical Density Actinometer Method

10.C.1. Phase One Experiments: UV–Visible Absorption Spectra

The procedure outlined in Section 10.B.1. is followed and the absorbance of the test chemical, $A_{\lambda c}$, is measured in triplicate as a function of wavelength in the range 290–800 nm. $A_{\lambda c}$ is measured in triplicate at 313.0 and 366.0 nm. The absorbance data at 313.0 and 366.0 nm will be used to calculate the molar absorptivities of the test chemical, ϵ_{313c} and ϵ_{366c}. Photoreaction experiments should be carried out at the wavelength corresponding to the highest value of the molar absorptivity.

10.C.2. Phase Two Experiments: Determination of the Reaction Quantum Yield of the Test Chemical

10.C.2.a. Preparation of Buffer Solutions

Buffer solutions are prepared according to the procedure outlined in Section 10.B.1. for a test chemical that ionizes or protonates.

10.C.2.b. Preparation of Test Chemical Solution

A homogeneous solution of test chemical is prepared according to the procedure outlined in Section 10.B.2.a.ii.

10.C.2.c. Preparation of Ferrioxalate Actinometer Solution

A ferrioxalate actinometer solution at 0.15 M is prepared as described in Section 9.D.2.a. following the procedure outlined by Murov (*3*), which has been modified slightly. Because the ferrioxalate actinometer solution is very sensitive to visible light, all actinometer experiments must be carried out in a dark room with photographic "safe lights."

10.C.2.d. Determination of the Reaction Quantum Yield of the Test Chemical

The reaction cell is completely filled with a solution of the ferrioxalate actinometer at high optical density (absorbance > 2) as just described. The reaction cell is placed in the photochemical optical bench (POB) with a 450-W medium-pressure mercury lamp and appropriate filters. The reaction cell should be controlled to a temperature of $t_i \pm 2$ °C in the range 20–30 °C. The filter system used should isolate the wavelength 313 or 366 nm corresponding to the larger value of ϵ_{313c} or ϵ_{366c} determined in the phase one experiments.

The actinometer solution is irradiated in the reaction cell and the concentration of Fe^{2+} formed is measured spectrophotometrically in duplicate at 1, 3, 6, 9, 12, and 15 min. No data should be taken when the optical density of the actinometer falls below 2. Only small volumes should be withdrawn for analysis so that the total volume of the actinometer solution in the reaction cell does not change appreciably. This procedure is repeated in a third set of experiments soon after the test chemical photoreaction experiments are performed.

In the second set of experiments, the initial molar concentration $[C_o]_c$ is measured in duplicate (Section 7.B.1.k.) by using a solution of test chemical at low optical density (absorbance > 0.02). The two reaction cells are filled with the same volume as that used in the actinometer experiments. One of these cells is wrapped with aluminum foil, placed close to the POB, and is temperature controlled to $t_i \pm 2$ °C. The second reaction cell (which is the same one used for the actinometry experiments) is placed in the POB and irradiated at the same temperature and wavelength used in the actinometry experiments.

The molar concentration of test chemical is measured in duplicate periodically between 10 and 80% transformation (There should be at least six data points at approximately equal intervals.) Only small volumes are withdrawn for analysis so that the total volume of the test chemical solution in the reaction cell does not change appreciably. The concentration of the control is determined in duplicate at each time point.

As an alternative procedure, the same series of experiments as just described can be carried out in a PMGRR by using individual reaction tubes for each data point.

References

1. Calvert, J. G.; Pitts, J. N., Jr. *Photochemistry*; Wiley: New York, 1966; Chapter 7.
2. de Mayo, P.; Shizuka, P. In *Creation and Detection of Excited States*; Ware, W. R., Ed.; Marcel Dekker: New York, 1976; Vol. 4, Chapter 4.

3. Murov, S. L. *Handbook of Photochemistry*; Marcel Dekker: New York, 1973; Sections 9–14.
4. Mill, T.; Mabey, W. R.; Bomberger, D. C.; Chou, T–W.; Hendry, D. G.; Smith, J. H. *Laboratory Protocols for Evaluation of the Fate of Organic Chemicals in Air and Water;* U.S. Environmental Protection Agency. U.S. Government Printing Office: Washington, DC, 1982; EPA 600/3–82–002.
5. Moses, F. G.; Liu, R. S. H.; Monroe, B. M. *Mol. Photochem.* **1969**, *1*, 245.
6. Andre, J. C.; Niclause, J.; Joussot-Dubien, J.; Deglise, X. *J. Chem. Educ.* **1977**, *54*, 387.
7. Barltrop, J. A.; Coyle, J. D. *Excited States in Organic Chemistry*; Wiley: New York, 1975; Chapter 3.
8. Dulin, D.; Mill, T. *Environ. Sci. Technol.* **1982**, *16*, 815.
9. Mill, T., SRI International, Menlo Park, CA, personal communication, 1984.
10. Zepp, R. G., U.S. Environmental Protection Agency, Office of Research and Development, Athens, GA, personal communication, 1984.
11. *Fed. Regist.* **1985**, *50*, 39472; 40 CFR 796.1050.
12. *Handbook of Chemistry and Physics*; Weast, R. C., Ed.; 57th ed.; Chemical Rubber Company: Cleveland, OH, 1976–1977.
13. U.S. Environmental Protection Agency. "Hydrolysis as a Function of pH"; CG, CS–5000; National Technical Information Service, U.S. Department of Commerce: Springfield, VA, 1982; EPA 560/6–82–003; NTIS PB82–233008.
14. *Fed. Regist.* **1985**, *50*, 39283; 40 CFR 796.3500.
15. *Fed. Regist.* **1987**, *52*, 36334; 40 CFR 796.3510.

Chapter 11

Illustrative Example

Determination of the Reaction Quantum Yield by the Low Optical Density Test Chemical and Actinometer Method and Sunlight Photoreaction Rates

A chemical plant just south of Peoria, IL, on the Illinois River, produces an organic chemical A that is neither an acid nor a base. The waste effluent passes through primary and secondary treatment plants. The waste, which still contains some chemical A, is then continuously discharged into the river. The plant is located at 40.7° N. latitude. Information is needed on direct photoreaction rates and half-lives of chemical A in aqueous media in the summer and winter seasons.

The company reseach laboratory is located in Peoria, IL, and the required photoreaction data were needed in January. Because the temperature outdoors was well below freezing during this month, the outdoor sunlight photoreaction experiments described in Chapters 6–9 could not be carried out. Thus, it was necessary to carry out photoreaction experiments in the laboratory. The research laboratory was equipped with an Ace-type PMGRR to carry out photoreaction experiments and, thus, procedure one was used.

Laboratory Experiments, Data, and Calculations

11.A. Phase One Experiments: Molar Absorptivity of the Test Chemical and Maximum Sunlight Photoreaction Rates

Chemical A had a saturated water solubility of 3.9×10^{-3} M at 25 °C. In the phase one procedure, the UV–vis absorption spectrum was obtained in triplicate for chemical A at a concentration of 1.00×10^{-5} M in a 10.0-cm quartz cell. By using the wavelength interval from

1464–6/88/0191$06.00/1

Table A-1 of Appendix A, the average absorbance of triplicate runs at λ(center) was obtained, and the results are summarized in Table 11-1. In addition, the average absorbance was measured in triplicate at 313.0 and 366.0 nm, and these results are summarized in Table 11-1. By using these data and the expression $A_\lambda = \epsilon_\lambda[C]l$ (equation 1.14), the average molar absorptivity is given by

$$\epsilon_{\lambda c} = 10^4 A_{\lambda c} \tag{11.1}$$

With the average value of $A_{\lambda c}$ at λ(center), the average molar absorptivity can be obtained from equation 11.1, and these results are summarized in Table 11-1. In addition, with $A_{\lambda c}$ at 313.0 and 366.0 nm in equation 11.1, the molar absorptivity ϵ_{313c} and ϵ_{366c} can be obtained, and these values are also given in Table 11-1.

The plant is located at 40.7° N. latitude, so the closest L_λ values are for 40 ° N. latitude. These values are obtained from Table A-8 of Appendix A and are summarized in Table 11-2 for the summer and winter solstices. By using the data from Table 11-2 and equations 9.5 and 9.6, the following results are obtained.

Table 11-1. Spectral Data for Chemical A

λ(center), nm	$A_{\lambda c}$	$\epsilon_{\lambda c}(M^{-1}cm^{-1})$
297.5	0.5904	504
300.0	0.5414	5414
302.5	0.4924	4924
305.0	0.4434	4434
307.5	0.3810	3810
310.0	0.3124	3124
312.5	0.2714	2714
315.0	0.2224	2224
317.5	0.2004	2004
320.0	0.1514	1514
323.1	0.1310	1310
330.1	0.1030	1030
340.0	0.0645	645
350.0	0.0473	473
360.0	0.0287	287
370.0	0.0170	170
380.0	0.0085	85
390.0	0.0030	30
400.0	0.0010	10
410.0	0.0000	0
313.0	0.2610	2610
366.0	0.0185	185

Table 11-2. Light Absorption Data for Chemical A

	Summer		*Winter*	
λ*(center), nm*	L_λ^a	$\epsilon_{\lambda c}L_\lambda(day^{-1})$	L_λ^a	$\epsilon_{\lambda c}L_\lambda(day^{-1})$
297.5	6.17(-5)	0.4	5.49(-7)	0.0
300.0	2.69(-4)	1.5	5.13(-6)	0.0
302.5	8.30(-4)	4.1	3.02(-5)	0.1
305.0	1.95(-3)	8.6	1.19(-4)	0.5
307.5	3.74(-3)	14.2	3.38(-4)	1.3
310.0	6.17(-3)	19.3	7.53(-4)	2.4
312.5	9.07(-3)	24.6	1.39(-3)	3.8
315.0	1.22(-2)	27.1	2.22(-3)	4.9
317.5	1.55(-2)	31.1	3.19(-3)	6.4
320.0	1.87(-2)	28.3	4.23(-3)	6.4
323.1	3.35(-2)	43.9	8.25(-3)	10.8
330.0	1.16(-1)	119.5	3.16(-2)	32.5
340.0	1.46(-1)	94.2	4.31(-2)	27.8
350.0	1.62(-1)	76.6	4.98(-2)	23.6
360.0	1.79(-1)	51.4	5.68(-2)	16.3
370.0	1.91(-1)	32.5	6.22(-2)	10.6
380.0	2.04(-1)	17.3	6.78(-2)	5.8
390.0	1.93(-1)	5.8	6.33(-2)	1.9
400.0	2.76(-1)	2.8	9.11(-2)	0.9
410.0	3.64(-1)	0.0	1.20(-1)	0.0
		$\Sigma_\lambda\ \epsilon_{\lambda c}L_\lambda = 603.2$		$\Sigma_\lambda\ \epsilon_{\lambda c}L_\lambda = 156.0$

[a]The units of L_λ are in millieinsteins cm^{-2}day^{-1}. The second number in the columns in parentheses is the power of 10 by which the first number is multiplied.

For the summer solstice:

$$(k_{dE})_{\max} = \sum_\lambda \epsilon_{\lambda c}L_\lambda = 603 \text{ days}^{-1} \quad (11.2)$$

$$[t_{(1/2)E}]_{\min} = 1.1 \times 10^{-3} \text{ days} \quad (11.3)$$

For the winter solstice:

$$(k_{dE})_{\max} = \sum_\lambda \epsilon_{\lambda c}L_\lambda = 156 \text{ days}^{-1} \quad (11.4)$$

$$[t_{(1/2)E}]_{\min} = 4.4 \times 10^{-3} \text{ days} \quad (11.5)$$

Because the chemical may transform rapidly on the summer and winter solstices, the reaction quantum yield of chemical A must be determined in the laboratory by using procedure one. Also, direct sunlight photoreaction rates and half-lives in the river must be obtained during the summer and winter seasons.

11.B. Phase Two Experiments: Determination of the Approximate Rate Constant of the Test Chemical and Chemical Actinometry Experiments

11.B.1. Test Chemical Photoreaction Experiments

Chemical A was dissolved directly in pure water, and a homogeneous solution was prepared. Analysis of duplicate samples indicated that the average concentration was 5.0×10^{-6} M. By using the UV spectral data obtained, the absorbance at 313 nm in a 1-cm absorption cell containing a solution at 5.0×10^{-6} M was determined to be less than 0.02 [i.e., $A_{313c} = 2610\ (5.0 \times 10^{-6})(1.00) = 0.0131$]. Hence, the test chemical solution in approximately 1-cm path length tubes was at low optical density.

A series of tubes (13- × 100-mm), with effective path lengths of 1.12 cm, were filled with the aqueous solution of test chemical and sealed. Because $\epsilon_{313c} > \epsilon_{366c}$ (Table 11-1), photoreaction experiments were carried out at 313 nm in a PMGRR. Duplicate irradiated and control samples were removed periodically, and the molar concentration of test chemical was determined. At $t = 39.0$ h, the average molar concentration of irradiated sample $[C_t]_c$ was 2.3×10^{-6} M, and the average molar concentration of control sample was 5.0×10^{-6} M. Thus, approximately 54% of the test chemical was transformed. Because no loss of chemical was observed in the control sample, no adventitious processes occurred and the loss of chemical was due only to photoreaction. By using equation 9.37 and these data:

$$\ln (5.0 \times 10^{-6}/2.3 \times 10^{-6}) = k_{dc}(39.0) \qquad (11.5)$$

$$k_{dc} = 0.020\ \text{h}^{-1} \qquad (11.6)$$

11.B.2. Chemical Actinometry Experiments

If equation 9.16 and the value of k_{dc} obtained in equation 11.6 are used, the half-life of the test chemical is

$$[t_{(1/2)}]_c = 0.693/0.020 = 35\ \text{h} \qquad (11.7)$$

Because $[t_{(1/2)}]_c > 12$ h, the PNAP–PYR actinometer is required for the photoreaction experiments.

By using equation 10.5 and the approximate rate constant k_{dc} from equation 11.6, the concentration of pyridine needed to make the rate constant of the actinometer approximately equal to the rate constant of the test chemical at 313 nm is calculated as follows:

$$[\text{PYR}] = 1.16k_{dc} = 1.16(0.020) \tag{11.8}$$

$$[\text{PYR}] = 2.32 \times 10^{-2} \tag{11.9}$$

An actinometer solution was then prepared according to the procedure given in Section 10.B.2.b.i.; the concentration of pyridine was 2.32×10^{-2} M. The concentration of PNAP was measured in duplicate and the average molar concentration $[C_o]_a$ was 0.90×10^{-5} M. By using the UV spectral data, the absorbance at 313 nm in a 1-cm absorption cell containing an actinometer solution at a concentration of 0.90×10^{-5} M was determined to be less than 0.02 [i.e., $A_{313a} = (2056)(0.90 \times 10^{-5})(1.00) = 0.019$]. Hence, the actinometer solution in approximately 1-cm path length tubes was at low optical density.

A series of tubes, identical to those used in the test chemical photoreaction experiments, were filled with the actinometer solution and sealed. The photoreaction experiments were carried out in the same PMGRR at 313 nm . Duplicate irradiated and control samples were periodically removed and analyzed for the concentration of PNAP. At 50.6 h, the average molar concentration of irradiated sample $[C_t]_a$ was 0.36×10^{-5} M and the average molar concentration of the control sample was 0.90×10^{-5} M. Thus, approximately 60% of the PNAP transformed. Because no loss of chemical was observed in the control sample, no adventitious processes occurred and the loss of PNAP was due only to photoreaction. By using equation 9.38 and the pertinent data:

$$\ln (0.90 \times 10^{-5}/0.36 \times 10^{-5}) = k_{da}(50.6) \tag{11.10}$$

$$k_{da} = 0.018 \ \text{h}^{-1} \tag{11.11}$$

Thus, at the molar concentration of pyridine given by equation 11.9, k_{da} is approximately the same as k_{dc} (within $\pm 50\%$).

11.C. Phase Three Experiments: Determination of the Reaction Quantum Yield of the Test Chemical

A solution of test chemical was prepared according to the procedure described in Section 10.B.2.a.ii. and the molar concentration $[C_o]_c$ was measured in duplicate. The average molar concentration was found to be 5.0×10^{-6} M. An actinometer solution (PNAP–PYR) was prepared according to the procedure described in Section 10.B.2.b.i. with a pyridine concentration of 2.32×10^{-2} M (equation 11.9). The average molar concentration of PNAP, $[C_o]_a$ from duplicate samples was found to be 9.0×10^{-6} M. These solutions were placed in identical tubes (13- × 100-mm), sealed, and irradiated at various times (t) at 313 nm in the PMGRR. The average concentrations of duplicate samples of test chemical, test chemical control, actinometer (PNAP), and actinometer control obtained in this photoreaction experiment, are summarized in Table 11-3.

Because no significant loss of PNAP or test chemical was observed in the control samples, no adventitious processes occurred and the loss of test chemical and PNAP was due only to photoreaction. By using the data in Table 11-3, ln $([C_o]/[C_t])$ for the test chemical and actinometer solution can be calculated at various times (t); the results are summarized in Table 11-4. By using all the data in equation 9.39 (including the time point $t = 0$) with ln $([C_o]/[C_t])_a$ as the independent variable and ln $([C_o]/[C_t])_c$ as the dependent variable, and linear regression analysis, the slope, which is equal to (k_{dc}/k_{da}), is found to be 1.09 with a correlation coefficient of 1.00.

The quantum yield of the PNAP–PYR actinometer is given by equation 9.24; $\phi_{da} = 0.0169[\text{PYR}]$. Because the pyridine concentration in the actinometer was 2.32×10^{-2} M, the quantum yield of the actinometer, ϕ_{da}, is 3.92×10^{-4}.

The reaction quantum yield of the test chemical is given by equation 9.36; $\phi_{dc} = (k_{dc}/k_{da})(\epsilon_{\lambda a}/\epsilon_{\lambda c})$. The pertinent data are summarized as follows: $\phi_{da} = 3.92 \times 10^{-4}$, $(k_{dc}/k_{da}) = 1.09$, $\epsilon_{313c} = 2610$, and $\epsilon_{313a} = 2060$. Substituting these data into equation 9.36 yields

$$\phi_{dc} = (1.09)(2060/2610)(3.92 \times 10^{-4}) \quad (11.12)$$

$$\phi_{dc} = 3.37 \times 10^{-4} \quad (11.13)$$

Table 11-3. Photoreaction Data for Test Chemical A and (PNAP–PYR) Actinometer

	Average Test Chemical Concentration × 10^6		*Average PNAP Concentration × 10^6*	
t (*hours*)	*Irradiated Samples*	*Controls*	*Irradiated Samples*	*Controls*
0	5.0	5.0	9.0	9.0
12	3.9	4.9	7.2	8.9
24	3.1	5.0	5.8	8.9
36	2.4	5.0	4.7	9.0
48	1.9	4.9	3.8	8.9
60	1.5	4.9	3.0	9.0
72	1.2	5.0	2.4	8.9

Table 11-4. Photoreaction Data for Test Chemical and (PNAP–PYR) Actinometer

	Test Chemical		*Actinometer (PNAP)*	
t (*hours*)	$[C_t]_c \times 10^6$ (*M*)	*ln* $([C_0]/[C_t])_c$	$[C_t]_a \times 10^6$ (*M*)	*ln* $([C_0]/[C_t])_a$
0	5.0	0.000	9.0	0.000
12	3.9	0.248	7.2	0.223
24	3.1	0.478	5.8	0.439
36	2.4	0.734	4.7	0.650
48	1.9	0.968	3.8	0.862
60	1.5	1.20	3.0	1.10
72	1.2	1.43	2.4	1.32

Thus, the reaction quantum yield of the test chemical is 3.4×10^{-4}.

The same set of experiments were carried out at 366 nm and the reaction quantum yield was found to be 3.3×10^{-4}. Thus, the reaction quantum yield of the chemical is independent of the wavelength. The average value of the reaction quantum yield of the test chemical is 3.4×10^{-4}.

11.D. Sunlight Photoreaction Rates of the Test Chemical

The rate constants and half-lives for direct photoreaction of test chemical in the river at Peoria, IL, under clear sky conditions for the winter and summer seasons can be calculated as follows: The values of $\sum_\lambda \epsilon_{\lambda c} L_\lambda$ have been calculated for the summer and winter solstices, Table 11-2. For summer $\sum_\lambda \epsilon_{\lambda c} L_\lambda = 603$ days^{-1}; for winter $\sum_\lambda \epsilon_{\lambda c} L_\lambda = 156$ days^{-1}. The reaction quantum yield for chemical A is 3.4×10^{-4}. Using these data in equation 9.4 yields

$$\text{summer: } k_{dE} = 3.4 \times 10^{-4}(603) = 0.205 \text{ days}^{-1} \qquad (11.14)$$

$$\text{winter: } k_{dE} = 3.4 \times 10^{-4}(156) = 0.0530 \text{ days}^{-1} \qquad (11.15)$$

and thus, k_{dE} values for the summer and winter seasons are 0.21 and 0.053 days^{-1}, respectively.

These values can be substituted into equation 9.2 to obtain the half-lives for these two seasons.

$$\text{summer: } [t_{(1/2)E}]_c = 0.693/0.21 = 3.3 \text{ days} \qquad (11.16)$$

$$\text{winter: } [t_{(1/2)E}]_c = 0.693/0.053 = 13 \text{ days} \qquad (11.17)$$

These results are valid for clear sky conditions and at shallow depths in the Illinois River at Peoria.

Part III. Indirect Photoreaction Test Methods

Chapter 12 is devoted to the theory and practice of measuring rates of indirect photoreaction in natural waters. It is based on the theory of indirect photoreaction described in Chapter 5 and Appendix B.

Chapters 13–15 describe a cost-effective screening test method that measures the indirect, or sensitized, aqueous photoreaction rate constant by carrying out sunlight photoreaction experiments with a chemical dissolved in humic acid water. The method yields a net photoreaction rate constant that represents the sum of the direct and indirect photoreaction rate constants. This test method describes the preparation and standardization of synthetic humic acid water (SHW).

For this method, a test chemical is exposed to sunlight in tubes containing solutions of SHW and in solutions containing buffered distilled water. Samples of SHW in tubes are irradiated simultaneously. To correct for variations in solar irradiance during photoreaction, and actinometer solution is simultaneously insolated. From a cocnentration-versus-time plot, the indirect and the net photoreaction rate constants are calculated that apply to clear sky, near-surface conditions in fresh water bodies in the environment.

Chapter 13 describes the theory and development of the test method. This test method was developed on the basis of the theory of the kinetics of aqueous photoreaction and the fundamental principles that guide the selection of experimental approaches. Procedures for obtaining the kinetic data are discussed in detail in Chapter 14 so that a scientist can carry out these experiments easily and precisely. Two examples are given in Chapter 15 to illustrate how the experimental data are used to calculate environmentally relevant indirect and net photoreaction rate constants.

Chapter 12

Indirect (Sensitized) Photoreaction Rates and Half-Lives in Humic Waters in the Environment

Some chemicals present in natural waters containing dissolved humic acids exhibit accelerated rates of photoreaction in comparison to the rates of photoreaction in pure water. These photoreactions with increased rates are called indirect (or sensitized) photoreactions. Chapter 12 is devoted to the theory and practice of measuring rates of indirect photoreaction in natural waters. It is based on the theory of indirect photoreaction described in Chapter 5 and Appendix B, primarily on the work of Zepp et al. (*1–3*), and also on the work of Haag et al. (*4–6*). Procedures are described for obtaining indirect photoreaction rates by carrying out photoreaction experiments either outdoors in sunlight or in the laboratory with monochromatic light. These studies lay the foundation for forecasting rates of indirect photoreaction in natural waters in the environment.

12.A. Introduction

Certain organic compounds dissolved in pure water do not undergo photoreaction (or do so only to a small extent) when exposed to sunlight; however, these compounds photoreact rapidly in water containing dissolved humic substances. For example, ethylene thiourea (ETU, Chart 12.1) dissolved in sterilized oxygenated agricultural drainage water decomposes rapidly in sunlight but is stable in pure water in sunlight (*7*). Because ETU has an absorption maximum at 240 nm and does not absorb light at $\lambda \geq 290$ nm, it cannot undergo direct photoreaction. However, because of its molecular structure, ETU undergoes rapid photoreaction in agricultural waters, presumably because of the pres-

1464–6/88/0201$07.25/1

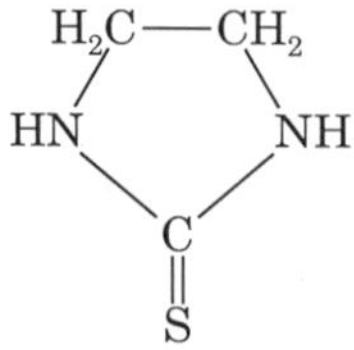

ethylenethiourea (ETU)

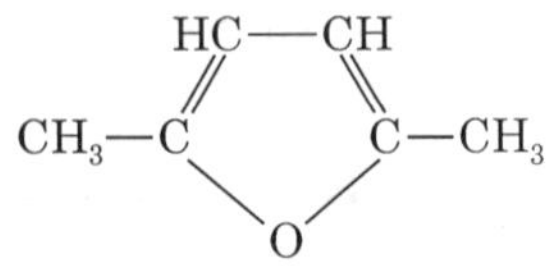

2,5-dimethylfuran (DMF)

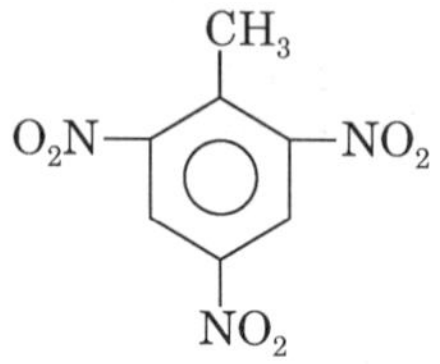

2,4,6-trinitrotoluene (TNT)

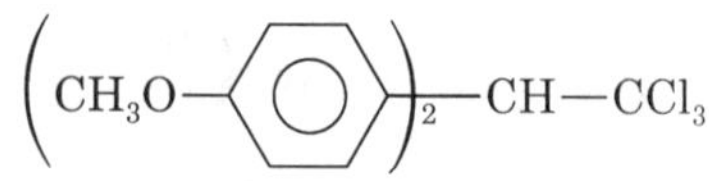

methoxychlor (MOC)

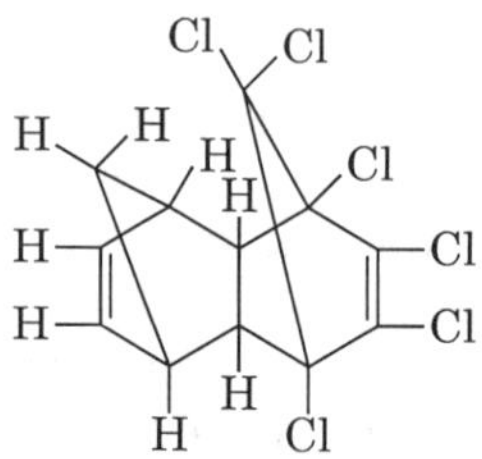

aldrin (A)

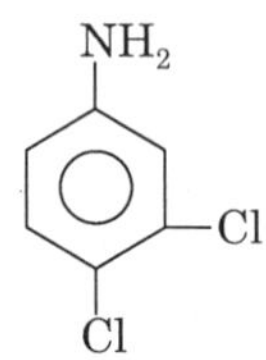

3,4-dichloroaniline (DCA)

disulfoton (D)

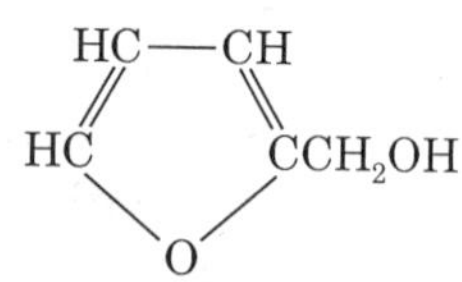

furfuryl alcohol (FFA)

Chart 12.1.

ence of natural sensitizers in these waters. The rapid photoreaction of ETU in agricultural water is called indirect (or sensitized) photoreaction.

2,5-Dimethylfuran (DMF, Chart 12.1) absorbs light well below 290 nm and thus cannot undergo direct photoreaction in sunlight in pure water. However, in the presence of dissolved humic substances and oxygen, DMF rapidly undergoes indirect photoreaction in sunlight (*1–3, 8, 9*). Aldrin* (A, Chart 12.1) is very stable when irradiated in summer sunlight in pure aqueous solution but undergoes rapid indirect photoreaction in agricultural water in summer sunlight (*10*). Other

* 1,2,3,4,10,10-hexachloro-1,4,4*a*,5,8,8*a*-hexahydro-1,4:5,8-dimethanonaphthalene

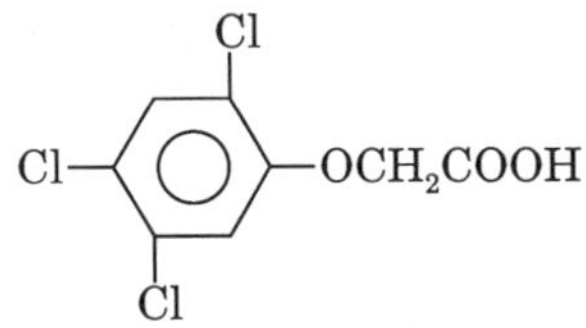

2,4,5-trichlorophenoxyacetic acid (2,4,5-T)

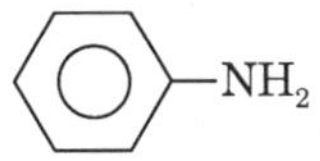

aniline

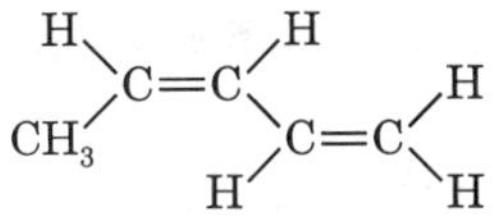

cis-1,3-pentadiene (cP)

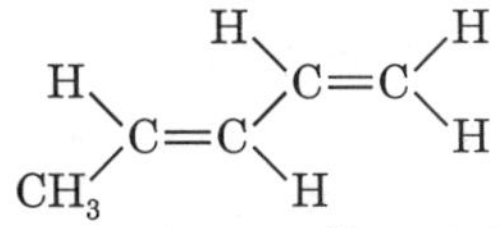

trans-1,3-pentadiene (tP)

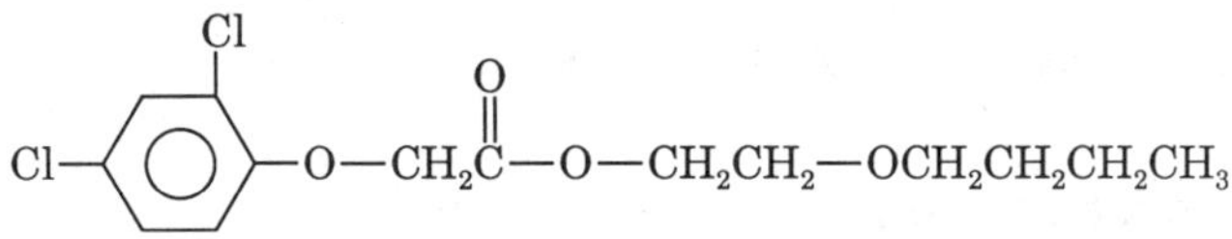

n-butoxyethyl ester of 2,4-dichlorophenoxyacetic acid (BEE)

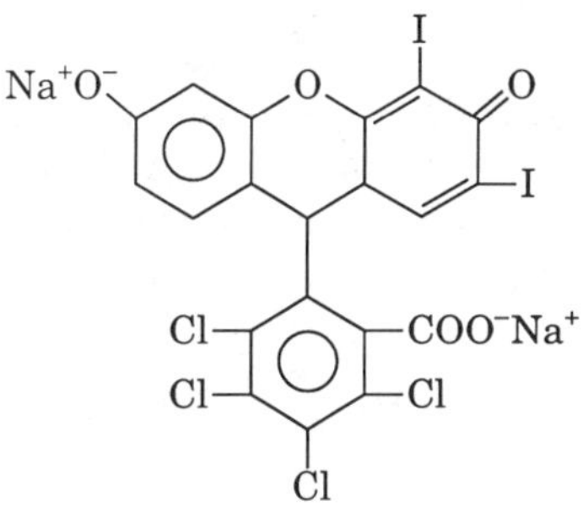

rose bengal, a sensitizer

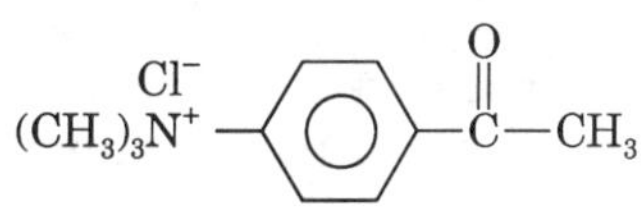

p-trimethylaminobenzophenone chloride (TMAB), a sensitizer

Chart 12.1. Continued.

examples of organic chemicals that undergo indirect photoreaction are given in Chart 12.1 and are: 2,4,6-trinitrotoluene [TNT] (*11*), methoxychlor [MOC] (*12*), *n*-butoxyethyl ester of 2,4-dichlorophenoxyacetic acid [BEE] (*13*), disulfoton [D] (*1*), *cis*- and *trans*-1,3-pentadiene [cP and tP, respectively] (*1, 3*), furfuryl alcohol [FFA] (*4–6*), 2,4,5-trichlorophenoxyacetic acid [2,4,5-T] (*14*), aniline (*1*), and 3,4-dichloroaniline [DCA] (*15*).

The photoreaction of the compounds in Chart 12.1 in the presence of dissolved humic acids is called indirect (or sensitized) photoreaction because the humic acids absorb the sunlight, form triplet excited-state species, transfer the energy to other species present in the water, and form reactive intermediates. These reactive intermediates are intimately involved in the decomposition of these organic compounds. These reactive intermediates include

- singlet oxygen ($^1O_2$) (*1–6, 8, 9*)
- alkylperoxy radicals ($RO_2\cdot$) (*16–19*)
- hydroxy radicals ($HO\cdot$) (*16, 18–20*)
- superoxide anion ($O_2^{\dot{-}}$) and hydroperoxy radicals ($HO_2\cdot$) (*21, 22*)
- triplet excited states of humic substances (*1, 3*)

In addition, some experimental data indicates that humic acids photoionize and liberate electrons that become hydrated (*23–25*).

12.B. Theoretical Aspects of Indirect Photoreaction

The theory of direct and indirect photoreaction has been described in detail in Chapter 5 and in Appendix B-2. Pertinent aspects of these photoreaction processes will be summarized in this section and developed further. These aspects are the theoretical basis for the measurement of indirect (or sensitized) rate constants in outdoor sunlight experiments or in laboratory experiments using monochromatic light.

12.B.1. Basic Equations for Photoreaction in Sunlight

In a well-mixed natural water system composed of dissolved humic acids and a chemical at a low concentration $[C]$ and at low absorbance (<0.02), both direct and indirect (sensitized) photoreaction can be described by first-order rate equations [equations 5.23 and 5.66]. The overall rate of disappearance of a chemical in sunlight is then the sum of these two equations and is given by

$$-(d[C]/dt) = k_p[C] \tag{12.1}$$

where

$$k_p = k_d + k_s \tag{12.2}$$

The net first-order rate constant k_p, in the units of reciprocal time, is defined by equation 12.2. The rate constant k_d is the first-order direct photoreaction rate constant, defined by equation 5.28; and k_s is the first-order indirect photoreaction rate constant (defined by equation 5.67).

Integrating equation 12.1 under the boundary conditions ($t = 0$, $[C_o]$) and (t, $[C_t]$) yields

$$\ln([C_o]/[C_t]) = k_p t \quad (12.3)$$

The associated half-life is given by

$$t_{(1/2)} = 0.693/k_p \quad (12.4)$$

Equations 12.1 through 12.4 apply to photoreaction rates in tubes; however, if photoreaction takes place in sunlight in a natural water body (e.g., a river or lake) in the environment, then the net photoreaction rate constant is called k_{pE} and the corresponding half-life is $t_{(1/2)E}$. The subscript E denotes that photoreaction is taking place in a natural water body in sunlight in the environment.

For the case when little or no direct photoreaction occurs, so that $k_s \gg k_d$, equation 12.3 reduces to

$$\ln([C_o]/[C_t]) = k_s t \quad (12.5)$$

12.B.2. Basic Equations for Photoreaction with Monochromatic Light

Consider a study of the indirect photoreaction of a chemical in the laboratory by performing experiments in a reaction cell using monochromatic light of wavelength λ. It will be assumed that the chemical does not absorb light at λ so that $k_d = 0$ and that only indirect photoreaction can occur. Consider that the chemical is present in a water system composed of dissolved humic acids and is at a low concentration such that its absorbance is very low (i.e., $A_{\lambda c} < 0.02$). The attenuation coefficient of the humic water system is α_λ. In indirect photoreaction, light is absorbed by humic acid sensitizers. The average rate of light absorption by humic acid sensitizers, $I^s_{a\lambda}$, is given by equation 5.56

$$I^s_{a\lambda} = 2.303 j^{-1} E^{av}_{o\lambda} \epsilon_{\lambda s}[S] \quad (12.6)$$

where $E^{av}_{o\lambda}$ is the average irradiance of the light source in the reaction cell at λ, $\epsilon_{\lambda s}$ is the molar absorptivity of the sensitizers at λ, $[S]$ is the molar concentration of the sensitizers, and j is a constant that makes the units of irradiance compatible with the units of molar concentration of the sensitizers.

At a fixed wavelength λ, the average rate of disappearance of a chemical is defined by equations 5.60, 5.61, and 5.62. These equations can be summarized as follows:

$$\text{average rate} = -(d[C]/dt)_{s\lambda} = k_{s\lambda}[C] \tag{12.7}$$

$$k_{s\lambda} = 2.303j^{-1}E^{\text{av}}_{o\lambda}\epsilon_{\lambda s}Q_{\lambda}[S] \tag{12.8}$$

where $k_{s\lambda}$ is the first-order rate constant for indirect (sensitized) photoreaction at wavelength λ in units of reciprocal time; Q_{λ} is a constant that is related to the photosensitized reaction quantum yield of the chemical [equation 5.59]; and the terms $E^{\text{av}}_{o\lambda}$, $\epsilon_{\lambda s}$, $[S]$, and j were defined for equation 12.6.

The average irradiance of the light source in the reaction cell at a fixed wavelength λ is given by equation 5.54 and is

$$E^{\text{av}}_{o\lambda} = I_{\lambda}(1 - 10^{-\alpha_{\lambda}l})(2.303\alpha_{\lambda})^{-1} \tag{12.9}$$

where l is the cell path length in cm, α_{λ} is the attenuation coefficient of the humic acid solution in cm^{-1}, and the light intensity term, I_{λ}, is the number of photons that enter the reaction cell per unit volume per unit time.

Equation 12.7 can be integrated to give

$$\ln([C_o]/[C_t])_{\lambda} = k_{s\lambda}t \tag{12.10}$$

12.B.3. Effects of Attenuation

Attenuation effects are extremely important to consider when evaluating indirect photoreaction kinetic data. If the light is weakly absorbed by the system, equation 5.72 applies and the rate constant $k_{s\lambda}$ is given by

$$k_{s\lambda} = 2.303j^{-1}I_{\lambda}l\epsilon_{\lambda s}Q_{\lambda}[S] \tag{12.11}$$

For these conditions, equation 12.11 indicates that $k_{s\lambda}$ is proportional to the sensitizer concentration $[S]$. At the other extreme, when the solution absorbs essentially all the light, $k_{s\lambda}$ is independent of the sensitizer concentration (equations 5.76 and 5.78). If the light is absorbed moderately (e.g., when the solution contains an appreciable concentration of sensitizer and the absorbance is on the order of 0.5–1.0), $k_{s\lambda}$ is not a linear function of sensitizer concentration because at higher sensitizer concentrations the enhancement of indirect photoreaction is partially offset by light attenuation effects.

Humic acids are probably mixtures of several components, only some of which actually sensitize reactions. To compare the photosensitizing abilities of these mixtures, Zepp et al. (*2*) had to correct for system-specific effects such as those of light intensity and attenuation on rate constants. Thus, Zepp introduced a "response function", $X_{s\lambda}$, defined by

$$X_{s\lambda} = k_{s\lambda}/E^{av}_{o\lambda} \quad (12.12)$$

Substituting equation 12.8 in equation 12.12 yields

$$X_{s\lambda} = 2.303j^{-1}\epsilon_{\lambda s}Q_{\lambda}[S] \quad (12.13)$$

The response function, $X_{s\lambda}$, defined by equation 12.13, is proportional to $\epsilon_{\lambda s}$, the strength of the light absorption of one or more sensitizers; proportional to $[S]$, the concentration of one or more sensitizers; and proportional to the photosensitized reaction quantum yield of the chemical (which is proportional to Q_{λ} in equation 5.59). If a single compound within the mixture called humic substances was the sensitizer, a plot of $X_{s\lambda}$ versus λ (i.e., the "action spectrum") would parallel the absorption spectrum of the sensitizer.

12.C. Determination of Indirect Photoreaction Rates in Sunlight

Aqueous sunlight and laboratory photoreaction experiments have been carried out with a variety of organic substrates in the presence of dissolved humic acids (*1–6, 12–15*). The results indicated that the rates of indirect photoreaction were first-order in substrate and obeyed equation 12.5.

Zepp et al. (*1*), presented quantitative kinetic data on the photosensitizing efficiency of several aquatic- and soil-derived humic substances from aqueous sunlight photoreaction experiments using 2,5-dimethylfuran, disulfoton, *cis*-1,3-pentadiene, and aniline as substrates. These substrates, (*See* Chart 12.1), underwent a variety of different indirect photoreactions. The photosensitizing efficiencies of these humic substances were remarkably similar, despite the diversity of their origins. These results indicated that several commercial humic or fulvic acids could serve as useful model sensitizers to aid in the estimation of the rate of transformation of chemicals that enter natural aquatic environments. Thus, the procedures developed by Zepp et al. (*1*), their results, and other pertinent indirect photoreaction data are described in detail in the following paragraphs.

12.C.1. Experimental Procedures

12.C.1.a. Materials

Commercial humic acids were obtained from Aldrich Chemical Co., Fluka AG (Switzerland), and Carl Roth (West Germany). Several lots of Fluka humic acids were obtained from Ciba–Geigy Ltd., Basel, Switzerland and from Columbia Organic Chemical Company. Natural water samples were obtained from the top meter of four water bodies in the United States (Aucilla River, GA; Okefenokee Swamp, GA; Quincy Bog, NH; and Williamson River, OR) and were refrigerated at 5 °C. A sample of fulvic acid isolated from Mattole soil, Petrolia, CA, was used. A commercial fulvic acid, was obtained from Contech E.T.C. Ltd., Ottawa, Ontario, Canada.

The organic substrates 2,5-dimethylfuran, disulfoton*, *cis*- and *trans*-1,3-pentadiene (cP and tP, respectively), and aniline were ≥99% pure). The sensitizer *p*-trimethylaminobenzophenone chloride was also very pure. The molecular structures of these compounds are shown in Chart 12.1.

12.C.1.b. Preparation of Solutions

In general, humic acids are insoluble in water; therefore, the following procedure was used to solubilize the humic acids. Humic acid (0.050 g) was added to 100 mL of 0.1% NaOH solution. The mixture was stirred for 12 h and centrifuged at 15,000 rpm in a Sorvall centrifuge at 25 °C for 1 h. The supernate was diluted 20-fold with distilled water and adjusted to pH 6 with 0.1 N H_2SO_4. The fulvic acids were dissolved directly in distilled water and adjusted to pH 6 with 0.1 N H_2SO_4. All the natural waters were obtained from "black water" systems as listed in Section 12.C.1.a. These natural waters were centrifuged at 25 °C, as just described, to remove particulates.

By using distilled water, the absorbances of the humic acid solutions, the fulvic acid solutions, and the natural water samples were adjusted to a value of 0.20 at 366 nm in a 1.00-cm cell. These optically matched solutions were used in all sunlight experiments.

12.C.1.c. Procedures for Photoreaction Studies

For each organic substrate, 1.0×10^{-5} M solutions were prepared in air-saturated, optically matched humic acid solutions and in distilled

* phosphorodithioic acid *O,O*-diethyl *S*-[2-(ethylthio)ethyl] ester

water. Solutions containing disulfoton and aniline were added to 13-mm o.d. borosilicate glass (Pyrex) cylindrical test tubes equipped with ground-glass stoppers. Because DMF and cP are volatile, solutions of these substrates were added to test tubes equipped with gas-tight Mininert caps (Section 7.B.1.e.). The tubes were filled to the top to ensure that the chemicals were completely in the aqueous phase.

Sunlight photoreaction experiments were conducted by placing the tubes in a rack with their axes pointing north–south and tilted at 20° from the horizontal. After exposure to sunlight, the concentration of the chemical $[C_t]$ was measured, and the photoreaction rate constant k_p was calculated by using equation 12.3. Dark controls were monitored during the course of the photoreaction experiments, and no reactions were observed for any of the substrates.

12.C.2. Discussion of Results

12.C.2.a. Photosensitized Oxygenations

2,5-Dimethylfuran and disulfoton were used to study photosensitized oxygenations. Both compounds photoreacted rapidly in sunlight in air-saturated solutions containing humic and fulvic acids. It was established (*8*) that DMF photoreacted in sunlight in humic natural waters to form diacetyl ethylene and H_2O_2, (*see* Scheme 12.1, part A) and that the reaction involved photochemically generated singlet oxygen. Disulfoton, a sulfide, photoreacted to the sulfoxide in natural agricultural waters (*26*). Zepp et al. (*1*) also studied the sunlight photoreaction of

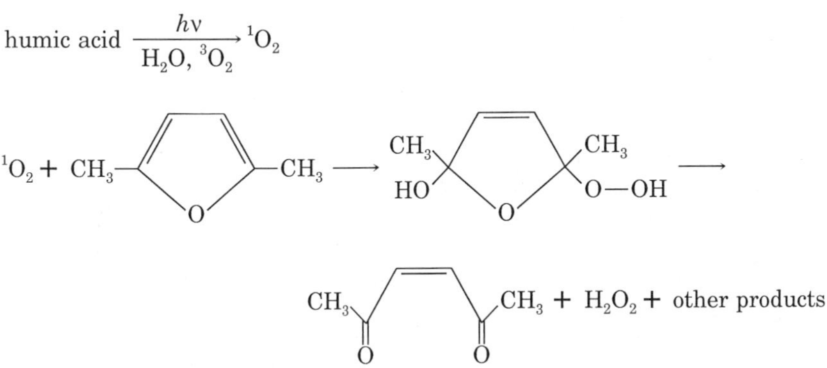

A

Scheme 12.1. Photooxygenations sensitized by dissolved humic acids in water. (A) Indirect photoreaction of 2,5-dimethylfuran. Continued on next page.

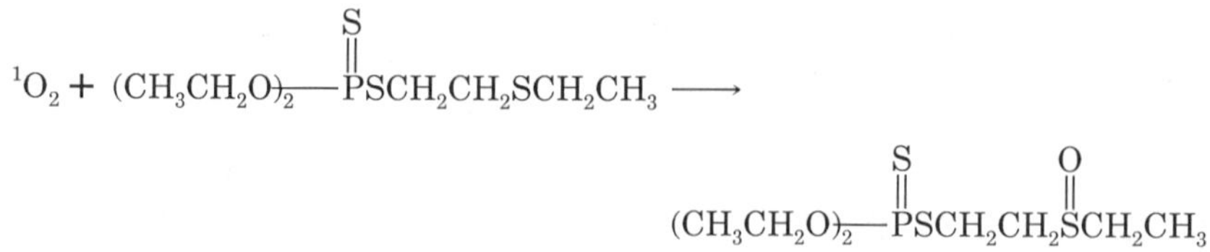

B

Scheme 12.1. Continued. (*B*) *Indirect photoreaction of disulfoton.*

this organic chemical in humic waters and found that disulfoton sulfoxide was the major product, Scheme 12.1, part B, and that the reaction was also mediated by photochemically generated singlet oxygen.

Haag et al. (*4–6*) studied the photosensitization of furfuryl alcohol (FFA, Chart 12.1) in the presence of humic acids and rose bengal by carrying out photoreaction experiments in sunlight and in the laboratory with a filtered xenon lamp. These studies focused on the intermediacy of singlet oxygen, $^1O_2$, generated by the humic acids in the presence of oxygen, $^3O_2$. The studies used FFA as a trapping agent. The theoretical aspects of this reaction are described in detail in Appendix B-2. The mechanism is depicted schematically in Scheme 12.2. The following experimental data supported this singlet oxygen mechanism. The results of high-performance liquid chromatographic (HPLC) studies clearly showed that the major product (I) was 1,6-hydroxy(2*H*)pyran-3(6*H*)-one. Hydrogen peroxide was formed in almost the same concentration as FFA was lost. The significantly reduced rates in both distilled and deoxygenated natural water demonstrated

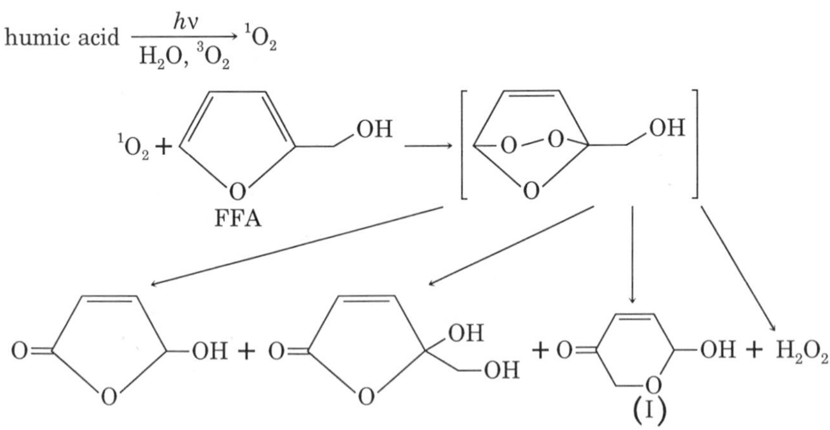

Scheme 12.2. Photooxygenation of furfuryl alcohol (FFA) in the presence of humic acids and oxygen in water. (Adapted from refs. 4 and 5.)

that both sensitizer and oxygen were required for the reaction to occur. The rate of photosensitization was increased in the presence of D_2O and decreased in the presence of azide ion. Rose bengal (Chart 12.1), a known singlet-oxygen sensitizer, gave the same photoreaction products as those produced with humic acids in the presence of $^1O_2$ and FFA.

In several photochemical studies, it had been found that the photooxidation of DMF (10 μM in humic acid solution), measured as the disappearance of DMF, was first-order and followed equation 12.3 or 12.5 (*1–3, 8*).

In the presence of a variety of humic acid sensitizers, DMF and disulfoton were irradiated outdoors in sunlight at Athens, GA, throughout the year under a wide variety of meteorological conditions. These substrates followed first-order kinetics (equation 12.3), and the calculated rate constants were directly proportional to the light intensity. In distilled water, these substrates showed no decomposition after 1 day of irradiation in summer sunlight. Thus, the first-order direct photoreaction rate constant k_d is zero under these photoreaction conditions and $k_p = k_s$.

The chemicals DMF and disulfoton in Aucilla River water were used as chemical actinometers to monitor the sunlight intensity during sunlight photoreaction. Controlled laboratory studies indicated that the photoefficiency of these substrates in Aucilla River water remained constant during these studies. Rate constants for these substrates in the various humic acid waters were calculated from the sunlight photoreaction data. The data were reported as the ratio of the rate constant for the photoreaction of the substrate in the various solutions of humic acids to the rate constant for the photoreaction of the substrate in Aucilla River water (Table 12-1). Inspection of the data indicated that all relative rates were within a factor of 2 for those of the Aucilla River water solution. These results, indeed, are remarkable considering that the humic acids used came from a wide variety of different origins, including several from commercial sources.

The half-life for DMF in Aucilla River water at midday, reported by Zepp et al. (*1*), ranged from 5 to 10 min in the summer to approximately 20 min in the winter. However, these half-lives correspond to values in tubes. For a natural water body in the environment, the half-life is approximately 7.5–15 min in summer and 30 min in the winter For disulfoton in Aucilla River water, the half-life at midday ranged from 5 h in the summer to 12 h in the winter in tubes; the corresponding half-life in a natural water body in the environment is 7.5 h in the summer to 18 h in the winter [*See* Zepp (*27*) in which he used $\gamma = 1.5$ in equation 6.14].

Table 12-1. Kinetic Data for Oxygenations Photosensitized by Various Humic Acids in Water Irradiated with Sunlight

Humic Acid Source	*2,5-Dimethylfuran*	*Disulfoton*
Aucilla River, FL	1.0	1.0
Okefenokee Swamp, GA	1.13 ± 0.32	
Quincy Bog, NH	1.52 ± 0.08	
Williamson River, OR	1.22 ± 0.40	
Fulvic acid, CA[a]	1.52 ± 0.40	1.2 ± 0.2
Contech fulvic acid[a,b]	0.98	1.1 ± 0.1
Aldrich humic acid[a,b]	1.21 ± 0.17	1.1 ± 0.4
Karl Roth humic acid[a,b]	0.50 ± 0.25	
Fluka humic acid[a,b]		
Lot A	0.81	1.2 ± 0.4
Lot B	0.68	1.15 ± 0.4
Lot C	0.80 ± 0.14	1.30 ± 0.53
Lot D	0.80 ± 0.15	1.2

NOTE: All results are relative rate constants, that is, the ratio of the sunlight photoreaction rate constant to that observed for Aucilla River water. All solutions were optically matched at pH 6.0 by adjusting the absorbance to 0.20 in a 1.00-cm cell at 366 nm.
[a]Isolated from soils.
[b]Obtained commercially.
SOURCE: Adapted from ref. 1.

12.C.2.b. Photosensitization Involving Energy Transfer

The photoisomerization of *cis-* or *trans*-1,3-pentadiene with various photosensitizers in organic solvents is a reversible reaction (*28, 29*). After prolonged irradiation in the presence of a photosensitizer, isomerization proceeds until the ratio of *cis-* to *trans*-diene reaches a constant value. When this photostationary state is reached, the rate of *cis–trans* isomerization equals the rate of *trans–cis* isomerization. The rate expression for this reversible reaction is derived in Appendix C, is given by equation C.29, and is

$$\ln \left[\frac{(t_\infty - t_0)}{(t_\infty - t_t)}\right] = k_i t \qquad (12.14)$$

where t_∞, t_0, and t_t are the fractions of the pentadiene that were *trans* at photostationary state, at time zero, and at time t, respectively. The rate constant for this reaction, k_i, represents the sum of the rate constants for the cis–trans conversion and for the trans–cis conversion. A typical plot of the kinetic data for this reaction in Aucilla River water at 10 μM is shown in Figure 12.1 (*3*). The same photostationary state was attained starting with either cP or tP.

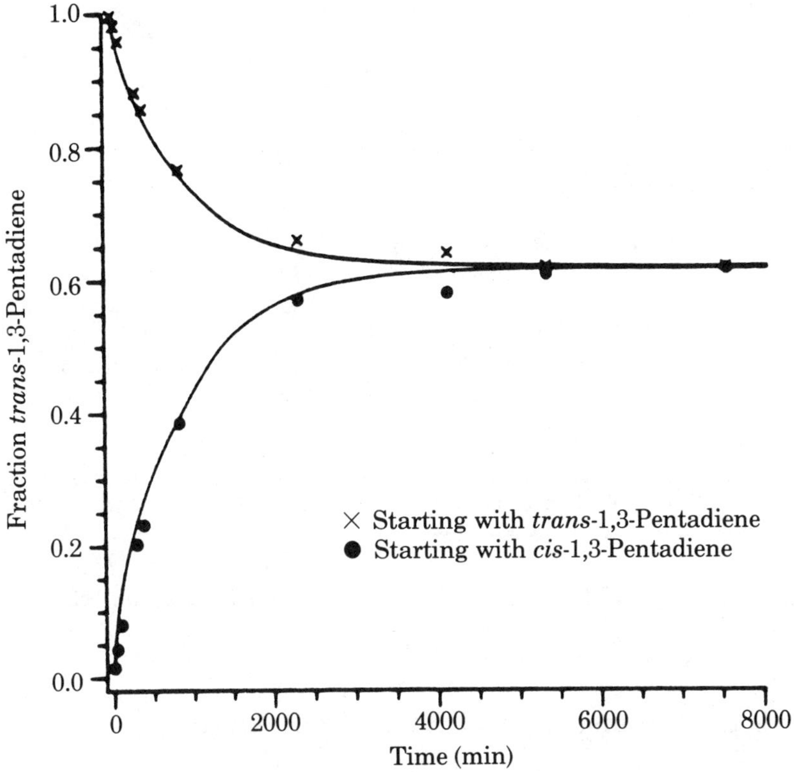

Figure 12.1. Kinetic data for the photosensitized isomerization of cis- *and* trans-*1,3-pentadiene (10 μM) in a water sample from the Aucilla River. (Reproduced from ref. 3. Copyright 1985 American Chemical Society.)*

In the presence of a variety of humic acids, the substrate cP was irradiated in sunlight at Athens, GA (*1*). The photoisomerization reaction is shown in Scheme 12.3, part A. After irradiation in sunlight for 1 day in distilled water, no isomerization was observed for cP. Thus, the direct photoreaction rate constant k_d was zero. In humic acid solution, isomerization proceeded to a photostationary state of 56% tP and 44% cP. The identical composition was observed when *p*-trimethylaminobenzophenone chloride (TMAB, Chart 12.1), a known triplet sensitizer for this reaction, was used as the photosensitizer (*1*).

By using the sunlight photoreaction data in equation 12.14, first-order plots were obtained, and are given in Figure 12.2. The data indicated that the Aucilla River water was the most efficient photosensitizer. By using the data in Figure 12.2, the half-life in tubes (midday, July, Athens, GA) was calculated to be approximately 20

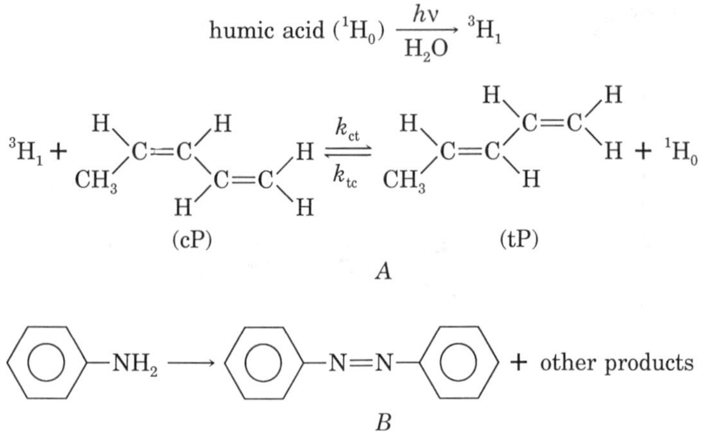

Scheme 12.3. Reactions photosensitized by humic acids in water. (A) Indirect photoreaction of cis-*1,3-pentadiene. (B) Indirect photoreaction of aniline.*

min. Details of further studies on the photoisomerization of cP are in ref. 3.

12.C.2.c. Photosensitized Reaction of Aniline

Aniline solutions containing four different humic acids were irradiated in sunlight at Athens, GA (*1*). Rate constants were calculated by using equation 12.3 and relative rates were computed by using the Aucilla River water sample as an actinometer. The relative rate data are summarized in Table 12-2. The half-lives in tubes ranged from 4 to 8 h in May sunlight in Athens, GA. Aniline in distilled water had a tube half-life of approximately 1 week, a result indicating that it had undergone some slow direct photoreaction. Thus, $k_s \gg k_d$ and $k_p = k_s$. Inspection of the data in Table 12-2 indicate that the rate constants were all within a factor of 2 of each other, despite the wide variety of origins of the humic acids. Azobenzene was a photoreaction product, presumably due to the coupling of anilino free radicals (*see* Scheme 12.3, part B). At the low concentrations of aniline used (1.0×10^{-5} M), the yield of azobenzene was very low in Aucilla River water (approximately 0.2% yield) and was even lower in the commercial humic acid solutions.

Miller et al. (*15*) carried out photoreaction experiments in Reno, NV, in the summer with the substrate 3,4-dichloroaniline (DCA, Chart 12.1) in the presence of humic acid sensitizers. The Fluka humic acid was obtained from Columbia Organic Chemicals Co. The natural humic acids were obtained from three rice fields and the Sacramento River

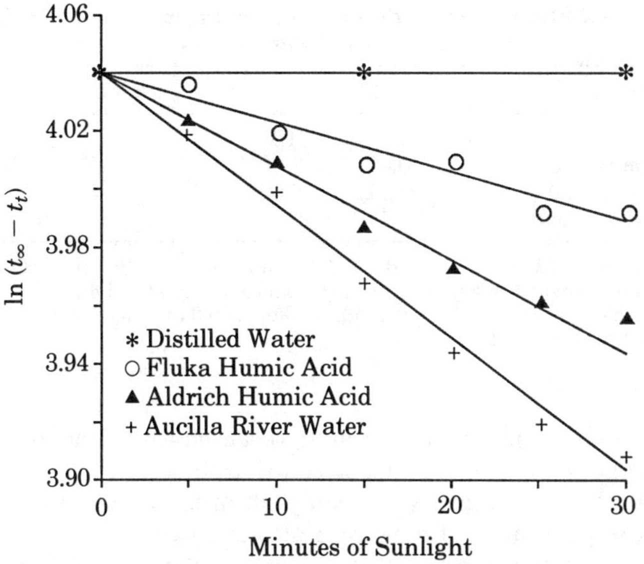

Figure 12.2. Kinetic data for isomerization of cis-*1,3-pentadiene photosensitized by various humic substances in water on an overcast day (midday, July, Athens, GA). Symbols:* t_∞ *and* t_t *represent the fraction of diene isomerized to* trans-*isomer at photostationary state and at time* t, *respectively. (Reproduced with permission from ref. 1. Copyright 1981 Pergamon.)*

near Davis, CA, and a Lovelock, NV, alfalfa irrigation ditch. Analysis of the kinetic data indicated that these humic acids photosensitized DCA with approximately the same efficiency as they did for aniline in Athens, GA.

The coupling product tetrachloroazobenzene was found to be a photoreaction product if 3×10^{-6} M solutions of DCA were irradiated in summer sunlight in the presence of the humic acids. The yield, however, was low (approximately 1%). In addition, the yield of the tetrachloroazobenzene coupling product was higher with the natural humic acid sensitizers in comparison to the commercial humic acids. This observation was similar to that observed with aniline.

12.D. Determination of Indirect Photoreaction Rates in the Laboratory and Sunlight Photoreaction Rates

Zepp et al. (*2*) used laboratory data to forecast rates of indirect photoreaction in natural waters in the environment. They introduced the

Table 12-2. Photosensitization of Aniline in Sunlight with Various Humic Acid Sensitizers

Humic Acid	k_s (h^{-1})	*Relative Rate Constant*
Aucilla River	0.17 ± 0.012	1.0
Aldrich humic acid	0.12 ± 0.0060	0.70
Fluka humic acid	0.091 ± 0.0080	0.54
Contech fulvic acid	0.11 ± 0.01	0.67

NOTE: Experiments were conducted during May at Athens, GA. Solutions of humic acids were optically matched by adjusting the absorbance to 0.20 at 366 nm in a 1.00-cm cell; the pH of the humic acid solutions was adjusted to pH 6.0 to match Aucilla River water.
SOURCE: Adapted from ref. 1.

theory of indirect photoreaction and presented equations that describe indirect photoreaction in aqueous solution. In very dilute solution, in which the incident light was weakly absorbed by the substrate, the rate of disappearance of substrate at wavelength λ, was first-order. With DMF as the substrate in laboratory experiments with monochromatic light, equation 12.10 was obeyed and the photosensitization efficiencies of several humic acids were compared. The data from these experiments were used to compute the rate of indirect photoreaction in sunlight in natural water bodies in the environment. A description of the procedures developed by Zepp et al. (*2*), their results, and other pertinent data used to forecast rates of indirect photoreaction in sunlight in natural water bodies in the environment follow.

12.D.1. Experimental Procedures

12.D.1.a. Materials

The organic substrate DMF was ≥99% pure. The sensitizer rose bengal was also very pure. The sources of the humic acids were described in Section 12.C.1.a.

12.D.1.b. Preparation of Solutions

The preparation of solutions was described in Section 12.C.1.b. For pH studies, the pH of the Aucilla River water or rose bengal solution was adjusted by the addition of 0.01 M phosphate buffer, HCl, or NaOH solution.

12.D.1.c. Procedures for Photoreaction Studies

Air-saturated, optically matched solutions of DMF at 1.0×10^{-4} M were prepared in both humic acid solution and distilled water. These solutions were put in borosilicate glass (Pyrex) cylindrical tubes with gas-tight Mininert caps (Section 7.B.1.e.). The tubes were filled to the top to ensure that all the DMF was in the aqueous phase. Photoreaction experiments were carried out outdoors in sunlight, as described in Section 12.C.1.c.

Laboratory photoreaction experiments were carried out using a Moses-type photochemical "merry-go-round" reactor (PMGRR) as described in Section 10.A.1.a.i. Monochromatic light was isolated from a 450-W medium-pressure mercury lamp by using the following glass filter combinations and filter solutions:

- 313 nm—Corning glass filter 7–54, 1-cm thick solution of 0.001 M K_2CrO_4 and 3% K_2CO_3, and 4 mm of Pyrex glass
- 366 nm—Corning glass filter combination 7–54 and 0–51
- 405 nm—Corning glass filter combination 3–75, 4–70, and 7–51
- 436 nm—Corning glass filter combination 3–73 and 5–58
- 546 nm—Corning glass filter combination 3–68, 4–72, and 1–60

Dark controls were monitored during the course of the photoreaction experiments and no reaction was observed. Rate constants $k_{s\lambda}$ were calculated at 313, 366, 405, 436, and 546 nm by using equation 12.10.

The response function $X_{s\lambda}$ was calculated using equation 12.12, where $k_{s\lambda}$ was the rate constant (in s^{-1}) for the photosensitized reaction at wavelength λ (in nm) and $E_{0\lambda}^{av}$ (in millieinsteins cm^{-2} s^{-1}) represented the average irradiance in the reaction cell. The average irradiance was computed using equation 12.9, where α_λ was the decadic attenuation coefficient in a natural water sample (in cm^{-1}) at wavelength λ, and l was the cell path length in cm, calibrated by the method described by Zepp (*30*) and the method described in Section 9.E.

The light intensity I_λ, which represents the number of photons that enter the reaction cell per unit volume and per unit time, was measured by using the ferrioxalate actinometer and procedures similar to those outlined in Sections 9.D.2.a and 10.C.2. Reaction cells identical to those used for the natural water photoreaction experiments were filled with 0.15 M ferrioxalate solution. The light intensity I_λ was computed in the following way. Equation 9.21 was modified to take into account the fraction of light absorbed by the ferrioxalate actinometer and is given by

$$(d[Fe^{2+}]/dt)_{d\lambda} = \phi_{d\lambda} I_\lambda F_\lambda \tag{12.15}$$

where $[Fe^{2+}]$ is the molar concentration of ferrous ion formed at time t, $\phi_{d\lambda}$ is the reaction quantum yield of the ferrioxalate actinometer at λ, I_λ is the light intensity at λ, and F_λ is the fraction of light absorbed at λ.

Integrating equation 12.15 under the boundary conditions ($[Fe^{2+}]_\lambda = 0$, $t = 0$) and ($[Fe^{2+}]_\lambda$, t) yields

$$[Fe^{2+}]_\lambda = \phi_{d\lambda} I_\lambda F_\lambda t = k_{d\lambda} t \tag{12.16}$$

Solving for I_λ in equation 12.16 yields

$$I_\lambda = \frac{[Fe^{2+}]_\lambda / t}{\phi_{d\lambda} F_\lambda} = k_{d\lambda} / \phi_{d\lambda} F_\lambda \tag{12.17}$$

The rate constant $k_{d\lambda}$ was calculated from the linear regression analysis of the experimental data for $[Fe^{2+}]_\lambda$ versus t. The values of $\phi_{d\lambda}$ and F_λ were obtained from Table 9-1. The fraction of light absorbed (F_λ) was essentially unity at all wavelengths except 546 nm. When $[Fe^{2+}]_\lambda$ is in the units of moles per liter, t is in seconds, I_λ is in the units of einsteins per liter per second, and the units of $E_{0\lambda}^{av}$ are in milliensteins per square centimeter per second. Under these conditions, the conversion factor j is equal to 1 (Section 1.C.4).

12.D.2. Discussion of Results

12.D.2.a. Introduction to Experimental Studies

The indirect photoreaction of DMF was used as a tool to learn how variation in reaction conditions affects the photosensitizing efficiencies of a few humic acids. This indirect photoreaction was an excellent a model system and was chosen for the following reasons:

1. The substrate DMF does not photoreact in distilled water when irradiated with light of $\lambda \geq 290$ nm. This is a result of the first law of photochemistry because DMF does not absorb light at $\lambda \geq 290$ nm (i.e., $\epsilon_\lambda = 0$ for $\lambda \geq 290$ nm and from equation 5.28 the direct photoreaction rate constant $k_d = 0$).

2. In the presence of dissolved humic acids, DMF undergoes only indirect photoreaction that is entirely via photosensitization by the dissolved humic acids.
3. The substrate photoreacts rapidly in sunlight (Section 12.C.2.a.) and when exposed to isolated monochromatic light from a medium-pressure mercury lamp.
4. The substrate DMF is stable in the dark.
5. The predominant mechanism for the photosensitization of DMF involves a rapid intermediate step in which sensitizers in their triplet state transfer energy to molecular oxygen to form singlet oxygen.

The mechanism of this reaction is given in Scheme 12.1, part A, and is described in more detail in Appendix B-2.

12.D.2.b. Effect of Sensitizer Concentration on the Rate Constant and Response Function

The rate constant $k_{s\lambda}$ and the response function $X_{s\lambda}$ were determined at 366 nm as functions of the concentration of humic acid by using the procedure outlined in Section 12.D.1. The results are summarized in Table 12-3. Linear regression analysis of $X_{s\lambda}$ as a function of concentration of humic acid (i.e., the sensitizer concentration [*S*] in milligrams of organic carbon per liter), including the point (0,0), gave a slope of 1.06 and a correlation coefficient of 0.993. Thus, the response function $X_{s\lambda}$ is a linear function of the concentration of sensitizer [*S*] as predicted by equation 12.13.

Table 12-3. Kinetic Parameters at 366 nm for the Indirect Photoreaction of DMF in Aucilla River Water Diluted with Distilled Water To Yield Solutions Containing Various Concentrations of Humic Acid

[S][a]	$k_{s(366)}$ (h^{-1})	$X_{s(366)}$
21	0.18 ± 0.02	23 ± 3
15	0.13 ± 0.01	15 ± 1
10	0.077 ± 0.007	8.6 ± 0.6
6.3	0.056 ± 0.005	6.0 ± 0.6
2.1	0.026 ± 0.002	2.7 ± 0.3

[a]Humic acid concentration in milligrams of organic carbon per liter.

SOURCE: Adapted from ref. 2.

Linear regression analysis with $k_{s\lambda}$ as a function of [S], including the point (0,0), gave a slope of 0.00838 and correlation coefficient of 0.996. Thus, the rate constant $k_{s\lambda}$ is also a linear function of the sensitizer concentration, as predicted by equation 12.11. The absorbance of these solutions was <0.1 at 366 nm (*27*) indicating weak light absorption. At higher concentrations of humic acids (>30 mg of organic carbon per liter), the absorbance is greater than 0.20 and $k_{s\lambda}$ is not a linear function of sensitizer (*27*). This is a direct result of the fact that at these sensitizer concentrations, light attenuation effects partially offset the enhancement of indirect photoreaction (Sections 5.B.3 and 12.B.3).

12.D.2.c. Effect of pH on Indirect Photoreaction Rates

The effects of pH on the indirect photoreaction of DMF were investigated by adjusting the pH of Aucilla River water with phosphate buffer, H_2SO_4, or NaOH. As the pH of the Aucilla River water was increased, the attenuation coefficient α_λ increased, indicating an increase in light absorption as a result of speciation changes. Thus, the increase in attenuation of light should decrease the rate of indirect photoreaction of DMF.

Sunlight photoreaction experiments were carried out with DMF in oxygenated Aucilla River water by varying the pH, and the results are depicted in Figure 12.3. In the pH range 5–9, the rate constant was essentially constant, a result that was unexpected. At pH < 5, the rate constant increased and at pH > 9, the rate constant decreased. The rate constants in unbuffered (point *) and in buffered phosphate (point ▲) Aucilla River water were essentially the same; hence, no buffer catalysis was observed in the pH range 5–9.

The pH effects for the Aucilla River water were possibly the results of speciation changes of the natural humic acid photosensitizers in the water. The pH effects could also be attributable to changes in the quantum efficiency. To investigate this point, the indirect photoreaction of DMF was studied as a function of pH in buffered water as just described but with rose bengal, a known sensitizer for singlet-oxygen reactions, in place of the Aucilla River water. The absorption spectrum of rose bengal did not change when the pH was raised from 5.5 to 11, indicating that no speciation occurred. Just as in the Aucilla River water (Figure 12.3), the rate constant for rose bengal sensitized photoreaction of DMF decreased at pH greater than 10. Zepp et al. (*2*) explained these results as follows

> *The component(s) of the humic substances that photosensitize the formation of singlet oxygen do* not *ionize in the pH 5 to*

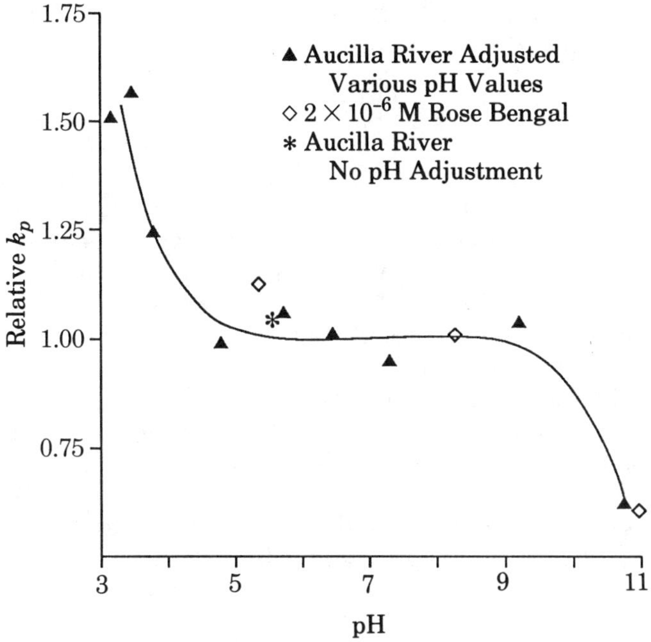

Figure 12.3. The photosensitized oxygenation of 2,5-dimethylfuran (DMF) as a function of pH. (Reproduced with permission from ref. 2. Copyright 1980 Pergamon.)

> *9 range. If so, this precludes the possibility that the photosensitizing component(s) are species such as phenols or carboxylic acids that are known to ionize in this pH range. That the decrease in rate constant at pH 10 is about the same with both rose bengal and humic substances as photosensitizers indicates that the same factor is responsible for the decrease with both photosensitizers. This decrease may be attributable to quenching of singlet oxygen by the hydroxide ion.*

Haag et al. (*6*) carried out photosensitization experiments with furfuryl alcohol as the substrate, as a function of pH, in a natural humic-rich water (Étang de la Gruére, Switzerland). The laboratory experiments were carried out with a high-pressure mercury lamp, filtered with a solidex borosilicate glass filter, in a PMGRR (Section 10.A.1.a.i.). In the pH region 4–9, the photosensitized rate constant was relatively constant but decreased rapidly above pH 9. These results were similar to those obtained by Zepp et al. (*2*) in the pH range 5–10. However, at pH less than 4, Haag et al. did not observe the sharp increase in the rate constant observed by Zepp.

12.D.2.d. Wavelength Studies: The Response Function as a Function of Wavelength

A series of photoreaction experiments was carried out with DMF in various humic waters in a PMGRR with a medium-pressure mercury lamp filtered to give monochromatic light at 313, 366, 405, 436, and 546 nm. The response function $X_{s\lambda}$ was evaluated at each wavelength as described in Section 12.D.1.c., and the results are depicted in Figure 12.4. The plot of $X_{s\lambda}$ versus λ is called the "action spectrum" for sensitized photoreactions. These action spectra were used to quantitatively compare the photosensitizing efficiencies of natural aquatic humic acid (Aucilla River water) to two commercial humic acids (Aldrich and Fluka humic acids). Inspection of Figure 12.4 indicates that in the near-UV and visible spectral regions (i.e., 336–546 nm), the action spectra for the reaction of DMF sensitized by the various humic sensitizers were the same within a factor of two. However, in the UV-B spectral region (313 nm), the Aucilla River water was found to be more efficient than the commercial humic acids.

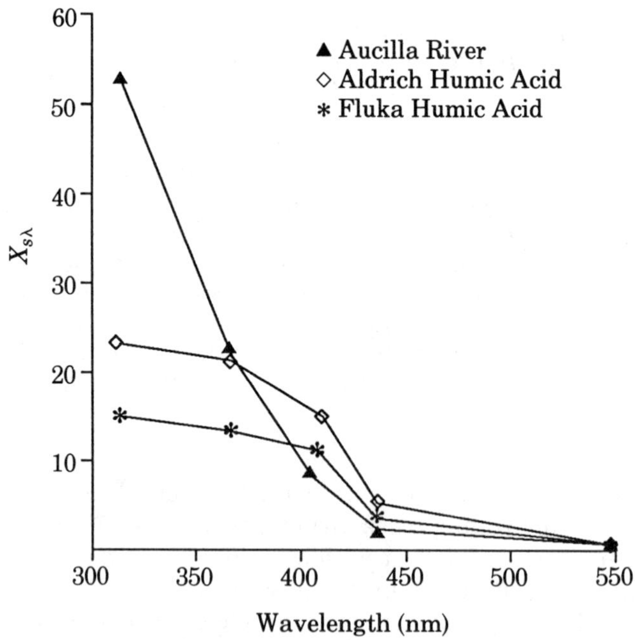

Figure 12.4. Action spectra for photosensitized oxygenation of 2,5-dimethylfuran (DMF). Reaction conditions: absorbance of each solution at 366 nm was adjusted to 0.20 in a 1.00-cm cell; pH was adjusted to 6.0. (Reproduced with permission from ref. 2. Copyright 1981 Pergamon.)

12.D.2.e. Computation of Sunlight Indirect Photoreaction Rate Constants in Natural Water Bodies in the Environment

For the indirect photoreaction of DMF in the presence of humic acids (Aucilla River water, Aldrich humic acid, and Fluka humic acid), near-surface sunlight rate constants, $k_{s\lambda}$, were computed as a function of λ at 40°N. latitude, at midday, for the summer season by using the SOLAR computer program of Zepp and Cline (*3*) [i.e., the average irradiance, $E_{o\lambda}^{av}$], the response function $X_{s\lambda}$ (obtained from Figure 12.4), and equation 12.12 in the form

$$k_{s\lambda} = X_{s\lambda}E_{o\lambda}^{av} \tag{12.18}$$

The results of these computations are depicted in Figure 12.5. The SOLAR computer program was then used to integrate $k_{s\lambda}$ over all wavelengths λ to obtain an average rate constant k_{sE} for each of the three humic acid solutions. The results were 5.4, 6.5, and 4.3 h^{-1} for

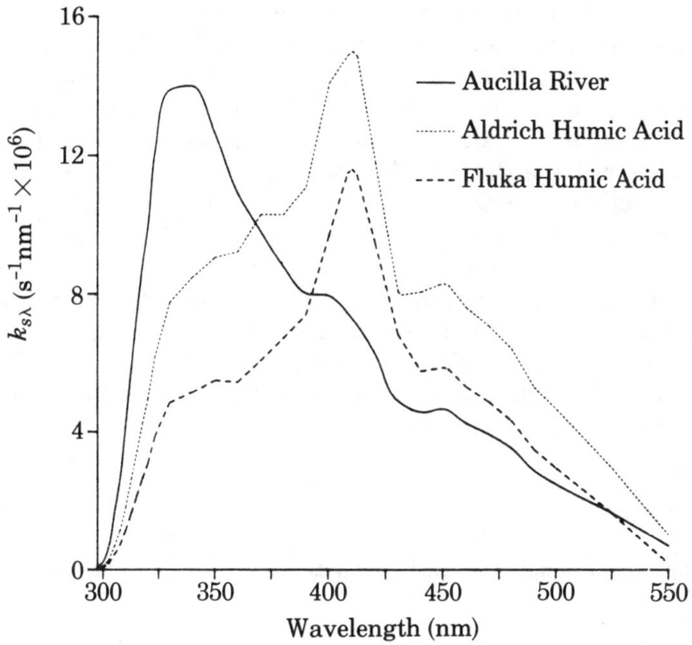

Figure 12.5. Wavelength dependence of rate constants computed for photosensitized reaction of 2,5-dimethylfuran (DMF) in sunlight (midday, summer, latitude 40°N). (Reproduced with permission from ref. 2. Copyright 1981 Pergamon.)

the Aucilla River water, Aldrich humic acid, and the Fluka humic acid, respectively.* The rates relative to Aucilla River water were 1 : 1.2 : 0.80 for Aucilla River water, Aldrich humic acid, and Fluka humic acid, respectively. These results agreed with those (from Table 12-1) obtained experimentally (1.0 : 1.21 : 0.77, respectively, for these three humic acids). By using equation 12.4 (in a form applicable to natural water bodies in the environment) with k_{sE} equal to 5.4 h^{-1} for Aucilla River water, $t_{(1/2)E}$ was equal to 0.13 h (7.8 min). This half-life agreed reasonably well with the experimental half-life of 7.5–15 min measured at Athens, GA (*see* Section 12.C.2.a.).

The data in Figure 12.5 indicate that the UV light was the most effective in promoting the indirect photoreaction of DMF in Aucilla River water; whereas, blue light was the most effective at promoting the photoreaction of DMF with commercial humic acids. Thus, there are differences between the photosensitization abilities of commercial and natural humic acids. Even though $X_{s\lambda}$ for the Aucilla River humic acids is much larger than for the commercial humic acids in the spectral region 300–330 nm, the influence of this difference on k_{sE} is negligible because the sunlight intensity is considerably lower in this spectral region.

Using furfuryl alcohol as a trapping agent, Haag et al. (*5*) measured the singlet-oxygen quantum yield as a function of wavelength from 366 to 546 nm for three humic acids [Fluka AG humic acid isolated from soil (a commercial grade humic acid); Black Lake humic acid isolated from Black Lake, NC; and Lake Baldegg humic acid water obtained from Lake Baldegg, NC, and used directly]. The singlet oxygen production rate was then calculated from the data and the SOLAR computer program of Zepp and Cline (*31*) for midday, summer, at 40° N. latitude from 300 nm to beyond 600 nm. The isolated humic acids exhibited peaks at approximately 420 nm while the Lake Baldegg humic acid exhibited at approximately 330 nm. These results were similar to those obtained by Zepp et al. (*2*). However, Haag et al. emphasized the differences in photosensitization ability of natural humic acids relative to isolated humic acids.

References

1. Zepp, R. G.; Baughman, G. L.; Schlotzhauer, P. F. *Chemosphere* **1981**, *10*, 109.
2. Ibid, p 119.
3. Zepp, R. G.; Schlotzhauer, P. F.; Sink, R. Merritt. *Environ. Sci. Technol.* **1985**, *19*, 74.

* The results reported by Zepp et al. (*2*) were 0.54, 0.65, and 0.43 h^{-1}. These results are in error by a factor of 10 from the actual calculated results (*27*).

4. Haag, W. R.; Hoigné, J.; Gassman, E.; Braun, A. M. *Chemosphere* **1984**, *13*, 631.
5. Haag, W. R.; Hoigné, J.; Gassman, E.; Braun, A. M. Ibid, p 641.
6. Haag, W. R.; Hoigné, J. *Environ. Sci. Technol.* **1986**, *20*, 341.
7. Ross, R. D.; Crosby, D. G. *J. Agric. Food Chem.* **1973**, *21*, 335.
8. Zepp, R. G.; Wolfe, N. L.; Baughman, G. L.; Hollis, R. C. *Nature* **1977**, *278*, 421.
9. Wolff, C. J. M.; Halmans, M. T. H.; Van der Heijde, H. B. *Chemosphere* **1981**, *10*, 59.
10. Ross, R. D.; Crosby, D. G. *Chemosphere* **1975**, *5*, 277.
11. Mabey, W. R.; Tse, D.; Baraze, A.; Mill, T. *Chemosphere* **1983**, *12*, 3.
12. Zepp, R. G.; Wolfe, N. L.; Gordon, J. A.; Fincher, R. C. *J. Agric. Food Chem.* **1976**, *24*, 727.
13. Zepp, R. G.; Wolfe, N. L.; Gordon, J. A.; Baughman, G. L. *Environ. Sci. Technol.* **1975**, *9*, 1144.
14. Skurlatov, Y. I.; Zepp, R. G.; Baughman, G. L. *J. Agric. Food Chem.* **1983**, *31*, 1065.
15. Miller, G. C.; Zisook, R.; Zepp, R. *J. Agric. Food Chem.* **1980**, *28*, 1953.
16. Mill, T.; Hendry, D. G.; Richardson, H. *Science* **1980**, *207*, 886.
17. Mill, T.; Mabey, W. R.; Bomberger, D. C.; Chou, T–W.; Hendry, D. G.; Smith, J. H. *Laboratory Protocols for Evaluation of the Fate of Organic Chemicals in Air and Water;* U.S. Environmental Protection Agency. U.S. Government Printing Office: Washington, DC, 1982; EPA 600/3–82–002.
18. Draper, W. M.; Crosby, D. G. *J. Agric. Food Chem.* **1981**, *29*, 699.
19. Draper, W. M.; Crosby, D. G. *J. Agric. Food Chem.* **1984**, *32*, 231.
20. Haag, W. R.; Hoigné, J. *Chemosphere* **1985**, *14*, 1659.
21. Cooper, W. J.; Zika, R. G. *Science* **1983**, *220*, 771.
22. Draper, W. M.; Crosby, D. G. *Arch. Environ. Contam. Toxicol.* **1983**, 121.
23. Fisher, A. M.; Kliger, D. S.; Winterle, J. S.; Mill, T. *Chemosphere* **1985**, *14*, 1299.
24. Zepp, R. G.; Braun, A. M.; Hoigné, J.; Leenheer, J. A. *Environ. Sci. Technol.* **1987**, *21*, 485.
25. Zafiriou, O. C.; Joussot-Dubien, J.; Zepp, R. G.; Zika, R. G. *Environ. Sci. Technol.* **1984**, *18*, 358A.
26. Draper, W. M.; Crosby, D. G. "Measurement of Photochemical Oxidants in Agriculture Field Water"; *Abstract of Papers*, 172nd National Meeting of the American Chemical Society, San Francisco, CA; American Chemical Society: Washington, DC, 1976.
27. Zepp, R. G., U.S. Environmental Protection Agency, Office of Research and Development, Athens, GA, personal communication, 1985.
28. Hammond, G. S.; Saltiel, J.; Lamola, A. A.; Turro, N. J.; Bradshaw, J. S.; Cowan, D. O.; Counsell, R. C.; Vogt, V.; Dalton, C. *J. Am. Chem. Soc.* **1964**, *86*, 3197.
29. Lamola, A. A.; Hammond, G. S. *J. Chem. Phys.* **1965**, *43*, 2129.
30. Zepp, R. G. *Environ. Sci. Technol.* **1978**, *12*, 327.
31. Zepp, R. G.; Cline, D. M. *Environ. Sci. Technol.* **1977**, *11*, 359.

Chapter 13

Theory and Development of the Screening Test Method

13.A. Introduction

Chapters 6–8 described a cost-effective screening test method to determine environmentally relevant direct photoreaction rate constants and half-lives by carrying out parallel sunlight photoreaction experiments with an actinometer solution and a test chemical solution containing pure water. Chapter 13 continues this work by describing a cost-effective screening test to measure environmentally relevant indirect (sensitized) photoreaction rate constants and half-lives by carrying out parallel sunlight photoreaction experiments with an actinometer solution, a test chemical solution containing buffered distilled water (W), and a test chemical solution containing humic acid dissolved in buffered distilled water. This method yields an environmentally relevant net photoreaction rate constant that is the sum of the direct and indirect aqueous photoreaction rate constants.

This screening test was developed on the basis of the four criteria listed in Section 6.A.: a comprehensive review of the literature on the experimental determination of indirect and direct aqueous photoreaction rate constants; the principles outlined by Zepp and Cline (*1*) and Zepp et al. (*2–4*); the method developed by the Office of Toxic Substances, U.S. Environmental Protection Agency (*5*); and by the research carried out by SRI International, sponsored by the Office of Toxic Substances (*6–10*). The following sections describe the development of this screening-level test method.

13.B. Theoretical Discussion

Consider a well-mixed natural water system composed of dissolved humic acids, with an attenuation coefficient α_λ (in cm^{-1}), and a chemical at a very low molar concentration $[C]$ with weak absorbance

1464–6/88/0227$06.00/1

(<0.02). The absorbance of the humic acid solution is much greater than the absorbance of the chemical. Furthermore, consider that the chemical, because of its molecular structure, absorbs light at $\lambda \geq 290$ nm so that it can undergo both direct and indirect photoreaction. When irradiated in sunlight, the overall rate of disappearance of the chemical is given by equation 5.82 and is

$$-(d[C]/dt)_p = k_p[C] \tag{13.1}$$

$$k_p = (k_d + k_s) \tag{13.2}$$

where k_d is the first-order direct photoreaction rate constant, k_s is the first-order indirect (sensitized) photoreaction rate constant, and k_p is the net first-order photoreaction rate constant in sunlight. All rate constants are in the units of reciprocal time.

Integrating equation 13.1 under the boundary conditions ($t = 0$, $[C_o]$) and (t, $[C_t]$) yields

$$\ln ([C_o]/[C_t]) = k_p t \tag{13.3}$$

and the associated half-life is given by

$$t_{(1/2)} = 0.693/k_p \tag{13.4}$$

Equations 13.3 through 13.4 apply to sunlight photoreaction in tubes. Because the rate of photoreaction in tubes is faster than the rate in a natural water body in the environment (e.g., a river or lake), equations 13.1 through 13.4 must be modified. Thus, based on equation 6.14

$$k_{pE} = 0.46k_p \tag{13.5}$$

If photoreaction takes place in sunlight in a natural water body in the environment, then the net photoreaction rate constant is called k_{pE} and the corresponding half-life is $t_{(1/2)E}$. The direct photoreaction rate constant is designated as k_{dE}; the indirect photoreaction rate constant is designated as k_{sE}. The subscript E denotes that photoreaction is in a natural water body in sunlight in the environment. The net rate of photoreaction in a natural water body in the environment can be described by equations 5.86 through 5.89.

13.C. Chemical Actinometry

To quantitatively measure rates of photoreaction of a chemical in sunlight, the integrated sunlight intensity incident on the sample must be measured. This measurement is accomplished by using the variable-rate low optical density actinometers *p*-nitroacetophenone–pyridine (PNAP–PYR) and *p*-nitroanisole–pyridine (PNA–PYR) developed by Dulin and Mill (*10*) and described completely in Section 6.B.4.b.ii. These actinometers have been used to measure rates of direct photoreaction in sunlight in the environment, and will be used extensively. In the screening test described in Chapters 13–15. In the screening tests described in Chapters 6–8, if the half-life of a chemical in tubes in sunlight is greater than 4 h, then the PNAP–PYR actinometer is used. If the half-life in tubes in sunlight is equal to or less than 4 h, then the PNA–PYR actinometer is used.

For the sunlight photoreaction experiments described in Chapters 6–8, the molar concentration of pyridine [PYR] required to adjust the rate of photoreaction of PNAP or PNA to equal the rate of photoreaction of the chemical in pure water is given by equations 7.9 and 7.11. These equations also apply when measuring rates of photoreaction of a chemical in humic acid solution. For the PNAP–PYR actinometer

$$[\mathrm{PYR}] = 26.9(k_{pc}/k_a) \tag{13.6}$$

For the PNA–PYR actinometer

$$[\mathrm{PYR}] = 1.04(k_{pc}/k_a) - 6.45 \times 10^{-4} \tag{13.7}$$

where k_{pc} is the net photoreaction rate constant of a chemical in days^{-1}, in tubes; k_a in equation 13.6 is the specific light absorption rate constant for PNAP in days^{-1} which is tabulated in Table A-3, Appendix A, as a function of latitude and season ; and k_a in equation 13.7 is the specific light absorption rate constant for PNA in days^{-1} and is tabulated in Table A-2, Appendix A, as a function of latitude and season.

The direct photoreaction rate constant for the PNAP–PYR actinometer, k_{da}, in the unit days^{-1} is given by equation 7.8 and is

$$k_{da} = 0.0372[\mathrm{PYR}]k_a \tag{13.8}$$

where [PYR] is the molar concentration of pyridine and k_a is the specific light absorption rate constant for PNAP in days^{-1}.

The direct photoreaction rate constant for the PNA–PYR actinometer in days^{-1} is given by equation 7.10 and is

$$k_{da} = 0.961[\mathrm{PYR}]k_a + 6.20 \times 10^{-4}k_a \tag{13.9}$$

where [PYR] is the molar concentration of pyridine and k_a is now the specific light absorption rate constant for the PNA actinometer in days^{-1}.

For 1 L of solution, the mass of pyridine required to adjust the rate of the actinometer is

$$\text{mass (grams)} = \text{fw[PYR]} \tag{13.10}$$

where fw = 79.10 g/mol, the formula weight of pyridine. Because the density of pyridine is 0.9810 g/mL at 20 °C, the volume (in mL) of pyridine, at 20 °C, required to equal this mass is

$$V = 80.6[\text{PYR}] \tag{13.11}$$

At low optical density, the actinometers PNAP–PYR and PNA–PYR undergo direct photoreaction in sunlight and follow the first-order differential equation

$$-d[C]_a/dt = k_{da}[C]_a \tag{13.12}$$

where k_{da} is the first-order direct photoreaction rate constant of the actinometer in the unit days^{-1} and $[C]_a$ is the molar concentration of the actinometer (Chapters 6–8 and Section 5.A.1.b.). Equation 13.12 is written in the form applicable to photoreaction in tubes. Integrating equation 13.12 under the boundary conditions ($t = 0$, $[C_o]$) and (t, $[C_t]$) yields

$$\ln([C_o]/[C_t])_a = k_{da}t \tag{13.13}$$

13.D. Physical and Photochemical Properties of Humic Acid Solutions

Spectroscopic properties were determined for a number of commercial and natural aquatic humic and fulvic acids (*11*). The sources of these humic and fulvic acids were described by Zepp et al. (*2*) and have also been described in Section 12.C.1.a. The electronic absorption spectra of all the humic and fulvic acids in aqueous solution were very similar. All spectra showed continuous absorption that increased as the wavelength decreased in the spectral region 600–300 nm with no absorption maxima. All aqueous humic and fulvic acid solutions obeyed the Beer–Lambert law (equation 1.14) in the humic and fulvic acid concentration range 1–40 mg of total organic carbon per liter.

Specific absorption coefficients, k_h, in the units liters per milligram of organic carbon per meter were computed using the Beer–Lambert law (equation 1.14) in the form

$$k_h = 2.303A_\lambda/lC_o \tag{13.14}$$

where A_λ is the absorbance at wavelength λ, l is the cell path length in meters, and C_o is the concentration of dissolved organic matter in milligrams of organic carbon per liter. Typical absorption curves of k_h versus λ are shown in Figure 13.1 for two commercial humic acids (Fluka and Aldrich); a natural aquatic humic acid from the Aucilla River, GA; and a fulvic acid isolated from California soil. The spectral data in the wavelength region 300–500 nm can be described by

$$k_h = A \exp[B(450 - \lambda)] \tag{13.15}$$

where A and B are constants and λ is the wavelength in nanometers. The mean values of the constants A and B for aquatic humic acids in 12 water bodies were $A = 0.6 \pm 0.3$; $B = 0.014 \pm 0.001$.

Mill et al. (*6*) obtained the electronic absorption spectra of four natural and synthetic humic acids in aqueous solution [Searsville Pond, Palo Alto, CA; Holston River North and South Forks, Kingsport, TN; and a commercial Aldrich humic acid] and the spectra were similar to those in Figure 13.1. The mean values of the constants A and B in equation 13.15 for these four humic acid solutions were $A = 0.26$ and $B = 0.016$. These values were similar to the values observed by Zepp and Schlotzhauer (*11*).

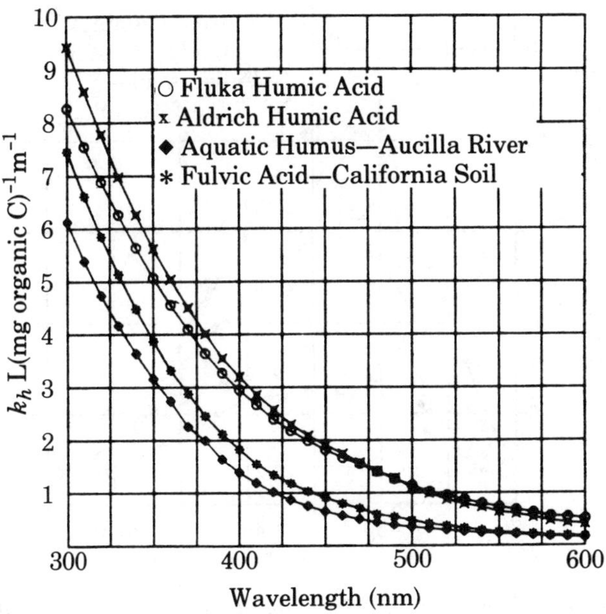

Figure 13.1. Absorption spectra of various humic acids in water at pH 6.0. (Reproduced with permission from ref. 11. Copyright 1981 Pergamon.)

Similar fluorescence spectra were obtained for the commercial humic acids, the fulvic acids, and the humic acid from a natural water. [*see* Figure 13.2 (*11*)]. With excitation at 350 nm, maximum emission occurred in the spectral region 430–470 nm with shoulders at 400–420 and 500–520 nm. Mill et al. (*6*) observed similar fluorescence spectra for their four humic acids described in the previous paragraph.

The fluorescence quantum efficiencies of the humic acids were measured in air-saturated solutions relative to the quantum efficiency of quinine sulfate in 0.1 N H_2SO_4 solution (*11*). All solutions were optically matched at 350 nm (absorbance was 0.25 in a 10-cm cell). All fluorescence quantum efficiencies were low and, remarkably, very similar. Removal of dissolved oxygen by degassing had no effect on the fluorescence efficiency of the aquatic humic acids in the Aucilla River water.

As a measure of the sensitizing efficiency of humic acids in the presence of furfuryl alcohol, Haag et al. (*12*) measured the singlet oxygen steady-state concentration in several Swiss natural waters and compared these results with those obtained from isolated humic acids (i.e., from a lake and a soil). In addition, these researchers compared

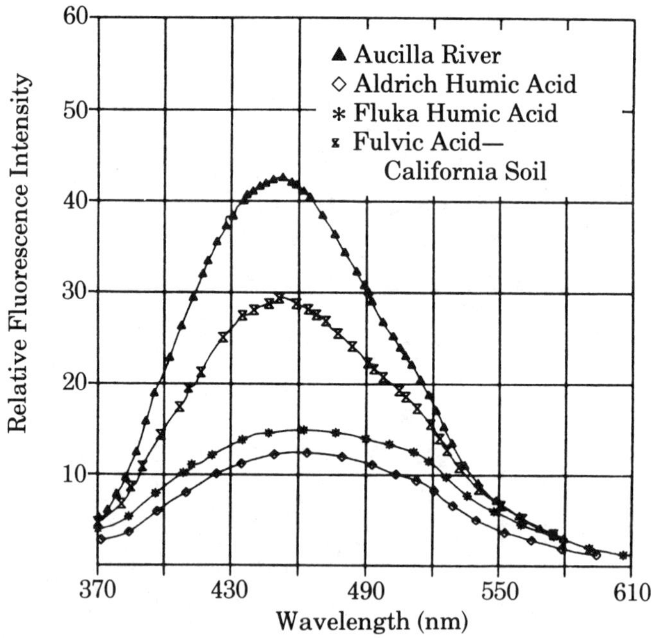

Figure 13.2. Fluorescence spectra of humic acids in water (excitation wavelength of 350 nm, pH 6.0). (Reproduced with permission from ref. 11. Copyright 1981 Pergamon.)

their results with those calculated from published data for other humic acid natural waters. The singlet oxygen steady-state concentration normalized to the absorptivity at 366 nm, $\{[^1O_2]/\alpha_{366}\} = \{M/m^{-1}\}$ Mm, is a measure of the quantum efficiency of the various humic acids. The values for the natural aquatic waters were in the range 1.0×10^{-14} to 8.3×10^{-14} Mm. Similarly, the three isolated humic acids had values from 2.5×10^{-14} to 7.1×10^{-14} Mm; the Fluka humic acid isolated from soil had a value of 2.5×10^{-14} Mm. Haag et al. (*12*) indicated that, in general, the natural aquatic humic waters had relatively low absorptivity and high quantum efficiency, but those isolated from a terrestrial source had a relatively high absorptivity and a low quantum efficiency.

All the electronic absorption and fluorescence spectra and the fluorescence quantum efficiencies of the humic and fulvic acids were remarkably similar, despite the wide diversity of origins. As indicated by Haag et al. (*12*), the quantum efficiency of the terrestrial humic acid was within a factor of 3 of the quantum efficiency of the natural humic acids. As described in Chapter 12, the photosensitizing efficiencies of natural humic acids from various rivers in the United States and those from a number of commercial humic acids were also remarkably similar (i.e., within a factor of 2). All of these studies have indicated that several commercial humic acids can serve as useful model sensitizers in aqueous photoreaction tests for forecasting rates of indirect (sensitized) photoreaction in natural waters in the environment.

Because the Aldrich humic acid has spectral and photochemical properties similar to natural aquatic humic acid waters; is readily available commercially to any laboratory, in contrast to a specific natural humic water; can be easily prepared and standardized; and can be diluted to the dissolved organic carbon content and spectral absorbance typical of most natural fresh waters, it was chosen as the model sensitizer for the indirect photoreaction screening test to be described in succeeding sections (*5, 8, 9*).

Mill et al. (*8*) and Winterle and Mill (*9*) found that the photoreactivity of the Aldrich humic acid aqueous solution (the synthetic humic acid water designated as SHW) depended upon pH and the history of sunlight exposure. The SHW solution initially photobleached with a time-dependent rate constant. After 7 days of irradiation with winter sunlight, however, SHW solutions photobleached at a nearly constant rate. Mill (*13*) indicated that SHW solutions can be photobleached with a nearly constant rate constant after 4 days of sunlight exposure in early spring or late fall or after 3 days of sunlight exposure in late spring, summer, or early fall. To control the photobleaching behavior, the pH was adjusted and maintained at a value of 7.0 with 0.01 M phosphate buffer. The preparation of pre-aged SHW solutions is described in Section 14.B.2.a.

On the basis of the indirect photoreaction screening tests developed by Mill et al. (*8*), Winterle and Mill (*9*), and the method developed by the Office of Toxic Substances, U.S. Environmental Protection Agency (*5*), the indirect (sensitized) photoreaction rate constant in tubes, k_s, after aging, was found to follow the mathematical expression

$$k_s = k_{so} \exp(-kt) \tag{13.16}$$

where k_{so} is the initial indirect (or sensitized) photoreaction rate constant and k is the SHW photobleaching rate constant. Equation 13.16 is a first-order bleaching equation.

Because SHW follows first-order kinetics, the rate of bleaching can be described by the equation

$$-(dC_{\mathrm{SHW}}/dt) = kC_{\mathrm{SHW}} \tag{13.17}$$

where k is the first-order bleaching rate constant defined in equation 13.16 and C_{SHW} is the concentration of the humic acid in SHW (e.g., in the units of milligrams of organic carbon per liter). From the Beer–Lambert law (equation 1.14), C_{SHW} is proportional to the absorbance at 370 nm, A_{370}, so that equation 13.17 becomes

$$-(dA_{370}/dt) = kA_{370} \tag{13.18}$$

Integrating equation 13.18 under the boundary conditions ($t = 0$, A^o_{370}) and (t, A_{370}) yields

$$\ln (A_{370}/A^o_{370}) = -kt \tag{13.19}$$

$$(A_{370}/A^o_{370}) = \exp(-kt) \tag{13.20}$$

The fraction of SHW photobleached is given by the equations

$$\text{fraction SHW photobleached} = [(A^o_{370} - A_{370})/A^o_{370}] \tag{13.21}$$

$$= [1 - (A_{370}/A^o_{370})] \tag{13.22}$$

Substituting equation 13.20 in 13.22 yields

$$\text{fraction SHW photobleached} = [1 - \exp(-kt)] \tag{13.23a}$$

$$= [1 - (A_{370}/A^o_{370})] \tag{13.23b}$$

13.E. Development of the Indirect Photoreaction Screening Test Method

The indirect photoreaction screening test is based on the research carried out by SRI International, sponsored by the Office of Toxic Substances, U.S. EPA, (*6–10*); by the research of Zepp et al. (*2–4*); and by the method developed by the Office of Toxic Substances (*5*).

This screening test allows the measurement of the indirect (or sensitized) aqueous photoreaction rate constant, k_{sE}, and the net aqueous photoreaction rate constant, k_{pE}, for a test chemical in dissolved humic acid waters (i.e., in synthetic humic acid water, SHW). It describes the preparation and standardization of SHW. In this method, a test chemical is exposed to sunlight in tubes containing solutions of SHW and solutions containing buffered distilled water (W). To correct for variations in solar irradiance during photoreaction, an actinometer is simultaneously insolated. From the concentration-versus-time data, k_{sE} and k_{pE} are calculated that apply to clear sky, near-surface conditions in fresh water bodies. Because this screening-level test does not require sophisticated and expensive equipment, it is cost-effective.

The screening-level test consists of two phases. In phase one, the test chemical is irradiated in both pre-aged SHW and W to obtain the approximate rate constants, $(k_{pc})_{\mathrm{SHW}}$ and $(k_{dc})_{\mathrm{W}}$; to determine whether direct and/or indirect photoprocesses are important; and to determine which actinometer (i.e., PNAP–PYR or PNA–PYR) should be used in the phase two experiments. In phase two, the test chemical is again irradiated in SHW and in W. To correct for photobleaching of SHW and for variations in solar intensity, tubes containing SHW and actinometer solutions are exposed simultaneously. From the time-versus-concentration data, k_{pE} (the sum of k_{sE} and k_{dE}) is calculated.

13.E.1. Phase One Test

Sunlight photoreaction rate constants are measured in tubes containing SHW and W. The molar concentration $[C]$ of the test chemical is adjusted so that the solution has low absorbance (≤ 0.02). In addition, the experimental conditions are adjusted so that the absorbance of SHW is 0.200 and is much greater than the absorbance of the test chemical. Thus, in a 1-cm cell, α_λ of SHW is much greater than $\epsilon_\lambda[C]$ of the test chemical, and equations 13.1 through 13.4 apply. Therefore, for the test chemical in SHW and W, equation 13.3 becomes

$$\ln([C_o]/[C_t])_{\mathrm{SHW}} = (k_{pc})_{\mathrm{SHW}} t \tag{13.24}$$

$$\ln([C_o]/[C_t])_{\mathrm{W}} = (k_{dc})_{\mathrm{W}} t \tag{13.25}$$

where k_{pc} is the net photoreaction rate constant and k_{dc} is the direct photoreaction rate constant of the test chemical in the unit days^{-1}. By measuring the molar concentration of the chemical in SHW and W in tubes as a function of the time t in days, k_{pc} and k_{dc} can be calculated with equations 13.24 and 13.25, respectively. The half-life can be calculated from these rate constants with equation 13.4.

The ratio of these rate constants, R, is defined by the equation

$$R = (k_{pc})_{\mathrm{SHW}}/(k_{dc})_{\mathrm{W}} \tag{13.26}$$

and is a measure of the effect of synthetic humic acid water on the photoreaction of the test chemical. If R is between 0 to 1, then the photoreaction of test chemical is inhibited by the synthetic humic acid water and phase two is not applicable. If R is between 1 to 2, the test chemical is marginally susceptible to indirect photoreaction. In this case, phase two studies are optional. If R is greater than 2, then phase two studies are necessary. Because both k_{pc} and k_{dc} are measured under the same conditions, R is a valid measure of the susceptibility of the test chemical to indirect photoreaction. However, because SHW is subject to photobleaching, k_{pc} will decrease with time because the indirect rate will decrease. Therefore, $R > 2$ is considered to be a conservative limit because k_{pc} will become systematically smaller with time.

Because the rate of photoreaction in tubes is faster than the rate in natural water bodies, values of near-surface photoreaction rate constants k_{pE} and k_{dE} in a natural body of water can be calculated from k_{pc} and k_{dc} by using equation 13.5.

13.E.2. Phase Two Test

13.E.2.a. Purpose

The purposes of the phase two test are

- to measure the initial indirect photoreaction rate constant of the test chemical, k_{soc} (defined by equation 13.16)
- to measure the direct photoreaction rate constant k_{dc}
- to calculate the net photoreaction rate constant k_{pc}
- to calculate k_{pE}, k_{sE}, and k_{dE} for the test chemical in a natural humic acid water body in the environment

If the approximate value of k_{pc} determined in phase one is not significantly greater then the value of k_{dc} measured for the experimental date, then phase two is not necessary because the test chemical is not subjected to indirect photoreaction. When k_{pc} in SHW is significantly larger than k_{dc}, then phase two studies are necessary. The rate constant k_{pc} in SHW is used to choose the appropriate actinometer and to choose an actinometer composition that makes the actinometer rate equal to the test chemical rate. Test chemical solutions in buffered SHW and in W are then irradiated in sunlight in parallel with actinometer solutions, in tubes.

13.E.2.b. Development of the Test Method

As described in Section 13.D., SHW is subject to photobleaching that decreases its ability to promote indirect photoreaction based on its ability to absorb sunlight. This effect will become significant when photoreaction exceeds a few days. To correct for photobleaching, tubes containing a test chemical in SHW and in W are irradiated simultaneously. At any time, the loss of test chemical is given by equation 13.1. Assuming actinometer correction to constant light flux, then equation 13.1 can be written as

$$-(dC/dt) = k_{sc}C + k_{dc}C \tag{13.27}$$

where C is the molar concentration of the test chemical; k_{sc} and k_{dc} are the first-order indirect and direct photoreaction rate constants of the test chemical, respectively; and t is the time.

The indirect photoreaction rate constant is actually time-dependent because SHW photobleaches. The rate constant k_{sc}, after preaging, obeys equation 13.16, where k_{soc} is the initial indirect photoreaction rate constant for the test chemical, and k is the SHW photobleaching rate constant. After substituting equation 13.16 in equation 13.27 and rearranging the result yields

$$-(dC/C) = k_{soc}[\exp(-kt)]dt + k_{dc}dt \tag{13.28}$$

Integrating equation 13.28 under the boundary conditions ($t = 0, C_o$) and (t, C_t) yields

$$\ln (C_o/C_t)_{\text{SHW}} = (k_{soc}/k)[1 - \exp(-kt)] + k_{dc}t \tag{13.29}$$

for the test chemical in SHW.

The term (k_{soc}/k) can be evaluated as follows. Because in W, $\ln (C_o/C_t)_{\text{W}} = k_{dc}t$, subtracting this equation from equation 13.29 yields

$$\ln (C_o/C_t)_{\text{SHW}} - \ln (C_o/C_t)_{\text{W}} = (k_{soc}/k)[1 - \exp(-kt)] \tag{13.30}$$

The photobleaching fraction $[1 - \exp(-kt)]$ is defined by equations 13.23a and 13.23b. Substituting these results in equation 13.30 yields

$$\ln (C_o/C_t)_{SHW} - \ln (C_o/C_t)_W = (k_{soc}/k)[1 - (A_{370}/A^o_{370})] \quad (13.31)$$

The terms A^o_{370} and A_{370} are the absorbance of SHW at 370 nm and are proportional to the humic acid sensitizer content at times zero and t, respectively. Samples of the test chemical in SHW and in W and samples of SHW (without test chemical) are irradiated simultaneously in the identical tubes as a function of the time. By using these data and linear regression analysis on equation 13.31 with $\{\ln (C_o/C_t)_{SHW} - \ln (C_o/C_t)_W\}$ as the dependent variable and $[1 - (A_{370}/A^o_{370})]$ as the independent variable, the slope is given by

$$S1 = (k_{soc}/k) \quad (13.32)$$

To evaluate k_{soc}, the parameter k has to be evaluated under standard sunlight conditions. This can be accomplished as follows: for the actinometer, equation 13.13 is applied in the form

$$\ln (C_o/C_t)_a = k_{da}t \quad (13.33)$$

For the SHW solution (without test chemical), equation 13.19 is used in the form

$$\ln (A^o_{370}/A_{370}) = kt \quad (13.34)$$

Because the SHW and actinometer solutions are irradiated simultaneously in the same size tubes, equation 13.33 can be divided into equation 13.34, to yield:

$$\ln (A^o_{370}/A_{370}) = (k/k_{da}) \ln (C_o/C_t)_a \quad (13.35)$$

Samples of SHW solutions and the appropriate actinometer solution (PNAP–PYR or PNA–PYR), chosen on the basis of the results of the phase one experiments, are irradiated simultaneously in the same size tubes as a function of the time. Linear regression analysis of these data using $\ln (A^o_{370}/A_{370})$ as the dependent variable and $\ln (C_o/C_t)_a$ as the independent variable gives the slope

$$S2 = (k/k_{da}) \quad (13.36)$$

The value of k_{da} for the appropriate actinometer can be obtained from equation 13.8 or 13.9. The value of k_{soc} can then be obtained from

$$k_{soc} = (k_{soc}/k)(k/k_{da})(k_{da}) \quad (13.37)$$

The initial indirect photoreaction rate constant of the test chemical, k_{soc}, can then be calculated by substituting the values of $S1$, $S2$, and k_{da} into equation 13.37.

The direct photoreaction rate constant for the test chemical, k_{dc}, is obtained under standard light conditions as follows. In W, the direct photoreaction of the test chemical can be described by the equation

$$\ln (C_o/C_t)_W = k_{dc}t \tag{13.38}$$

The photoreaction of the actinometer is described by equation 13.33. Because the actinometer and test chemical solutions are irradiated simultaneously in the same size tubes, equation 13.33 can be divided into equation 13.38, yielding

$$\ln (C_o/C_t)_W = (k_{dc}/k_{da}) \ln (C_o/C_t)_a \tag{13.39}$$

Samples of the test chemical in W and the actinometer are simultaneously irradiated as a function of time. Linear regression analysis using $\ln (C_o/C_t)_a$ as the independent variable and $\ln (C_o/C_t)_W$ as the dependent variable gives a slope

$$S3 = (k_{dc}/k_{da}) \tag{13.40}$$

The direct photoreaction rate constant k_{dc} can be obtained from

$$k_{dc} = (k_{dc}/k_{da})k_{da} \tag{13.41}$$

using the slope $S3$ and the value of k_{da} for the appropriate actinometer from equation 13.8 or 13.9.

Because the net photoreaction rate constant corresponds to the sum of the direct and indirect photoreaction rate constants, k_{pc} in SHW can be determined by taking the sum of k_{dc} and k_{soc} according to

$$(k_{pc})_{SHW} = k_{soc} + k_{dc} \tag{13.42}$$

Finally, k_{pE} for the test chemical in humic acid water can be calculated from k_{pc} by using equation 13.5.

13.E.2.c. Test Procedure

On the basis of the test chemical rate constant in SHW, $(k_{pc})_{SHW}$, determined in the phase one experiments, a half-life can be calculated with equation 13.4. If the half-life is equal to or less than 4 h, then the PNA–PYR actinometer is used. If the half-life is greater than 4 h, then the PNAP–PYR actinometer is used. After the appropriate actinometer is chosen, the rate constant of the actinometer is adjusted to match the rate constant of the test chemical (±50%) by varying the concentration of pyridine (equation 13.6 or 13.7).

The following solutions are prepared and added to a large number of tubes: actinometer solution, test chemical in pH 7.0 phosphate buffer solution, test chemical in pH 7.0 phosphate buffer/SHW solution, and pH 7.0 phosphate buffer/SHW solution. An appropriate number of samples in tubes, including dark controls, are placed in a rack in sunlight and irradiated. The concentration of each of these solutions, including the appropriate dark control, is measured as a function of time. From these data, the following quantities are calculated and arranged in tabular form in the order of increasing time: t, $\ln (C_o/C_t)_{\mathrm{SHW}}$, $\ln (C_o/C_t)_{\mathrm{W}}$, $[1 - (A_{370}/A_{370}^{o})]$, $\ln (A_{370}^{o}/A_{370})$, $\ln (C_o/C_t)_a$. By using the appropriate data and the equations derived in Section 13.E.2.b., k_{sE}, k_{dE}, and k_{pE} can be calculated.

This screening-level procedure is based on the theory of the kinetics of photoreaction in aquatic humic media and is in its preliminary stages of development. More research is needed to verify the method.

13.F. Applicability and Specificity

The screening level test is applicable to those chemicals with unique molecular structures that may or may not have UV–vis absorptions in the range of 290–800 nm and do have the potential to undergo indirect photoreaction in humic acid waters in sunlight (e.g., phenols, aromatic amines, hydroxylamines, mercaptans, sulfides, disulfides, olefins, dienes, conjugated carbon–carbon double-bonded systems, trialkyleneamines, and imidazoles).

This test method is applicable only to pure and not to technical grade chemicals. Impurities present can complicate the interpretation of the test results.

This screening test has been designed to use synthetic humic water. However, the method is easily adapted so that a natural humic acid water can be used.

For first-order rate equations to be applicable, the test chemical must be at a very low concentration with low absorbance (i.e., absorbance ≤ 0.02 at $\lambda \geq 290$ nm) and the SHW absorbance must be much greater than the absorbance of the test chemical. Thus, for this screening test, the absorbance of the synthetic humic acid solution has been set at 0.200 at 370 nm.

The test method is simple and inexpensive, but does require that the chemical dissolve in water at sufficient concentration to be measured by some analytical technique. Phase one tests should be performed during a clear sky period, if possible, to obtain the best test results. Phase two tests should be started as soon as possible after completion of the phase one test to ensure that the rate constant (k_{pc}) estimate in SHW remains valid.

To correct for variations in solar irradiance during photoreaction in sunlight, an actinometer is simultaneously insolated. In this way, all rate constants and half-lives are normalized to clear sky, near-surface conditions in fresh water bodies.

In contrast to k_{dE}, which, once measured, can be calculated for different seasons and latitudes in the northern hemisphere, k_{sE} only applies to the season and latitude for which it is measured. This condition exists because the solar action spectrum for indirect photoreaction in humic acid waters is generally not known and can be expected to change for different test chemicals. Thus, k_{pE} , which contains k_{sE}, is likewise valid only for the experimental date and latitude at which the experiments were carried out.

The values of k_{pE} and k_{sE} are atypical quantities because their values will change from water body to water body as the amount and quality of dissolved aquatic humic acids change. To standardize the screening-level indirect photoreaction test, a synthetic humic acid water is prepared by water extracting commercial humic acid material (Aldrich). The humic acid material is inexpensive and available to any laboratory, in contrast to a specific natural humic acid water. The standard SHW can be diluted to the dissolved organic carbon content and UV–vis absorbance typical of most surface fresh waters. For optically-matched natural and commercial humic acid solutions, the differences in photoefficiency are usually within a factor of 2 [Section 12.C.]. However, more research is needed to establish that Aldrich humic acid does, indeed, mimic naturally occurring humic acids in photosensitizing efficiency.

References

1. Zepp, R. G.; Cline, D. M. *Environ. Sci. Technol.* **1977**, *11*, 359.
2. Zepp, R. G.; Baughman, G. L.; Schlotzhauer, P. F. *Chemosphere* **1981**, *10*, 109.
3. Zepp, R. G.; Baughman, G. L.; Schlotzhauer, P. F. *Chemosphere* **1981**, *10*, 119.
4. Zepp, R. G.; Schlotzhauer, P. F.; Sink, R. Merritt. *Environ. Sci. Technol.* **1985**, *19*, 74.
5. *Fed. Regist.* **1986**, *51*, 483; 40 CFR 796.3765.
6. Mill, T.; Mabey, W. R.; Winterle, J. S.; Davenport, J. E.; Barich, V. P.; Dulin, D. E.; Tse, D.; Lee, G. "Design and Validation of Screening and Detailed Methods for Environmental Processes"; Final Report; U.S. Environmental Protection Agency, Office of Toxic Substances: Washington, DC, 1982; EPA Contract No. 68–01–6325.

7. Mill, T.; Davenport, J.; Winterle, J. S.; Mabey, W. R.; Drossman, H.; Tse, D.; Liu, A. "Toxic Substances Process Data Generation and Protocol Development"; Work assignment 12; Test standard development; Final Report, Washington, DC, 1983; EPA Contract No. 68–03–2981.
8. Mill, T., SRI International, Menlo Park, CA, personal communication, 1984.
9. Winterle, J. S.; Mill, T. "Toxic Substances Process Data Generation and Protocol Development"; Work assignment 18, EPA Special Report; *Indirect Photoreaction Protocol*; U.S. Environmental Protection Agency: Washington, DC, 1985; EPA Contract No. 68–03–2981.
10. Dulin, D.; Mill, T. *Environ. Sci. Technol.* **1982**, *16*, 815.
11. Zepp, R. G.; Schlotzhauer, P. F. *Chemosphere* **1981**, *10*, 479.
12. Haag, W. R.; Hoigné, J. *Environ. Sci. Technol.* **1986**, ***20***, **341.**
13. Mill, T., SRI International, Menlo Park, CA, personal communication, 1985.

Chapter 14

Experimental Procedures

14.A. UV–Visible Absorption Spectra

The absorbance of the synthetic humic acid solutions at 370 nm and the absorbance of solutions of test chemical at $\lambda \geq 290$ nm can be measured by standard spectroscopic procedures (*1*).

14.B. Indirect Photoreaction Screening Test

14.B.1. Experimental Conditions

Most of the experimental conditions described in Chapter 7 are applicable to these experiments. Thus, the reader is referred to Sections 7.B.1.a. Special Laboratory Equipment; 7.B.1.b. Purity of Water; 7.B.1.c. Sterilization; 7.B.1.e. Volatile Chemical Substances; 7.B.1.f. Control Solution; 7.B.1.h. Sunlight Actinometers; and 7.B.1.k. Chemical Analysis of Solutions.

14.B.2. Preparations

14.B.2.a. Standard Synthetic Humic Acid Solution

Aldrich humic acid (20 g) is added to a 2-L Pyrex Erlenmeyer flask. The flask is filled with 2 L of 0.1% NaOH. A stir bar is added to the flask, the flask is capped, and the solution is stirred for 1 h at room temperature. After stirring, the dark brown supernate is decanted and centrifuged at 15,000 rpm for 1 h. The supernate is then decanted, the pH is adjusted to 7.0 with dilute H_2SO_4 and the solution is filter-sterilized through a 0.2-μm filter into a 2-L Erlenmeyer flask. This solution contains approximately 60 ppm of dissolved organic carbon and the absorbance (in a 1-cm path length cell) is approximately 1.7 at 313 nm and 0.7 at 370 nm.

1464–6/88/0243$06.00/1

Pre-aging is accomplished by exposing the concentrated solution in the 2-L flask to direct sunlight for 4 days in early spring or late fall or for 3 days in late spring, summer, or early fall. The absorbance of the solution is then measured at 370 nm, a dilution factor is calculated, and the absorbance is decreased to 0.50 in a 1-cm path length cell. If necessary, the pH is readjusted to 7.0. The solution is brought, by exact dilution to an absorbance of 0.500 in a 1-cm path length cell with a precalculated volume of 0.010 M phosphate buffer solution. Finally, the solution is air-saturated by allowing it to equilibrate in the dark in a vessel plugged with sterile cotton. The flask is tightly capped and refrigerated at approximately 5 °C. This solution is standard synthetic humic acid water (SHW) stock solution. Before use in the phase one and two procedures, it is diluted with a precalculated volume of 0.010 M phosphate buffer to produce a pH 7.0 solution with an absorbance of 0.200 at 370 nm in a 1-cm cell.

14.B.2.b. Preparation of Test Chemical Solution

Solutions of test chemical should be prepared with sterile, air-saturated, 0.010 M pH 7.0 phosphate buffer and reagent grade (or purer) test chemical. Reaction solutions should be prepared with chemicals at concentrations less than one-half their solubility in pure water and at a concentration such that at any wavelength greater than 290 nm, the absorbance in a standard quartz 1-cm path length cell is equal to or less than 0.033. For very hydrophobic chemicals, it is difficult and time consuming to prepare aqueous solutions. To facilitate the preparation of aqueous solutions containing very hydrophobic chemicals and to allow for easier analytical measurement procedures, the following procedure may be used to aid in the dissolution of the chemical. The pure chemical is dissolved in reagent grade acetonitrile. The 0.010 M phosphate buffer solution is added to an aliquot of the acetonitrile solution. One volume percent of acetonitrile must not be exceeded in the final solution.

The test chemical solution should be mixed 3.00:2.00 by volume with 0.010 M phosphate buffered distilled water or SHW stock solution to prepare working solutions. SHW gives a 2:5 dilution with an absorbance of 0.200 at 370 nm in a 1-cm cell. The test chemical gives a 3:5 dilution to give a test chemical absorbance of 0.0198 or less at $\lambda \geq 290$ nm. Place the reaction solutions in the appropriate photoreaction tubes as described in Section 7.B.1.a. Tubes containing the synthetic humic acid solution should be labeled SHW and those containing buffered distilled water should be labeled W.

14.B.2.c. Preparation of Actinometer Solutions

The appropriate actinometer solution for the phase two test is a function of the half-life of the test chemical as determined in the phase one test. If the half-life of the test chemical is greater than 4 h, the PNAP–PYR actinometer solution should be prepared. If the half-life of the test chemical is equal to or less than 4 h, the PNA–PYR actinometer solution should be prepared.

14.B.2.c.i. The PNAP–PYR Actinometer Solution

By using the test chemical photoreaction rate constant, $(k_{pc})_{SHW}$, determined in the phase one test, and the variable k_a listed in Table A-3, Appendix A, the molar concentration of pyridine needed to adjust the rate of disappearance of the test chemical can be obtained from equation 13.6. The variable k_a is equal to the day-averaged rate constant for sunlight absorption by PNAP which is a function of latitude and season. The value of k_a is selected from Table A-3 for the season nearest the midexperiment date of the phase one studies and the decadic latitude nearest the latitude of the experimental site.

Once the molar concentration of pyridine [PYR] has been determined, an actinometer solution can be prepared as follows: PNAP (0.0165 g) is dissolved in 10 mL of acetonitrile to yield a 0.01 M solution. One milliliter of this solution is added to a 1-L volumetric flask. To the volumetric flask is added the mass in grams, or the volume of pyridine in milliliters at 20 °C, obtained from equations 13.10 or 13.11. The volumetric flask is filled with pure water to give 1 L of solution and shaken vigorously to ensure homogeneity. The resulting concentration of PNAP is 1.00×10^{-5} M. The flask should be wrapped with aluminum foil and kept from bright light.

14.B.2.C.ii. The PNA–PYR Actinometer Solution

By using the test chemical photoreaction rate constant, $(k_{pc})_{SHW}$, determined in the phase one test, and the variable k_a listed in Table A-2, Appendix A, the molar concentration of pyridine needed to adjust the rate of disappearance of *p*-nitroanisole (PNA) to match the rate of disappearance of the test chemical can be obtained from equation 13.7. The value of k_a is selected from Table A-2 for the season nearest the midexperiment date of the phase one studies and the decadic latitude nearest the latitude of the experimental site.

Once the molar concentration of pyridine [PYR] has been determined, an actinometer solution can be prepared as follows: PNA (0.0153 g) is dissolved in 10 mL of acetonitrile to yield a 0.01 M solution. One milliliter of this solution is added to 1-L volumetric flask. To the volumetric flask is added the mass in grams, or the volume of pyridine in milliliters at 20 °C, obtained from equations 13.10 or 13.11. The volumetric flask is filled with pure water to give 1 L of solution and shaken vigorously to ensure homogeneity. The resulting concentration of PNA is 1.00×10^{-5} M. The flask should be wrapped with aluminum foil and kept from bright light.

14.B.3. Performance of the Tests

14.B.3.a. Phase One Experiments

For all experiments, an aqueous solution of the test chemical in buffered distilled water (W) and in SHW, as well as a sufficient number of samples to perform all the required tests are prepared in quartz or borosilicate glass tubes (*see* Section 14.B.2.b.). The tubes are filled as completely as possible and sealed. Control samples are prepared in the absence of UV light and light is totally excluded by wrapping the tubes with aluminum foil or by other suitable methods. These samples are analyzed for the chemical after each time point to measure the loss of chemical in the absence of light. The samples and the controls are placed outdoors in an area free of shade and reflections of sunlight from windows and buildings. The samples are placed on a black, nonreflective background and are inclined at approximately 30° from the horizontal with the upper end pointing due north (in the northern hemisphere). The photoreaction experiments should be conducted during a frost-free time of year (i.e., May, June, July, August, or September in the northern hemisphere—weather permitting) and started initially at noon (1200). The following should be recorded:

- the date and time the experiment was begun
- the date and time the experiment was completed
- the time of sunrise and sunset on all days when photoreaction experiments were performed
- the times exposure was stopped and restarted for intermittent exposure
- the weather conditions during the period
- the latitude of the site

If a significant loss of test chemical has occurred in the control samples, the cause should be determined and eliminated or minimized. If hydrolysis is found to be significant, hydrolysis studies should be carried out first (*see* Section 7.B.1.f.).

One of the following procedures should be used depending on how fast the test chemical photoreacts in SHW.

14.B.3.a.i. Procedure 1

If the chemical transforms 50–80% in SHW within 28 days, the concentration of the chemical should be measured in SHW and in buffered distilled water (W), in triplicate, at time $t = 0$ and periodically at noon until at least 50% of the chemical has been consumed in SHW. To simplify, the sampling times can be estimated as the photoreaction experiments are in progress. The concentration of test chemical is determined in SHW and in W from three freshly opened reaction tubes for each time point. The concentration of test chemical should be determined in the control SHW and W solutions from three freshly opened reaction tubes for each time point and as soon as photoreaction experiments are completed.

14.B.3.a.ii. Procedure 2

If the test chemical transforms in SHW in the range of 20–50% in 28 days, the concentration of the chemical should be determined in SHW and W from three freshly opened reaction tubes at time $t = 0$. The concentration of the chemical in the three separate reaction tubes and the three control tubes is determined for SHW and W after 28 days of photoreaction.

14.B.3.a.iii. Procedure 3

For test chemicals that transform in SHW in sunlight 50–80% within 1 or 2 days, the samples are placed outside at noon (1200) and three samples are analyzed for the concentration of the chemical in SHW and W at $t = 0$, and three, freshly opened, reaction tubes are analyzed at noon the next day, and again, three, freshly opened, reaction tubes are analyzed at noon the second day. The concentration of the test chemical is determined in each of the three control solutions for SHW and W after the first day of irradiation and as soon as the photoreaction experiments have been completed on the second day.

14.B.3.a.iv. Analytical Methodology

An analytical method that is most applicable to the analysis of the specific chemical being tested (Section 7.B.1.k.).

14.B.3.b. Phase Two Experiments

By using the test chemical photoreaction rate constant, $(k_{pc})_{\mathrm{SHW}}$, determined in the phase one test, the appropriate actinometer solution, as well as a sufficient number of samples to perform all the required tests, are prepared in quartz or borosilicate tubes as described in Section 14.B.2.c. The tubes are filled as completely as possible after preparation. Aqueous solutions of test chemical in SHW and W are prepared as described in Section 14.B.2.b., as well as a sufficient number of samples in quartz or borosilicate tubes to perform all the required tests. These tubes are filled as completely as possible, sealed, and covered with aluminum foil as soon as possible after preparation.

All the samples are placed outdoors in an area free of shade and reflections of sunlight from windows and buildings. The samples are placed on a black, nonreflective background and inclined at approximately 30° from the horizontal with the upper end pointing due north (in the northern hemisphere). The foil should be removed from all samples except for the test chemical control samples and the actinometer control samples at noon (1200).

On the basis of the results of the phase one experiments, the concentration of test chemical in SHW and W and actinometer (PNAP or PNA) is determined in triplicate (from three separate tubes) at time $t = 0$ and periodically (at least five data points at approximately equal time intervals). The concentration of PNAP or PNA in the three actinometer control samples and the concentration of test chemical in the three control samples SHW and W are determined for each time point.

An analytical method should be selected that is most applicable to the analysis of the specific chemical tested (*see* Section 7.B.1.k) and the procedure given in that section should be followed for the analysis of PNAP or PNA.

Reference

1. *Fed. Regist.* **1985**, *50*, 39472; 40 CFR 796.1050.

Chapter 15

Illustrative Examples

15.A. Phase One Test: Illustrative Example 1

15.A.1. Description

Consider a chemical plant located in Cuyahoga Falls, OH (slightly northeast of Akron), on the Cuyahoga River which produces an organic chemical, a sulfide, continuously every day of the year. The waste effluent passes through a primary and secondary treatment plant and is discharged directly into the river which contains dissolved humic acids. The plant is located at 41° 9′ N. latitude. Using the phase one procedure, carry out experiments to estimate the net photoreaction rate constant of the sulfide in the river, the half-life, and the value of R. R is the ratio of the net photoreaction rate constant of a chemical in buffered humic acid water to the direct photoreaction rate constant of the chemical in buffered distilled water (defined in equation 13.26).

15.A.2. Photoreaction Experiments and Calculations

UV–vis absorption spectroscopic data indicated that the chemical absorbed light appreciably below 340 nm so that 11-mm i.d. quartz tubes were used (for these tubes, the path length was 1.00 cm). The chemical had a water solubility of 5.60×10^{-3} M at 25 °C. The chemical was dissolved in air-saturated 0.010 M pH 7.0 phosphate buffer at 25 °C. Analytical data indicated that the concentration of the chemical in this solution was 5.2×10^{-5} M, much less than one-half its water solubility.

The UV–vis absorption spectrum in a 1.00-cm cell showed no absorbance greater than 0.02 in the wavelength interval 290–800 nm, a condition required for the test method. Exactly 300 mL of the test chemical solution (5.2×10^{-5} M) was added to exactly 200 mL of air-saturated 0.010 M phosphate buffer. The test chemical was analyzed, and the concentration was found to be 3.1×10^{-5} M. Exactly 300 mL of the test chemical solution was added to exactly 200 mL of standard

1464–6/88/0249$06.00/1

SHW. The test chemical concentration was 3.1×10^{-5} M and the absorbance at 370 nm was 0.200. Absorbance must be ≥0.200 for this method.

The photoreaction experiments were carried out according to the phase one Procedure and the experiments were started on September 20 at noon (1200). The weather conditions are summarized and the analytical data given represent the mean of triplicate determinations.

1. Noon September 20 ($t = 0$): $[C_o]_{SHW} = [C_o]_W = 3.1 \times 10^{-5}$ M
2. Noon September 20 to noon September 21: weather was sunny and clear

   Noon September 21: $[C_t]_{SHW} = 2.6 \times 10^{-5}$ M

   $[C_t]_W = 2.9 \times 10^{-5}$ M

   control solutions: $[C_t]_{SHW} = [C_t]_W = 3.1 \times 10^{-5}$ M
3. Noon September 21 to noon September 23: weather was sunny and clear

   Noon September 23: $[C_t]_{SHW} = 1.8 \times 10^{-5}$ M

   $[C_t]_W = 2.6 \times 10^{-5}$ M

   control solutions: $[C_t]_{SHW} = [C_t]_W = 3.0 \times 10^{-5}$ M
4. At 0100 September 24, rain started and continued to 1010. The rain stopped at 1010 and the weather cleared so that it was sunny at 1110.
5. From 1110 September 24 to 1241 on September 25, the weather was clear and sunny.

   1241 September 24: $[C_t]_{SHW} = 1.3 \times 10^{-5}$ M

   $[C_t]_W = 2.3 \times 10^{-5}$ M

   control solutions: $[C_t]_{SHW} = 3.1 \times 10^{-5}$ M

   $[C_t]_W = 3.1 \times 10^{-5}$ M

Because 58% of the test chemical in SHW transformed, the photoreaction experiments were terminated. The average concentrations of the control samples for SHW and W were essentially the same as the initial concentrations, so that no adventitious processes occurred and the loss of chemical was only due to sunlight photoreaction.

Listed in Table 15-1 are the times of sunrise and sunset for the dates sunlight photoreaction experiments were carried out along with total number of hours of sunlight.

Table 15-2 summarizes the dates irradiated, the times exposed to sunlight, the total sunlight irradiation time for each date in days, and the total number of days of sunlight irradiated.

Table 15-1. Summary of Times for Sunrise and Sunset for the Period September 20 to September 25

Date	*Sunrise*	*Sunset*	*Total Sunlight* (*h*)
9/20	0554	2016	14.4
9/21	0555	2015	14.3
9/22	0556	2014	14.3
9/23	0557	2013	14.3
9/24	0558	2012	14.2
9/25	0559	2011	14.2

The calculations are as follows:

$$t = 4.7 \text{ days:} \tag{15.1}$$

$$[C_o]_{\mathrm{SHW}} = [C_o]_{\mathrm{W}} = 3.1 \times 10^{-5} \text{ M} \tag{15.2}$$

$$[C_o]_{\mathrm{SHW}} = 1.3 \times 10^{-5} \text{ M}; [C_t]_{\mathrm{W}} = 2.3 \times 10^{-5} \tag{15.3}$$

Using equations 13.24, 13.25, and 13.26 yields

$$(k_{pc})_{\mathrm{SHW}} = (1/4.7) \ln (3.1 \times 10^{-5}/1.3 \times 10^{-5}) = 0.18 \text{ days}^{-1} \tag{15.4}$$

$$(k_{dc})_{\mathrm{W}} = (1/4.7) \ln (3.1 \times 10^{-5}/2.3 \times 10^{-5}) = 0.064 \text{ days}^{-1} \tag{15.5}$$

$$R = (0.18/0.064) = 2.8 \tag{15.6}$$

Using the appropriate data in equations 13.4 and 13.5 yields

$$k_{pE} = 0.46(0.18) = 0.083 \text{ days}^{-1} \tag{15.7}$$

$$t_{(1/2)E} = (0.693/0.083) = 8.3 \text{ days} \tag{15.8}$$

Table 15-2. Summary of Irradiation Time

Date	*Times Irradiated*	*Time Irradiated/ Total Sunlight*	*Sunlight Irradiation Time for Each Date* (*days*)
9/20	1200 to 2016	(8.3/14.4)	0.58
9/21	0555 to 2015	(14.3/14.3)	1.00
9/22	0556 to 2014	(14.3/14.3)	1.00
9/23	0557 to 2013	(14.3/14.3)	1.00
9/24	1110 to 2012	(9.0/14.2)	0.63
9/25	0559 to 1241	(6.7/14.2)	0.47
			Total 4.68

Because R is greater than 2, indirect photoreaction is significant and the phase two experiments were started.

15.B. Phase Two Test: Illustrative Example 2

15.B.1. Description

Consider the same scenario as described in illustrative example 1. Using the phase two procedure, experiments are carried out to determine k_{sE}, k_{dE}, k_{pE}, and $t_{(1/2)E}$ for the test chemical in the Cuyahoga River for the fall season at 41° N. latitude.

15.B.2. Photoreaction Experiments and Calculations

On the basis of the results of the phase one experiments, the test chemical was found to have a photoreaction rate constant $(k_{pc})_{SHW}$ equal to 0.18 days^{-1} in the fall season in round tubes at 41° N. latitude. Using this rate constant in equation 13.4 yields a half-life of 3.9 days in tubes. Because the half-life is greater than 4 h, the PNAP–PYR actinometer was chosen for use in the phase two experiments. The value of k_a for the PNAP–PYR actinometer (245 days^{-1}) was chosen from Table A-3, Appendix A, which corresponds to the fall season at 40° N. latitude. Using equation 13.6, the molar concentration of pyridine required to adjust the actinometer rate to match the rate of disappearance of the test chemical is

$$[\text{PYR}] = 26.9(0.18/245) = 0.0198\text{M} \qquad (15.9)$$

The actinometer solution was prepared by adding 1.60 mL of pyridine (calculated from equation 13.11) to a 1-L volumetric flask containing 1.00 mL of 1.00×10^{-2} M PNAP in acetonitrile. The flask was filled to the mark with pure water to give a pyridine concentration of 0.0198 M and [PNAP] = 1.00×10^{-5} M. The PNAP was analyzed in triplicate, and the mean value of $[C_o]_{\text{PNAP}} = 1.0 \times 10^{-5}$ M.

Solutions of SHW, test chemical in standard SHW, and test chemical in phosphate buffer were prepared as described in illustrative example 1.

The following solutions were added to a sufficient number of tubes to carry out the phase two experiments:

1. test chemical (3.1×10^{-5} M) in standard SHW (0.010 M, pH 7.0 phosphate buffer)

2. test chemical (3.1×10^{-5} M) in 0.010 M, pH 7.0 phosphate buffer
3. SHW solution (absorbance = 0.200) in 0.010 M, pH 7.0 phosphate buffer
4. PNAP–PYR actinometer solution, [PNAP] = 1.0×10^{-5} M, ([PYR] = 0.0198 M).

Half of the tubes for all solutions were covered with foil and were used as controls. All the tubes were placed outdoors in sunlight on September 27, and the phase two experiments were started at 1200. Three tubes from each set of irradiated solutions and three tubes from each set of controls were removed at noon on days 0, 1, 2, 3, 4, and 5. The average analytical results are given in Table 15-3 for the irradiated samples. No significant loss of chemical or actinometer was found in any of the control samples.

From the data in Table 15-3 the functions ln $(C_o/C_t)_{SHW}$, ln $(C_o/C_t)_W$, $[1 - (A/A^o)_{370}]$, ln $(A^o/A)_{370}$, and ln $(C_o/C_t)_{PNAP}$ were calculated and are presented in Table 15-4.

Using the data in columns 2, 3, and 4 of Table 15-4, the slope $S1 = (k_{soc}/k)$ was calculated according to equation 13.31 and linear regression analysis. The slope was 4.6 with a correlation coefficient of 0.992.

By using the data in columns 5 and 6 of Table 15-4, the slope $S2 = (k/k_{da})$ was calculated according to equation 13.35 and linear regression analysis. The slope was 0.30 with a correlation coefficient of 0.999.

By using the data in columns 3 and 6 of Table 15-3, the slope $S3 = (k_{dc}/k_{da})$ was calculated according to equation 13.39 and was 0.45 with a correlation coefficient of 0.987.

By using values for the molar concentration of pyridine from equation 15.9 and k_a = 245 days^{-1} (from Table A-3 of Appendix A) for the fall season at 40° N. latitude in equation 13.8 yields

$$k_{da} = 0.0372(0.0198)(245) = 0.181 \text{ days}^{-1} \tag{15.10}$$

Table 15-3. Analytical Results for Illustrative Example 2, Phase Two

t (*days*)	*Solution 1,* $10^5[C]_{SHW}$, *M*	*Solution 2,* $10^5[C]_W$, *M*	*Solution 3,* A_{370}^{SHW}	*Solution 4,* 10^5[PNAP], *M*
0	3.1	3.1	0.200	1.0
1	2.4	2.9	0.189	0.84
2	1.6	2.7	0.179	0.68
3	1.3	2.4	0.169	0.58
4	0.88	2.3	0.159	0.46
5	0.66	2.0	0.150	0.39

Table 15-4. Photoreactive Functions for Illustrative Example 2, Phase Two

t (*days*)	$ln\ (C_o/C)_{SHW}$	$ln\ (C_o/C)_W$	$[1 - (A/A^o)_{370})]$	$ln\ (A^o/A)_{370}$	$ln\ (C_o/C)_{PNAP}$
0	0	0	0	0	0
1	0.256	0.0667	0.055	0.0566	0.174
2	0.661	0.138	0.105	0.111	0.386
3	0.869	0.256	0.155	0.168	0.545
4	1.26	0.298	0.205	0.229	0.777
5	1.55	0.438	0.250	0.288	0.942

The values of $S1$, $S2$, and k_{da}, just calculated, were combined in equation 13.37 to yield k_{soc} as follows:

$$k_{soc} = (4.6)(0.30)(0.181) = 0.25 \text{ days}^{-1} \qquad (15.11)$$

The rate constant k_{dc} was calculated from the product of $S3$ and k_{da} according to equation 13.41 as follows:

$$k_{dc} = (0.45)(0.181) = 0.081 \text{ days}^{-1} \qquad (15.12)$$

Substituting k_{soc} and k_{dc} in equation 13.42 yields

$$(k_{pc})_{SHW} = 0.25 + 0.081 = 0.33 \text{ days}^{-1} \qquad (15.13)$$

Substituting the values of k_{dc}, k_{soc}, and $(k_{pc})_{SHW}$, in equation 15.5 yields

$$k_{dE} = 0.46(0.081) = 0.037 \text{ days}^{-1} \qquad (15.14)$$

$$k_{sE} = 0.46(0.25) = 0.12 \text{ days}^{-1} \qquad (15.15)$$

$$k_{pE} = 0.46(0.33) = 0.15 \text{ days}^{-1} \qquad (15.16)$$

Finally, substituting the values of k_{pE} in equation 13.4, applicable to a natural water body in the environment, yields

$$t_{(1/2)E} = (0.693/0.15) = 4.6 \text{ days} \qquad (15.17)$$

The values of k_{sE}, k_{pE}, and $t_{(1/2)E}$ are only applicable to the fall season at 41° N. latitude and are only valid for the fall season at the location of the manufacturing plant.

Appendix A

Tables of L_λ Solar Irradiance, Wavelength Centers for L_λ, and Specific Light Absorption Rate Constants for *p*-Nitroanisole and *p*-Nitroacetophenone

Table A-1. Wavelength Centers and Intervals for L_λ (in nm)

	Interval Range		
λ(*center*)	*From*	*To*	$\Delta\lambda$
297.5	296.2	298.7	2.5
300.0	298.7	301.2	2.5
302.5	301.2	303.7	2.5
305.0	303.7	306.2	2.5
307.5	306.2	308.7	2.5
310.0	308.7	311.2	2.5
312.5	311.2	313.7	2.5
315.0	313.7	316.2	2.5
317.5	316.2	318.7	2.5
320.0	318.7	321.2	2.5
323.1	321.2	325.0	3.8
330.0	325.0	335.0	10.0
340.0	335.0	345.0	10.0
350.0	345.0	355.0	10.0
360.0	355.0	365.0	10.0
370.0	365.0	375.0	10.0
380.0	375.0	385.0	10.0
390.0	385.0	395.0	10.0
400.0	395.0	405.0	10.0
410.0	405.0	415.0	10.0
420.0	415.0	425.0	10.0
430.0	425.0	435.0	10.0
440.0	435.0	445.0	10.0
450.0	445.0	455.0	10.0
460.0	455.0	465.0	10.0
470.0	465.0	475.0	10.0
480.0	475.0	485.0	10.0

1464–6/88/0255$06.00/1

Table A-1. Continued

	Interval Range		
λ(*center*)	*From*	*To*	Δλ
490.0	485.0	495.0	10.0
500.0	495.0	505.0	10.0
525.0	512.5	537.5	25.0
550.0	537.5	562.5	25.0
575.0	562.5	587.5	25.0
600.0	587.5	612.5	25.0
625.0	612.5	637.5	25.0
650.0	637.5	662.5	25.0
675.0	662.5	687.5	25.0
700.0	687.5	712.5	25.0
750.0	725.0	775.0	50.0
800.0	775.0	825.0	50.0

Table A-2. Day-Averaged Specific Light Absorption Rate Constant (k_a in days^{-1}) for Sunlight Absorption by *p*-Nitroanisole (PNA) as a Function of Season and Decadic Latitude

Latitude (°N.)	*Spring*	*Summer*	*Fall*	*Winter*
0	4660	4420	4700	4440
10	4850	4830	4350	3900
20	4790	5100	3830	3100
30	4570	5140	3170	2270
40	4150	5030	2390	1400
50	3570	4770	1540	668
60	2890	4350	789	215
70	2120	3880	271	0

Table A-3. Day-Averaged Specific Light Absorption Rate Constant (k_a in days^{-1}) for Sunlight Absorption by *p*-Nitroacetophenone (PNAP) as a Function of Season and Decadic Latitude

Latitude (°N.)	*Spring*	*Summer*	*Fall*	*Winter*
0	505	477	513	482
10	525	523	470	417
20	514	551	409	327
30	483	551	333	233
40	431	532	245	139
50	362	496	154	63.7
60	286	445	76.1	19.7
70	200	390	25.1	0

Table A-4. L_λ at 0° N. Latitude as a Function of Season

λ(*center*), *nm*	*Spring*	*Summer*	*Fall*	*Winter*
297.5	1.60 (-4)	1.34 (-4)	2.04 (-4)	1.69 (-4)
300.0	5.30 (-4)	4.58 (-4)	6.36 (-4)	5.46 (-4)
302.5	1.33 (-3)	1.18 (-3)	1.52 (-3)	1.34 (-3)
305.0	2.68 (-3)	2.41 (-3)	2.96 (-3)	2.65 (-3)
307.5	4.60 (-3)	4.19 (-3)	4.94 (-3)	4.48 (-3)
310.0	6.98 (-3)	6.42 (-3)	7.35 (-3)	6.74 (-3)
312.5	9.66 (-3)	8.95 (-3)	1.00 (-2)	9.27 (-3)
315.0	1.25 (-2)	1.16 (-2)	1.28 (-2)	1.19 (-2)
317.5	1.53 (-2)	1.43 (-2)	1.56 (-2)	1.45 (-2)
320.0	1.80 (-2)	1.69 (-2)	1.82 (-2)	1.71 (-2)
323.1	3.16 (-2)	2.97 (-2)	3.19 (-2)	2.99 (-2)
330.0	1.06 (-1)	1.01 (-1)	1.07 (-1)	1.01 (-1)
340.0	1.32 (-1)	1.26 (-1)	1.33 (-1)	1.26 (-1)
350.0	1.46 (-1)	1.39 (-1)	1.46 (-1)	1.39 (-1)
360.0	1.61 (-1)	1.53 (-1)	1.61 (-1)	1.53 (-1)
370.0	1.71 (-1)	1.63 (-1)	1.71 (-1)	1.63 (-1)
380.0	1.82 (-1)	1.74 (-1)	1.82 (-1)	1.74 (-1)
390.0	1.73 (-1)	1.65 (-1)	1.73 (-1)	1.65 (-1)
400.0	2.48 (-1)	2.36 (-1)	2.48 (-1)	2.36 (-1)
410.0	3.26 (-1)	3.11 (-1)	3.26 (-1)	3.11 (-1)
420.0	3.35 (-1)	3.20 (-1)	3.35 (-1)	3.20 (-1)
430.0	3.23 (-1)	3.08 (-1)	3.23 (-1)	3.08 (-1)
440.0	3.81 (-1)	3.65 (-1)	3.81 (-1)	3.64 (-1)
450.0	4.29 (-1)	4.10 (-1)	4.29 (-1)	4.10 (-1)
460.0	4.34 (-1)	4.15 (-1)	4.34 (-1)	4.15 (-1)
470.0	4.48 (-1)	4.29 (-1)	4.49 (-1)	4.29 (-1)
480.0	4.59 (-1)	4.40 (-1)	4.59 (-1)	4.40 (-1)
490.0	4.34 (-1)	4.16 (-1)	4.35 (-1)	4.16 (-1)
500.0	4.43 (-1)	4.25 (-1)	4.44 (-1)	4.25 (-1)
525.0	1.17	1.12	1.17	1.12
550.0	1.22	1.17	1.22	1.17
575.0	1.23	1.18	1.24	1.18
600.0	1.24	1.20	1.25	1.20
625.0	1.25	1.20	1.25	1.20
650.0	1.26	1.21	1.26	1.21
675.0	1.25	1.21	1.26	1.21
700.0	1.24	1.20	1.24	1.20
750.0	2.39	2.30	2.39	2.30
800.0	2.27	2.19	2.27	2.19

NOTES: L_λ is in units of millieinsteins cm^{-2} day^{-1}. Numbers in parentheses are powers of 10.

Table A-5. L_λ at 10° N. Latitude as a Function of Season

λ(center), nm	*Spring*	*Summer*	*Fall*	*Winter*
297.5	1.56 (-4)	1.62 (-4)	1.39 (-4)	9.75 (-5)
300.0	5.28 (-4)	5.42 (-4)	4.69 (-4)	3.49 (-4)
302.5	1.34 (-3)	1.36 (-3)	1.19 (-3)	9.25 (-4)
305.0	2.73 (-3)	2.67 (-3)	2.43 (-3)	1.95 (-3)
307.5	4.71 (-3)	4.74 (-3)	4.19 (-3)	3.45 (-3)
310.0	7.18 (-3)	7.20 (-3)	6.39 (-3)	5.36 (-3)
312.5	9.97 (-3)	9.97 (-3)	8.88 (-3)	7.55 (-3)
315.0	1.29 (-2)	1.29 (-2)	1.15 (-2)	9.88 (-3)
317.5	1.58 (-2)	1.58 (-2)	1.41 (-2)	1.22 (-2)
320.0	1.87 (-2)	1.86 (-2)	1.67 (-2)	1.45 (-2)
323.1	3.28 (-2)	3.27 (-2)	2.93 (-2)	2.57 (-2)
330.0	1.11 (-1)	1.10 (-1)	9.91 (-2)	8.75 (-2)
340.0	1.38 (-1)	1.37 (-1)	1.24 (-1)	1.10 (-1)
350.0	1.52 (-1)	1.52 (-1)	1.37 (-1)	1.22 (-1)
360.0	1.67 (-1)	1.66 (-1)	1.51 (-1)	1.53 (-1)
370.0	1.78 (-1)	1.77 (-1)	1.61 (-1)	1.45 (-1)
380.0	1.89 (-1)	1.89 (-1)	1.71 (-1)	1.54 (-1)
390.0	1.79 (-1)	1.79 (-1)	1.62 (-1)	1.46 (-1)
400.0	2.57 (-1)	2.57 (-1)	2.32 (-1)	2.09 (-1)
410.0	3.38 (-1)	3.38 (-1)	3.06 (-1)	2.75 (-1)
420.0	3.47 (-1)	3.47 (-1)	3.14 (-1)	2.83 (-1)
430.0	3.35 (-1)	3.34 (-1)	3.03 (-1)	2.73 (-1)
440.0	3.96 (-1)	3.95 (-1)	3.59 (-1)	3.24 (-1)
450.0	4.45 (-1)	4.44 (-1)	4.03 (-1)	3.64 (-1)
460.0	4.50 (-1)	4.49 (-1)	4.08 (-1)	3.69 (-1)
470.0	4.65 (-1)	4.65 (-1)	4.22 (-1)	3.82 (-1)
480.0	4.76 (-1)	4.76 (-1)	4.32 (-1)	3.92 (-1)
490.0	4.51 (-1)	4.50 (-1)	4.09 (-1)	3.71 (-1)
500.0	4.60 (-1)	4.59 (-1)	4.18 (-1)	3.79 (-1)
525.0	1.21	1.21	1.10	1.00
550.0	1.26	1.26	1.15	1.04
575.0	1.28	1.28	1.16	1.06
600.0	1.29	1.29	1.18	1.07
625.0	1.30	1.30	1.18	1.08
650.0	1.30	1.30	1.19	1.08
675.0	1.30	1.30	1.19	1.08
700.0	1.29	1.29	1.18	1.07
750.0	2.47	2.47	2.26	2.07
800.0	2.35	2.35	2.15	1.97

NOTES: L_λ is in units of millieinsteins cm^{-2} day^{-1}. Numbers in parentheses are powers of 10.

Table A-6. L_λ at 20° N. Latitude as a Function of Season

λ(*center*), *nm*	*Spring*	*Summer*	*Fall*	*Winter*
297.5	1.09 (-4)	1.52 (-4)	7.76 (-5)	3.71 (-5)
300.0	4.06 (-4)	5.26 (-4)	2.96 (-4)	1.62 (-4)
302.5	1.10 (-3)	1.35 (-3)	8.21 (-3)	4.99 (-4)
305.0	2.37 (-3)	2.79 (-3)	1.79 (-3)	1.17 (-3)
307.5	4.24 (-3)	4.86 (-3)	3.24 (-3)	2.25 (-3)
310.0	6.65 (-3)	7.45 (-3)	5.13 (-3)	3.72 (-3)
312.5	9.42 (-3)	1.04 (-2)	7.33 (-3)	5.47 (-3)
315.0	1.24 (-2)	1.35 (-2)	9.68 (-3)	7.40 (-3)
317.5	1.54 (-2)	1.66 (-2)	1.21 (-2)	9.38 (-3)
320.0	1.82 (-2)	1.96 (-2)	1.44 (-2)	1.13 (-2)
323.1	3.22 (-2)	3.45 (-2)	2.55 (-2)	2.03 (-2)
330.0	1.10 (-1)	1.17 (-1)	8.75 (-2)	7.08 (-2)
340.0	1.37 (-1)	1.45 (-1)	1.10 (-1)	9.02 (-2)
350.0	1.52 (-1)	1.60 (-1)	1.22 (-1)	1.01 (-1)
360.0	1.67 (-1)	1.76 (-1)	1.35 (-1)	1.12 (-1)
370.0	1.78 (-1)	1.88 (-1)	1.45 (-1)	1.21 (-1)
380.0	1.89 (-1)	2.00 (-1)	1.55 (-1)	1.30 (-1)
390.0	1.79 (-1)	1.89 (-1)	1.46 (-1)	1.22 (-1)
400.0	2.57 (-1)	2.71 (-1)	2.09 (-1)	1.75 (-1)
410.0	3.38 (-1)	3.57 (-1)	2.76 (-1)	2.31 (-1)
420.0	3.47 (-1)	3.67 (-1)	2.84 (-1)	2.38 (-1)
430.0	3.35 (-1)	3.54 (-1)	2.74 (-1)	2.30 (-1)
440.0	3.95 (-1)	4.18 (-1)	3.25 (-1)	2.72 (-1)
450.0	4.45 (-1)	4.70 (-1)	3.65 (-1)	3.07 (-1)
460.0	4.50 (-1)	4.75 (-1)	3.70 (-1)	3.11 (-1)
470.0	4.65 (-1)	4.91 (-1)	3.83 (-1)	3.22 (-1)
480.0	4.76 (-1)	5.03 (-1)	3.92 (-1)	3.31 (-1)
490.0	4.50 (-1)	4.76 (-1)	3.72 (-1)	3.13 (-1)
500.0	4.59 (-1)	4.85 (-1)	3.80 (-1)	3.20 (-1)
525.0	1.21	1.28	1.00	8.48 (-1)
550.0	1.26	1.33	1.05	8.83 (-1)
575.0	1.27	1.35	1.06	8.92 (-1)
600.0	1.29	1.36	1.07	9.05 (-1)
625.0	1.29	1.37	1.08	9.15 (-1)
650.0	1.30	1.38	1.09	9.24 (-1)
675.0	1.30	1.38	1.09	9.27 (-1)
700.0	1.29	1.36	1.08	9.21 (-1)
750.0	2.48	2.62	2.08	1.78
800.0	2.38	2.51	2.00	1.71

NOTES: L_λ is in units of millieinsteins cm^{-2} day^{-1}. Numbers in parentheses are powers of 10.

Table A-7. L_λ at 30° N. Latitude as a Function of Season

λ(center), nm	*Spring*	*Summer*	*Fall*	*Winter*
297.5	5.73 (-5)	1.09 (-4)	3.18 (-5)	6.78 (-6)
300.0	2.50 (-4)	4.11 (-4)	1.46 (-4)	4.23 (-5)
302.5	7.65 (-4)	1.14 (-3)	4.64 (-4)	1.71 (-4)
305.0	1.79 (-3)	2.46 (-3)	1.12 (-3)	4.95 (-4)
307.5	3.43 (-3)	4.45 (-3)	2.19 (-3)	1.11 (-3)
310.0	5.64 (-3)	7.02 (-3)	3.67 (-3)	2.04 (-3)
312.5	8.27 (-3)	1.00 (-2)	5.46 (-3)	3.26 (-3)
315.0	1.12 (-2)	1.32 (-2)	7.43 (-3)	4.69 (-3)
317.5	1.41 (-2)	1.64 (-2)	9.48 (-3)	6.21 (-3)
320.0	1.70 (-2)	1.95 (-2)	1.15 (-2)	7.76 (-3)
323.1	3.04 (-2)	3.46 (-2)	2.07 (-2)	1.43 (-2)
330.0	1.05 (-1)	1.18 (-1)	7.23 (-2)	5.17 (-2)
340.0	1.33 (-1)	1.48 (-1)	9.23 (-2)	6.75 (-2)
350.0	1.47 (-1)	1.63 (-1)	1.03 (-1)	7.65 (-2)
360.0	1.62 (-1)	1.80 (-1)	1.15 (-1)	8.60 (-2)
370.0	1.73 (-1)	1.91 (-1)	1.24 (-1)	9.31 (-2)
380.0	1.84 (-1)	2.04 (-1)	1.33 (-1)	1.01 (-1)
390.0	1.74 (-1)	1.93 (-1)	1.25 (-1)	9.39 (-2)
400.0	2.50 (-1)	2.77 (-1)	1.79 (-1)	1.35 (-1)
410.0	3.29 (-1)	3.64 (-1)	2.36 (-1)	1.79 (-1)
420.0	3.38 (-1)	3.74 (-1)	2.43 (-1)	1.84 (-1)
430.0	3.26 (-1)	3.61 (-1)	2.35 (-1)	1.78 (-1)
440.0	3.86 (-1)	4.26 (-1)	2.79 (-1)	2.12 (-1)
450.0	4.34 (-1)	4.79 (-1)	3.14 (-1)	2.39 (-1)
460.0	4.39 (-1)	4.85 (-1)	3.19 (-1)	2.42 (-1)
470.0	4.54 (-1)	5.01 (-1)	3.30 (-1)	2.51 (-1)
480.0	4.65 (-1)	5.13 (-1)	3.38 (-1)	2.58 (-1)
490.0	4.40 (-1)	4.85 (-1)	3.20 (-1)	2.44 (-1)
500.0	4.49 (-1)	4.95 (-1)	3.27 (-1)	2.50 (-1)
525.0	1.18	1.31	8.67 (-1)	6.61 (-1)
550.0	1.23	1.36	9.03 (-1)	6.87 (-1)
575.0	1.24	1.37	9.11 (-1)	6.93 (-1)
600.0	1.25	1.38	9.24 (-1)	7.04 (-1)
625.0	1.26	1.39	9.34 (-1)	7.15 (-1)
650.0	1.27	1.40	9.45 (-1)	7.27 (-1)
675.0	1.28	1.40	9.48 (-1)	7.32 (-1)
700.0	1.27	1.39	9.42 (-1)	7.31 (-1)
750.0	2.44	2.67	1.82	1.41
800.0	2.34	2.57	1.75	1.37

NOTES: L_λ is in units of millieinsteins cm^{-2} day^{-1}. Numbers in parentheses are powers of 10.

Table A-8. L_λ at 40° N. Latitude as a Function of Season

λ(*center*), *nm*	*Spring*	*Summer*	*Fall*	*Winter*
297.5	1.85 (-5)	6.17 (-5)	7.83 (-6)	5.49 (-7)
300.0	1.06 (-4)	2.69 (-4)	4.76 (-5)	5.13 (-6)
302.5	3.99 (-4)	8.30 (-4)	1.89 (-4)	3.02 (-5)
305.0	1.09 (-3)	1.95 (-3)	5.40 (-4)	1.19 (-4)
307.5	2.34 (-3)	3.74 (-3)	1.19 (-3)	3.38 (-4)
310.0	4.17 (-3)	6.17 (-3)	2.19 (-3)	7.53 (-4)
312.5	6.51 (-3)	9.07 (-3)	3.47 (-3)	1.39 (-3)
315.0	9.18 (-3)	1.22 (-2)	4.97 (-3)	2.22 (-3)
317.5	1.20 (-2)	1.55 (-2)	6.57 (-3)	3.19 (-3)
320.0	1.48 (-2)	1.87 (-2)	8.18 (-3)	4.23 (-3)
323.1	2.71 (-2)	3.35 (-2)	1.51 (-2)	8.25 (-3)
330.0	9.59 (-2)	1.16 (-1)	5.44 (-2)	3.16 (-2)
340.0	1.23 (-1)	1.46 (-1)	7.09 (-2)	4.31 (-2)
350.0	1.37 (-1)	1.62 (-1)	8.04 (-2)	4.98 (-2)
360.0	1.52 (-1)	1.79 (-1)	9.02 (-2)	5.68 (-2)
370.0	1.63 (-1)	1.91 (-1)	9.77 (-2)	6.22 (-2)
380.0	1.74 (-1)	2.04 (-1)	1.05 (-1)	6.78 (-2)
390.0	1.64 (-1)	1.93 (-1)	9.86 (-2)	6.33 (-2)
400.0	2.36 (-1)	2.76 (-1)	1.42 (-1)	9.11 (-2)
410.0	3.10 (-1)	3.64 (-1)	1.87 (-1)	1.20 (-1)
420.0	3.19 (-1)	3.74 (-1)	1.93 (-1)	1.24 (-1)
430.0	3.08 (-1)	3.61 (-1)	1.87 (-1)	1.20 (-1)
440.0	3.65 (-1)	4.26 (-1)	2.22 (-1)	1.43 (-1)
450.0	4.11 (-1)	4.80 (-1)	2.51 (-1)	1.61 (-1)
460.0	4.16 (-1)	4.85 (-1)	2.54 (-1)	1.64 (-1)
470.0	4.30 (-1)	5.02 (-1)	2.63 (-1)	1.69 (-1)
480.0	4.40 (-1)	5.14 (-1)	2.70 (-1)	1.74 (-1)
490.0	4.16 (-1)	4.86 (-1)	2.56 (-1)	1.65 (-1)
500.0	4.25 (-1)	4.96 (-1)	2.62 (-1)	1.68 (-1)
525.0	1.12	1.31	6.93 (-1)	4.45 (-1)
550.0	1.16	1.36	7.21 (-1)	4.61 (-1)
575.0	1.17	1.37	7.22 (-1)	4.61 (-1)
600.0	1.18	1.38	7.39 (-1)	4.69 (-1)
625.0	1.20	1.40	7.50 (-1)	4.82 (-1)
650.0	1.21	1.41	7.62 (-1)	4.95 (-1)
675.0	1.22	1.41	7.68 (-1)	5.03 (-1)
700.0	1.21	1.40	7.66 (-1)	5.05 (-1)
750.0	2.33	2.69	1.48	9.84 (-1)
800.0	2.25	2.59	1.43	9.56 (-1)

NOTES: L_λ is in units of millieinsteins cm^{-2} day^{-1}. Numbers in parentheses are powers of 10.

Table A-9. L_λ at 50° N. Latitude as a Function of Season

λ(*center*), *nm*	*Spring*	*Summer*	*Fall*	*Winter*
297.5	3.61 (-6)	2.86 (-5)	9.58 (-7)	5.47 (-8)
300.0	3.05 (-5)	1.50 (-4)	8.27 (-6)	4.17 (-7)
302.5	1.54 (-4)	5.33 (-4)	4.47 (-5)	2.62 (-6)
305.0	5.24 (-4)	1.39 (-3)	1.63 (-4)	1.34 (-5)
307.5	1.32 (-3)	2.89 (-3)	4.39 (-4)	5.14 (-5)
310.0	2.66 (-3)	5.05 (-3)	9.32 (-4)	1.49 (-4)
312.5	4.53 (-3)	7.75 (-3)	1.66 (-3)	3.43 (-4)
315.0	6.82 (-3)	1.08 (-2)	2.58 (-3)	6.52 (-4)
317.5	9.34 (-3)	1.40 (-2)	3.64 (-3)	1.07 (-3)
320.0	1.19 (-2)	1.71 (-2)	4.76 (-3)	1.57 (-3)
323.1	2.25 (-2)	3.12 (-2)	9.19 (-3)	3.39 (-3)
330.0	8.26 (-2)	1.10 (-1)	3.48 (-2)	1.45 (-2)
340.0	1.08 (-1)	1.40 (-1)	4.70 (-2)	2.12 (-2)
350.0	1.22 (-1)	1.57 (-1)	5.43 (-2)	2.53 (-2)
360.0	1.36 (-1)	1.74 (-1)	6.18 (-2)	2.96 (-2)
370.0	1.47 (-1)	1.86 (-1)	6.76 (-2)	3.30 (-2)
380.0	1.57 (-1)	1.99 (-1)	7.37 (-2)	3.65 (-2)
390.0	1.48 (-1)	1.87 (-1)	6.89 (-2)	3.49 (-2)
400.0	2.12 (-1)	2.69 (-1)	9.90 (-2)	4.98 (-2)
410.0	2.80 (-1)	3.55 (-1)	1.31 (-1)	6.54 (-2)
420.0	2.89 (-1)	3.65 (-1)	1.35 (-1)	6.71 (-2)
430.0	2.79 (-1)	3.52 (-1)	1.31 (-1)	6.47 (-2)
440.0	3.31 (-1)	4.17 (-1)	1.55 (-1)	7.66 (-2)
450.0	3.73 (-1)	4.69 (-1)	1.75 (-1)	8.62 (-2)
460.0	3.78 (-1)	4.75 (-1)	1.78 (-1)	8.74 (-2)
470.0	3.90 (-1)	4.91 (-1)	1.84 (-1)	8.95 (-2)
480.0	4.00 (-1)	5.03 (-1)	1.89 (-1)	9.15 (-2)
490.0	3.78 (-1)	4.76 (-1)	1.79 (-1)	8.62 (-2)
500.0	3.86 (-1)	4.85 (-1)	1.83 (-1)	8.77 (-2)
525.0	1.10	1.28	4.84 (-1)	2.28 (-1)
550.0	1.05	1.33	5.03 (-1)	2.32 (-1)
575.0	1.05	1.34	5.04 (-1)	2.28 (-1)
600.0	1.06	1.35	5.13 (-1)	2.32 (-1)
625.0	1.08	1.37	5.26 (-1)	2.42 (-1)
650.0	1.10	1.38	5.39 (-1)	2.53 (-1)
675.0	1.11	1.39	5.47 (-1)	2.61 (-1)
700.0	1.11	1.38	5.49 (-1)	2.66 (-1)
750.0	2.15	2.66	1.07	5.22 (-1)
800.0	2.08	2.57	1.04	5.11 (-1)

NOTES: L_λ is in units of millieinsteins cm^{-2} day^{-1}. Numbers in parentheses are powers of 10.

Table A-10. L_λ at 60° N. Latitude as a Function of Season

λ(center), nm	Spring	Summer	Fall	Winter
297.5	5.33 (-7)	1.07 (-5)	1.41 (-7)	8.93 (-9)
300.0	6.76 (-6)	7.04 (-5)	9.91 (-7)	6.66 (-8)
302.5	4.76 (-5)	2.96 (-4)	6.00 (-6)	3.37 (-7)
305.0	2.09 (-4)	8.83 (-4)	2.81 (-5)	1.31 (-6)
307.5	6.41 (-4)	2.02 (-3)	9.62 (-5)	4.40 (-6)
310.0	1.49 (-3)	3.80 (-3)	2.51 (-4)	1.38 (-5)
312.5	2.82 (-3)	6.15 (-3)	5.25 (-4)	3.86 (-5)
315.0	4.58 (-3)	8.92 (-3)	9.27 (-4)	9.17 (-5)
317.5	6.64 (-3)	1.19 (-2)	1.44 (-3)	1.84 (-4)
320.0	8.84 (-3)	1.49 (-2)	2.03 (-3)	3.18 (-4)
323.1	1.73 (-2)	2.77 (-2)	4.19 (-3)	8.09 (-4)
330.0	6.63 (-2)	1.00 (-1)	1.72 (-2)	4.15 (-3)
340.0	8.96 (-2)	1.30 (-1)	2.47 (-2)	6.89 (-3)
350.0	1.02 (-1)	1.47 (-1)	2.93 (-2)	8.81 (-3)
360.0	1.15 (-1)	1.64 (-1)	3.42 (-2)	1.08 (-2)
370.0	1.25 (-1)	1.76 (-1)	3.81 (-2)	1.25 (-2)
380.0	1.35 (-1)	1.89 (-1)	4.21 (-2)	1.43 (-2)
390.0	1.26 (-1)	1.78 (-1)	3.99 (-2)	1.30 (-2)
400.0	1.81 (-1)	2.55 (-1)	5.70 (-2)	1.83 (-2)
410.0	2.39 (-1)	3.37 (-1)	7.49 (-2)	2.38 (-2)
420.0	2.74 (-1)	3.47 (-1)	7.70 (-2)	2.42 (-2)
430.0	2.39 (-1)	3.35 (-1)	7.43 (-2)	2.31 (-2)
440.0	2.84 (-1)	3.97 (-1)	8.80 (-2)	2.70 (-2)
450.0	3.20 (-1)	4.48 (-1)	9.91 (-2)	3.01 (-2)
460.0	3.25 (-1)	4.54 (-1)	1.01 (-1)	3.02 (-2)
470.0	3.35 (-1)	4.69 (-1)	1.03 (-1)	3.02 (-2)
480.0	3.43 (-1)	4.80 (-1)	1.06 (-1)	3.04 (-2)
490.0	3.25 (-1)	4.54 (-1)	9.98 (-2)	2.81 (-2)
500.0	3.31 (-1)	4.64 (-1)	1.02 (-1)	2.81 (-2)
525.0	8.71 (-1)	1.22 (-1)	2.66 (-1)	6.97 (-2)
550.0	8.98 (-1)	1.27	2.73 (-1)	6.71 (-2)
575.0	8.95 (-1)	1.27	2.70 (-1)	6.19 (-2)
600.0	9.07 (-1)	1.29	2.74 (-1)	6.18 (-2)
625.0	9.29 (-1)	1.31	2.85 (-1)	6.68 (-2)
650.0	9.52 (-1)	1.33	2.95 (-1)	7.23 (-2)
675.0	9.66 (-1)	1.34	3.03 (-1)	7.69 (-2)
700.0	9.68 (-1)	1.33	3.06 (-1)	7.79 (-2)
750.0	1.88	2.58	6.01 (-1)	1.58 (-1)
800.0	1.81	2.46	5.80 (-1)	1.54 (-1)

NOTES: L_λ is in units of millieinsteins cm^{-2} day^{-1}. Numbers in parentheses are powers of 10.

Table A-11. L_λ at 70° N. Latitude as a Function of Season

λ(*center*), *nm*	*Spring*	*Summer*	*Fall*	*Winter*
297.5	8.73 (-8)	3.08 (-6)	3.11 (-8)	0
300.0	1.24 (-6)	2.67 (-5)	1.78 (-7)	0
302.5	1.12 (-5)	1.39 (-4)	7.47 (-7)	0
305.0	6.31 (-5)	4.89 (-4)	2.66 (-6)	0
307.5	2.39 (-4)	1.27 (-3)	8.90 (-6)	0
310.0	6.58 (-4)	2.61 (-3)	2.74 (-5)	0
312.5	1.42 (-3)	4.53 (-3)	7.18 (-5)	0
315.0	2.56 (-3)	6.92 (-3)	1.57 (-4)	0
317.5	4.01 (-3)	9.61 (-3)	2.90 (-4)	0
320.0	5.66 (-3)	1.24 (-2)	4.68 (-4)	0
323.1	1.17 (-2)	2.37 (-2)	1.11 (-3)	0
330.0	4.76 (-2)	8.85 (-2)	5.34 (-3)	0
340.0	6.71 (-2)	1.18 (-1)	8.55 (-3)	0
350.0	7.79 (-2)	1.35 (-1)	1.09 (-2)	0
360.0	8.92 (-2)	1.52 (-1)	1.33 (-2)	0
370.0	9.78 (-2)	1.65 (-1)	1.53 (-2)	0
380.0	1.07 (-1)	1.79 (-1)	1.75 (-2)	0
390.0	9.94 (-2)	1.67 (-1)	1.60 (-2)	0
400.0	1.43 (-1)	2.40 (-1)	2.66 (-2)	0
410.0	1.89 (-1)	3.16 (-1)	2.93 (-2)	0
420.0	1.95 (-1)	3.26 (-1)	2.98 (-2)	0
430.0	1.89 (-1)	3.16 (-1)	2.84 (-2)	0
440.0	2.24 (-1)	3.74 (-1)	3.33 (-2)	0
450.0	2.53 (-1)	4.22 (-1)	3.72 (-2)	0
460.0	2.57 (-1)	4.29 (-1)	3.73 (-2)	0
470.0	2.64 (-1)	4.43 (-1)	3.76 (-2)	0
480.0	2.71 (-1)	4.54 (-1)	3.79 (-2)	0
490.0	2.56 (-1)	4.29 (-1)	3.53 (-2)	0
500.0	2.61 (-1)	4.38 (-1)	3.54 (-2)	0
525.0	6.83 (-1)	1.15	8.91 (-2)	0
550.0	7.01 (-1)	1.19	8.74 (-2)	0
575.0	6.93 (-1)	1.20	8.24 (-2)	0
600.0	7.04 (-1)	1.21	8.39 (-2)	0
625.0	7.29 (-1)	1.24	8.78 (-2)	0
650.0	7.55 (-1)	1.26	9.35 (-2)	0
675.0	7.73 (-1)	1.28	9.80 (-2)	0
700.0	7.80 (-1)	1.28	1.01 (-1)	0
750.0	1.53	2.49	1.99 (-1)	0
800.0	1.47	2.38	1.93 (-1)	0

NOTES: L_λ is in units of millieinsteins cm^{-2} day^{-1}. Numbers in parentheses are powers of 10.

Appendix B

Indirect Photoreaction (Photosensitization) in Humic Acid Waters

Proof That $\phi_{s\lambda} = Q_\lambda[C]$

B.1. Introduction

In Section 5.B. a general theoretical scheme was developed for determining the kinetics of indirect photoreaction (or photosensitization) of a chemical in water containing dissolved humic acids. In the derivation, it was postulated that if the concentration of the chemical is low, as is generally the case under environmental conditions, the presence of the chemical has only a negligible effect on the lifetime of the electronically excited sensitizer. Under these conditions, the photosensitized reaction quantum yield of the chemical ($\phi_{s\lambda}$), at a fixed wavelength λ, is given by equation 5.59 and is

$$\phi_{s\lambda} = Q_\lambda[C] \tag{B.1}$$

where $[C]$ is the molar concentration of the chemical and Q_λ is a proportionality constant at λ. The following two photosensitized processes will be described in this Appendix to prove equation B.1: singlet oxygen photosensitization, and electronic energy transfer photosensitization of the cis–trans isomerization of 1,3-pentadiene.

B.2. Singlet Oxygen Photosensitization

Consider the photoreaction of a solution containing humic acids, dissolved oxygen ($^3O_2$), and an acceptor molecule A such as dimethylfuran

1464–6/88/0265$06.00/1

at wavelength λ. Using conventional spectroscopic notation, let the sensitizer (or sensitizers) in the humic acid solution be denoted by the symbol S_0 with S denoting the singlet state and the subscript 0 denoting the ground state. The photoreaction of this aqueous humic solution can be described by the following general scheme:

$$S_0 \xrightarrow{h\nu(I^{s}_{a\lambda})} S_1 \tag{B.2}$$

$$S_1 \xrightarrow{{}^1k_{ic}} S_0 + \text{heat} \tag{B.3}$$

$$S_1 \xrightarrow{{}^1k_{isc}} T_1 \tag{B.4}$$

$$T_1 \xrightarrow{{}^3k_{ic}} S_0 + \text{heat} \tag{B.5}$$

$$T_1 + {}^3O_2 \xrightarrow{k_{ox}} {}^1O_2 + S_0 \tag{B.6}$$

$$^1O_2 + H_2O \xrightarrow{k_d} {}^3O_2 + H_2O^* \tag{B.7}$$

$$^1O_2 + A \xrightarrow{k_q} {}^3O_2 + A^* \tag{B.8}$$

$$^1O_2 + A \xrightarrow{k_r} AO_2 \tag{B.9}$$

The ground-state sensitizer S_0 absorbs light, $I^s_{a\lambda}$, to form the first excited singlet state S_1 (equation B.2). Singlet excited states with lifetimes on the order of 10 ns or less are far too short-lived to interact significantly with oxygen present at the very low concentrations found in the environment. The singlet excited state, S_1, can decay back to the ground state by internal conversion (ic), a nonradiative decay process, with rate constant ${}^1k_{ic}$ and liberate heat (equation B.3) or it can decay to the first excited triplet state (T_1) via intersystem crossing (isc) with rate constant ${}^1k_{isc}$ (equation B.4).

The long-lived triplet state can interact with ground-state oxygen, ${}^3O_2$, to form singlet oxygen, ${}^1O_2$, with rate constant k_{ox} (equation B.6) or it can be deactivated by internal conversion to the ground state with rate constant ${}^3k_{ic}$ and liberate heat (equation B.5). Singlet oxygen is a more powerful oxidant than ground-state oxygen and it can react rapidly with certain types of acceptor molecules, A, with rate constant k_r (equation B.9). Reactions with singlet oxygen must compete with its decay back to the ground state by quenching with water with rate constant k_d (equation B.7) and by quenching (q) with acceptor molecules A with rate constant k_q (equation B.8).

Dimethylfuran is one of the most reactive singlet oxygen acceptors and has been studied in detail in humic acid waters (*1–4*). Furfuryl alcohol is also a very strong acceptor molecule and has also been studied in detail in humic acid waters (*5–7*). The kinetic scheme for singlet oxygen photosensitization, based on the preceding discussion and on the work of Haag et al. (*5–7*), will now be described.

The kinetic scheme of Haag et al. (*5*) is summarized in Figure B.1 for the photooxidation of an acceptor molecule A [e.g., furfuryl alcohol (FFA)—structure in Chart 12.1] with a sensitizer S_0 [e.g., dissolved humic acids or rose bengal, structure in Chart 12.1] in water. The kinetic scheme involves the absorption of light, $I^s_{a\lambda}$, by the ground-state sensitizer S_0 in the singlet state to form an excited singlet-state sensitizer S_1.

By intersystem crossing, S_1 is converted to a long-lived triplet-state sensitizer T_1 that can interact with ground-state oxygen, $^3O_2$, to form the energetic singlet oxygen, $^1O_2$. The quantum yield for the production of $^1O_2$, $\Phi(^1O_2)$, is defined as the fraction of all photons absorbed by sensitizer S_0 at λ that ultimately yield $^1O_2$ in the bulk water solution. Thus, the quantum yield for formation of $^1O_2$ is given by

$$\Phi(^1O_2) = \Phi_{isc}\gamma(O_2) \quad \text{(B.10)}$$

where Φ_{isc} is the quantum yield of intersystem crossing of S_1 to the triplet state T_1 and $\gamma(O_2)$ is the efficiency of capture of triplet sensitizers by ground-state oxygen, $^3O_2$. This efficiency is high for dissolved humic acids in water and close to 1 for rose bengal, a very efficient singlet-oxygen dye sensitizer.

The singlet oxygen is quenched rapidly by water with rate constant k_d or it can interact with acceptor molecule A with rate constant k_A. After the interception of $^1O_2$ by A, the intermediate can undergo reaction to form AO_2 with rate constant k_r or it can decay by quenching with rate constant k_q. From the kinetic scheme given in Figure B.1 and using the steady-state approximation (*8, 9*), it is seen that

$$\Phi(AO_2) = \Phi(^1O_2) \prod_x \gamma(x) \quad \text{(B.11)}$$

where the pathway for reaction with $^1O_2$ is x, $\gamma(x)$ is the fraction of precursor molecules following pathway x with associated rate constant k_x (i.e., the quantum efficiency), and $\prod_x$ denotes the product of $\gamma(x)$ for all pathways x. For example, $\gamma(\mathrm{r})$ corresponds to the pathway r for reaction with associated rated constant k_r and involves quenching or reaction. The term $\prod_x \gamma(x)$ will be evaluated for the kinetic scheme as follows.

Consider the pathway for x = A in which the acceptor molecule A has been introduced to intercept $^1O_2$. For this pathway, the interception efficiency, $\gamma(\mathrm{A})$, involves the capture of $^1O_2$ by A or the quenching by water. Thus, the fraction of precursor molecules captured by $^1O_2$ with rate constant k_A is given by the relationship

$$\gamma(A) = k_A[A]/(k_A[A] + k_d) \qquad (B.12)$$

Rearranging equation B.12 yields

$$\gamma(A) = [A]/([A] + \beta) \qquad (B.13)$$

where

$$\beta = k_d/k_A \qquad (B.14)$$

The kinetic scheme in Figure B.1 shows that not every interaction of A with $^1O_2$ leads to a chemical reaction that forms AO_2. The acceptor molecule also quenches $^1O_2$ with rate constant k_q. Therefore, the efficiency of reaction, $\gamma(r)$, for pathway r, with associated rate constant k_r, is given by

$$\gamma(r) = k_r[A]/(k_r[A] + k_q[A]) \qquad (B.15)$$

$$\gamma(r) = k_r/(k_r + k_q) \qquad (B.16)$$

Furthermore, from the kinetic scheme

$$k_A[A] = k_r[A] + k_q[A] \qquad (B.17)$$

$$k_A = k_r + k_q \qquad (B.18)$$

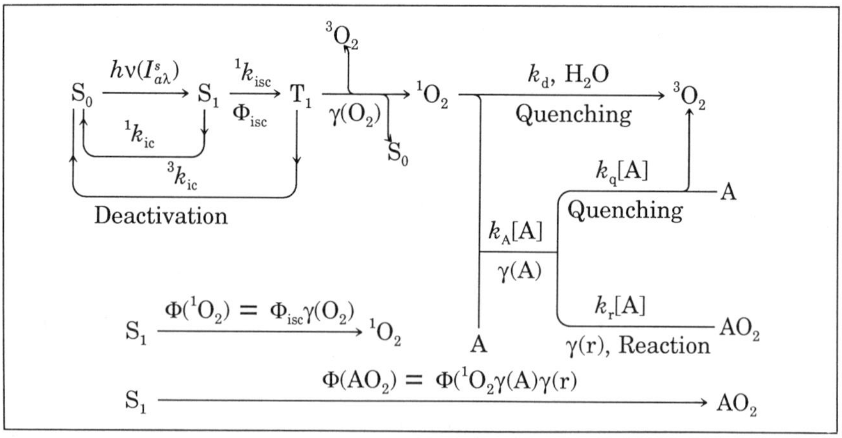

Figure B.1. Photochemical pathways for singlet oxygen formation and consumption in water by an acceptor molecule. (Adapted from ref. 5.)

so that equations B.14 and B.16 can be written as

$$\beta = k_d/(k_r + k_q) \tag{B.19}$$

$$\gamma(r) = k_r/k_A \tag{B.20}$$

From the kinetic scheme in Figure B.1 and equation B.11, $\Phi(AO_2)$ is given by

$$\Phi(AO_2) = \Phi(^1O_2)\gamma(A)\gamma(r) \tag{B.21}$$

Substituting equation B.10, B.12, and B.16 in equation B.21 yields

$$\Phi(AO_2) = \Phi_{isc}\gamma(O_2)\{k_A[A]/(k_A[A] + k_d)\}\{k_r/(k_r + k_q)\} \tag{B.22}$$

For the boundary condition in which [A] is very low, which is generally the case in the environment, $k_d >> k_A[A]$; and equation B.22 becomes

$$\Phi(AO_2) = \{\Phi_{isc}\gamma(O_2)k_A k_r/[k_d(k_r + k_q)]\}[A] \tag{B.23}$$

However, $\Phi(AO_2)$ corresponds to the reaction quantum yield of A, $\Phi_{A\lambda}$, so that equation B.23 becomes

$$\Phi_{A\lambda} = \{\Phi_{isc}\ \gamma(O_2)k_A k_r/[k_d(k_r + k_q)]\}[A] \tag{B.24}$$

If the acceptor molecule A is the chemical, so that $[A] = [C]$ and $\Phi_{A\lambda} = \phi_{s\lambda}$ [i.e., the photosensitized reaction quantum yield of the chemical], then equation B.24 becomes

$$\phi_{s\lambda} = Q_\lambda[C] \tag{B.25}$$

where

$$Q_\lambda = \{\Phi_{isc}\gamma(O_2)k_A k_r/[k_d(k_r + k_q)]\} \tag{B.26}$$

Because all the terms in the brackets { } of equation B.26 are constants at a fixed wavelength λ, Q_λ is a constant and the photosensitized reaction quantum yield of the chemical, $\phi_{s\lambda}$, is proportional to $[C]$. This verifies equation B.1 (or equation 5.59).

B.3. Electronic Energy Transfer for the Photosensitized cis–trans Isomerization of 1,3-Pentadiene

The chemicals *cis*-1,3-pentadiene (cP) or *trans*-1,3-pentadiene (tP) [structures depicted in Chart 12.1] undergo reversible photoisomerization in the presence of a wide range of photosensitizers (*10, 11*). After prolonged radiation, the ratio of cp to tp reaches a constant value. This value is independent of starting material (i.e., pure cP or pure tP). Details of this isomerization reaction are in Section 12.C.2.b. and Appendix C.

Zepp et al. (*4*) developed the following general scheme for the electronic energy transfer for the photosensitized cis–trans isomerization of 1,3-pentadiene in humic acid solution.

$$^1H_0 \xrightarrow{h\nu(I^s_{a\lambda})} {}^1H_1 \tag{B.27}$$

$$^1H_1 \xrightarrow{^1k_{isc}} {}^3H_1 \tag{B.28}$$

$$^1H_1 \xrightarrow{^1k_{ic}} {}^1H_0 + \text{heat} \tag{B.29}$$

$$^3H_1 \xrightarrow{k_d} {}^1H_0 + \text{heat} \tag{B.30}$$

$$^3H_1 + {}^3O_2 \xrightarrow{k_{ox}} {}^1H_0 + {}^1O_2 \tag{B.31}$$

$$^3H_1 + Q \xrightarrow{k_q} {}^1H_0 + Q^* \tag{B.32}$$

$$^3H_1 + cP \xrightarrow{k_c} T_1 + {}^1H_0 \tag{B.33}$$

$$^3H_1 + tP \xrightarrow{k_t} T_1 + {}^1H_0 \tag{B.34}$$

$$T_1 \xrightarrow{k_1} cP \tag{B.35}$$

$$T_1 \xrightarrow{k_2} tP \tag{B.36}$$

In this kinetic scheme, $^1H_0$ is the singlet ground state of the humic acid sensitizer with the subscript 0 denoting the ground state and the superscript 1 denoting the singlet state. The first step in this scheme is the absorption of light, $I^s_{a\lambda}$, at a fixed wavelength λ, by the ground-state humic acid sensitizer, $^1H_0$, to form the first excited singlet state $^1H_1$ (equation B.27).

Singlet excited states with lifetimes $\leq$10 ns are far too short-lived to interact with the *cis*- or *trans*-pentadiene that is present at very low concentration found in the environment or in the research of Zepp et al. (*4*). Singlet excited-state molecules decay to the excited triplet state $^3H_1$ via intersystem crossing with rate constant $^1k_{isc}$ (equation B.28) or they can decay to the ground singlet state $^1H_0$ by internal conversion (nonradiative) with rate constant $^1k_{ic}$ (equation B.29).

The excited-state triplets are considerably longer-lived than excited-state singlets and can be deactivated via a variety of pathways (equations B.30 to B.34). These pathways include the nonradiative decay (internal conversion) to the ground state with rate constant k_d (equation B.30), energy transfer to $^3O_2$ with rate constant k_{ox} (equation

B.31), energy transfer to *cis*- or *trans*-pentadiene to form its triplet state T_1 with rate constant k_c or k_t, respectively (equations B.33 and B.34); or interaction with Q that quenches $^3H_1$ with rate constant k_q (equation B.32). The triplet excited state of the diene then decays to the ground state cP and tP with rate constants k_1 and k_2, respectively (equations B.35 and B.36). This kinetic scheme is summarized in Figure B.2.

By using the steady state approximation for the *k*inetic scheme for the photoisomerization of diene from cis- to trans-isomer (*8,9*), the reaction quantum yield ϕ_{ct} is given by the mathematical relationship

$$\phi_{ct} = \prod_{x} \gamma(x) \tag{B.37}$$

where the pathways for this kinetic scheme are designated by x, $\gamma(x)$ is the fraction of precursor molecules following pathway x with associated rate constant k_x (the quantum efficiency), and $\prod_{x}$ denotes the product of $\gamma(x)$ for all pathways x.

There are three distinct pathways for the photoisomerization from cis to trans so that equation B.37 becomes

$$\phi_{ct} = [\gamma(\text{isc})][\gamma(^3\text{H}_1)(\text{cP})][\gamma(\text{tP})] \tag{B.38}$$

where $\gamma(\text{isc})$, $\gamma(^3\text{H}_1)(\text{cP})$, and $\gamma(\text{tP})$ are defined as follows: $\gamma(\text{isc})$ is the fraction of singlet excited states $^1H_1$ that form the triplet state $^3H_1$. It is also the quantum yield for intersystem crossing $\phi(\text{isc})$, and is given by the expression

$$\gamma(\text{isc}) = \phi_{\text{isc}} = {}^1k_{\text{isc}}/({}^1k_{\text{isc}} + {}^1k_{\text{ic}}) \tag{B.39}$$

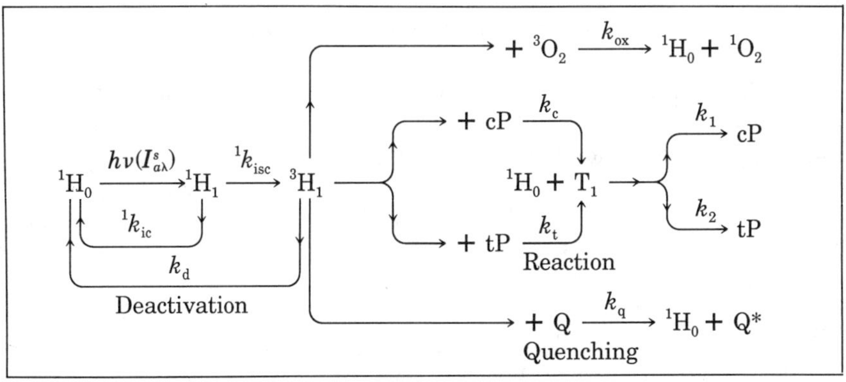

Figure B.2. Photochemical pathways for the cis–trans *isomerization of 1,3-pentadiene in humic acid solution.*

The second pathway involves the fraction of triplet-state molecules $^3H_1$ that react with cP to form T_1 [$(\gamma(^3H_1)(cP)$]. This is given by the expression

$$\gamma(^3H_1)(cP) = k_c[^3H_1][cP]\{k_d[^3H_1] + k_{ox}[^3O_2][^3H_1] + k_q[Q][^3H_1] + k_c[^3H_1][cP]\}^{-1} \quad (B.40)$$

$$\gamma(^3H_1)(cP) = k_c[cP]\{k_d + k_{ox}[^3O_2] + k_q[Q] + k_c[cP]\}^{-1}$$

The third pathway involves $\gamma(tP)$, the fraction of triplet diene molecules T_1 that form tP, and is given by the expression

$$\gamma(tP) = k_2[T_1]/(k_1[T_1] + k_2[T_1]) \quad (B.41)$$

$$\gamma(tP) = k_2/(k_1 + k_2) \quad (B.42)$$

If $k_c[cP]$ in the denominator of equation B.40 is considerably smaller than the other terms, which is generally the case in the environment or in the research by Zepp et al. (*4*), equation B.40 reduces to

$$\gamma(^3H_1)(cP) = k_c[cP]\{k_d + k_{ox}[^3O_2] + k_q[Q]\}^{-1} \quad (B.43)$$

$$\gamma(^3H_1)(cP) = k_c\tau[cP] \quad (B.44)$$

where the lifetime of the photosensitizers triplet state τ is

$$\tau = \{k_d + k_{ox}[^3O_2] + k_q[Q]\}^{-1} \quad (B.45)$$

Combining equations B.39, B.42, and B.44 in equation B.38 yields

$$\phi_{ct} = \phi_{isc}\{k_c\tau[cP]\}\{k_2/(k_1 + k_2)\} \quad (B.46)$$

A similar expression can be written for the trans to cis isomerization.

If we now let $[cP] = [C]$ and $\phi_{ct} = \phi_{s\lambda}$, the photosensitized reaction quantum yield of the chemical, then equation B.44 becomes

$$\phi_{s\lambda} = Q_\lambda[C] \quad (B.47)$$

where

$$Q_\lambda = \phi_{isc}k_c\tau[k_2/(k_1 + k_2)] \quad (B.48)$$

Because all the terms ϕ_{isc}, k_c, τ, k_1, and k_2 are constants at a fixed wavelength λ, Q_λ is a constant and the photoisomerization reaction quantum yield of the chemical, $\phi_{s\lambda}$, is proportional to $[C]$, thus verifying equation B.1 (or equation 5.59).

References

1. Zepp, R. G.; Wolfe, N. L.; Baughman, G. L.; Hollis, R. C. *Nature* **1977,** *278,* 421. 327.
2. Zepp, R. G.; Baughman, G. L.; Schlotzhauer, P. F. *Chemosphere* **1981**, *10*, 109.
3. Ibid, p 119.
4. Zepp, R. G.; Schlutzhauer, P. F.; Sink, R. Merrit. *Environ. Sci. Technol.* **1985,** *19,* 74.
5. Haag, W. R.; Hoigné, J.; Gassman, E.; Braun, A. M. *Chemosphere* **1984,** *13*, 631.
6. Haag, W. R.; Hoigné, J.; Gassman, E.; Braun, A. M. *Chemosphere* **1984,** *13*, 641.
7. Haag, W. R.; Hoigné, J. *Environ. Sci. Technol.* **1986,** *20,* 341.
8. Balzani, V.; Carassiti, V. *Photochemistry of Coordination Compounds*; Academic: New York, 1970; Chapter 2.
9. Barltrop, J. A.; Coyle, J. D. *Excited States in Organic Chemistry*; Wiley: New York, 1975; Chapter 3.
10. Hammond, G. S.; Saltiel, J.; Lamola, A. A.; Turro, N. J.; Bradshaw, J. S.; Cowan, D. O.; Counsell, R. C.; Vogt, V.; Dalton, C. *J. Am. Chem. Soc.* **1964,** *86*, 3197.
11. Lamola, A. A.; Hammond, G. S. *J. Chem. Phys.* **1965,** *43*, 2129.

Appendix C

Rate Expression for the Indirect (Photosensitized) Isomerization of *cis*- and *trans*-1,3-Pentadiene

The photosensitized isomerization of *cis*- and *trans*-1,3-pentadiene (cP and tP, respectively) is reversible and can be represented by the following reaction

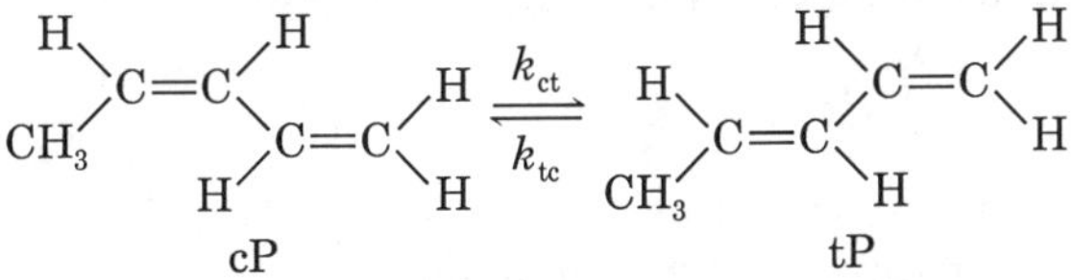

(C.1)

where k_{ct} is the first-order rate constant for isomerization of cP to tP and k_{tc} is the first-order rate constant for isomerization of tP to cP. Because photoisomerization involves a reversible equilibrium, the kinetics are more complex than usually encountered. The solution of this kinetic problem, however, is straightforward, as shown by the mathematical analysis that follows (*1*).

The isomerization of *cis*- and *trans*-1,3-pentadiene can be described by the equilibrium reaction

$$\mathrm{A} \underset{k'}{\overset{k}{\rightleftharpoons}} \mathrm{B} \tag{C.2}$$

where A and B represent cP and tP, respectively, and $k = k_{ct}$ and $k' = k_{tc}$. Let A and B represent the molar concentration of cP and tP, respectively, and let A_0 and B_0 represent their corresponding initial molar concentrations. Kinetic analysis of reaction C.2 indicates that

$$-(dA/dt) = kA - k'B \tag{C.3}$$

1464–6/88/0275$06.00/1

If only A is present at the start of the reaction, then

$$A_0 - A = B \tag{C.4}$$

where A_0 is the initial concentration of A. Substituting equation C.4 in equation C.3 yields

$$-(dA/dt) = kA - k'(A_0 - A) \tag{C.5}$$

This first-order differential equation can be solved in the following manner.

$$-(dA/dt) = (k + k')A - k'A_0 \tag{C.6}$$

$$\alpha = k + k' \tag{C.7}$$

$$-(dA/dt) = \alpha A - k'A_0 \tag{C.8}$$

$$\int [dA/(\alpha A - k'A_0)] = -\int dt \tag{C.9}$$

By using Peirce's integral No. 26 (*2*), given by

$$\int dx/(a + bx) = (1/b) \ln (a + bx) \tag{C.10}$$

equation C.9 can be solved as follows: let $A = x$, $dA = dx$, $b = \alpha$, $a = -k'A_0$ so that integrating C.9 yields

$$(1/\alpha) \ln (\alpha A - k'A_0) = -t + C \tag{C.11}$$

where C is a constant of integration. When $t = 0$, and $A = A_0$,

$$C = (1/\alpha) \ln (\alpha A_0 - k'A_0)$$

$$C = (1/\alpha) \ln \{(k + k')A_0 - k'A_0\}$$

$$C = (1/\alpha) \ln (kA_0) \tag{C.12}$$

Substituting equation C.12 in equation C.11 yields

$$(1/\alpha) \ln (\alpha A - k'A_0) = -t + (1/\alpha) \ln (kA_0) \tag{C.13}$$

$$\ln \{kA_0/[(k + k')A - kA_0]\} = (k + k')t \tag{C.14}$$

At equilibrium $(dA/dt) = 0$ and $A = A_e$; A_e is the equilibrium concentration of A. Using these results in equation C.6 yields

$$0 = (k + k')A_e - k'A_0 \tag{C.15}$$

$$A_e = k'A_0/(k + k') \tag{C.16}$$

Using equation C.16, one can write

$$A_0 - A_e = A_0 - \{k'A_0/(k + k')\} \qquad (C.17)$$

$$A_0 - A_e = [A_0(k + k') - k'A_0]/(k + k') \qquad (C.18)$$

$$A_0 - A_e = kA_0/(k + k') \qquad (C.19)$$

$$kA_0 = (k + k')(A_0 - A_e) \qquad (C.20)$$

Substituting equation C.20 in equation C.14 yields

$$\ln\left[\frac{(k + k')[A_0 - A_e]}{(k + k')A - k'A_0}\right] = (k + k')t \qquad (C.21)$$

$$\ln\left[\frac{[A_0 - A_e]}{[A - k'A_0/(k + k')]}\right] = (k + k')t \qquad (C.22)$$

Substituting equation C.16 in equation C.22 yields

$$\ln\left[\frac{[A_0 - A_e]}{[A - A_e]}\right] = k_it \qquad (C.23)$$

where

$$k_i = k + k' \qquad (C.24)$$

These results indicate that the approach to equilibrium is a first-order process with an effective rate constant, k_i, equal to the sum of the forward and reverse rate constants.

The photoisomerization of *cis*- and *trans*-1,3-pentadiene is given by reaction C.1, so equation C.23 applies. If only the cis isomer is present, $[cP]_0$, $[cP]_\infty$, and $[cP]_t$ represent the concentration of cP at time zero, at equilibrium (photostationary state), and at time *t*; and $k_i = k_{ct} + k_{tc}$. Using these results in equation C.23 yields

$$\ln\left[\frac{([cP]_0 - [cP]_\infty)}{([cP]_t - [cP]_\infty)}\right] = k_it \qquad (C.25)$$

If *P* is the total concentration of *cis*- and *trans*-1,3-pentadiene, then

$$[P] = [cP] + [tP] \qquad (C.26)$$

$$t = \text{fraction of trans isomer} = [tP]/[P] \qquad (C.27)$$

$$1 - t = \text{fraction of cis isomer} = [cP]/[P] \qquad (C.28)$$

$$1 - t_0 = [cP]_0/[P] \text{ at time } (t) = 0 \qquad (C.29)$$

$$[cP]_0 = (1 - t_0)[P] \qquad (C.30)$$

and in a similar way

$$[\mathrm{cP}]_\infty = (1 - t_\infty)[P] \text{ at time}(t) = \infty \tag{C.31}$$

$$[\mathrm{cP}]_t = (1 - t_t)[P] \text{ at time}(t) = t \tag{C.32}$$

Substituting equations C.30, C.31, and C.32 in equation C.25 yields

$$\ln\left\{\frac{(1 - t_0)[P] - (1 - t_\infty)[P]}{(1 - t_t)[P] - (1 - t_\infty)[P]}\right\} = k_i t \tag{C.33}$$

$$\ln\left\{\frac{(1 - t_0) - (1 - t_\infty)}{(1 - t_t) - (1 - t_\infty)}\right\} = k_i t \tag{C.34}$$

$$\ln\left\{\frac{(t_\infty - t_0)}{(t_\infty - t_t)}\right\} = k_i t \tag{C.35}$$

If no trans isomer is present at $t = 0$, then $t_0 = 0$ and equation C.35 becomes

$$\ln\{t_\infty/(t_\infty - t_t)\} = k_i t \tag{C.36}$$

References

1. Frost, A. A.; Pearson, R. G. *Kinetics and Mechanism*; Wiley: New York, 1961; p 186.
2. Peirce, B. O. *A Short Table of Integrals;* Ginn: New York, 1929; Integral No. 26, p 5.

Epilogue

This text provides, in one source, the detailed theory and practice for determining environmentally relevant rates of photoreaction in aquatic media. The important literature published in the last two decades in environmental photochemistry has been described by using a historical approach.

I have given detailed discussions on the nature of light emanating from the sun, the absorption and the transmission of sunlight through the atmosphere and through the air–water interface, and the absorption and transmission of sunlight in a natural water body. These four processes are critical to understanding the fundamentals of measuring photoreaction rates.

Using the first and second laws of photochemistry, I developed a general theory of the kinetics of direct and indirect photochemical reactions in aquatic media in the environment (Chapter 5). This theoretical treatment provided the foundation for the development of experimental photochemical procedures described in Chapters 6 through 15.

Many papers have been published that verify the theory of direct photoreaction in aquatic media in the environment. Significant advances have been made in chemical actinometry for very low chemical concentrations in aqueous solution and in environmental photochemistry.

On the other hand, the theory of indirect, or sensitized, photoreaction in aquatic media is still being developed. Nevertheless, general rate equations have been derived and used to analyze the kinetic data for a number of photosensitized reactions in aquatic humic media. On the basis of this theory, procedures have been developed to measure rates of indirect photoreaction in the laboratory or outdoors in sunlight. However, more research is needed to verify these methods and, in particular, to understand the role of natural and synthetic humic acids in promoting indirect photoreactions.

I hope that this book will contribute to a better understanding of environmental photochemistry and will stimulate others to carry on the work that remains. I share the task of every scientist and researcher whose work, in some small way, will lead to a better environment for all who inhabit this earth.

Symbols

a_λ	Absorption coefficient of an aqueous solution at wavelength λ in the Napierian (natural) logarithm system. It corresponds to 2.303 times the absorbance per unit path length. The units are in length^{-1} (e.g., cm^{-1}). (*See* pages 85–86.)
$(A^\circ_{370} - A_{370})/A^\circ_{370}$, $1 - (A_{370}/A^\circ_{370})$, $1 - \exp(-kt)$	Fraction of synthetic humic water photobleached at time t. (*See* pages 4–5, 87–88, 103, 118, 192.)
A_λ	Absorbance at wavelength λ. (*See* pages 4–5, 87–88, 103, 118, 192.)
$A_{\lambda c}$	Absorbance of a chemical at wavelength λ. (*See* pages 132, 182, 187, 192.)
A_{313a}, A_{366a}	Absorbance of an actinometer at 313 and 366 nm, respectively. (*See* pages 183, 194.)
A°_{370}, A_{370}	Absorbance of synthetic humic acid water at 370 nm at time $t = 0$ and t, respectively. (*See* pages 234, 238, 240, 249, 253.)
α_λ	Attenuation coefficient of the medium (e.g., air, ozone, or water) at wavelength λ in the decadic logarithm system. The units are length^{-1} (e.g., cm^{-1}). (*See* pages 4–5, 8, 19, 25–26, 38–42, 49, 79, 84–89, 91, 144, 152, 164.)
A_p	Aperture in a Moses-type photochemical merry-go-round reactor (cm^2). (*See* pages 179–180.)
c	Velocity of light (cm/s). (*See* pages 4–5.)
$[C]$, $[C_o]$, $[C_t]$	Molar concentration of a chemical, the molar concentration at time $t = 0$, and at time t, respectively.
$[C_t]_r$, $[C_t]_p$	Molar concentration of a reactant (r) and product (p) at time t. (*See* pages 146, 158–159, 161–162.)
D	Depth in a natural water body in the environment. (*See* pages 33–41, 53–54.)
DMF	2,5-Dimethylfuran. (*See* pages 202, 209–212, 216–223, 265–266.)
$-(d[C]_r/dt)_{d\lambda}$, $-(d[C]_p/dt)_{d\lambda}$	Rate of direct photoreaction of reactant (r) and product (p), respectively, at wavelength λ.

1464–6/88/0281$06.00/1

$-(d[C]dt)_{d\lambda}$, $-(d[C]dt)_{s\lambda}$	Rate of direct and indirect (sensitized) photoreaction, respectively, at wavelength λ).
$-(d[C]/dt)_{d}$, $-(d[C]/dt)_{s}$, $-(d[C]/dt)_{p}$	Rate of direct, indirect (sensitized), and net photoreaction, respectively, integrated over all wavelengths).
$-(d[C]/dt)_{dE}$, $-(d[C]/dt)_{sE}$, $-(d[C]/dt)_{pE}$	Rate of direct photoreaction, indirect (sensitized) photoreaction, and net photoreaction, respectively, in a natural water body in the environment).
ϵ_{λ}	Molar absorptivity at wavelength λ (M^{-1} cm^{-1}). (*See* pages 4–5, 8, 25–28, 39–41, 49–50, 54, 55, 66, 79–84, 87, 101–104.)
ϵ_{a}, $\epsilon_{\lambda c}$, $\epsilon_{\lambda s}$	Molar absorptivity of an actinometer, chemical, and sensitizer, respectively, at wavelength λ (M^{-1} cm^{-1}). ($\epsilon_{\lambda a}$, *See* pages 108, 110–111, 121, 125, 137, 156–156; $\epsilon_{\lambda c}$, *See* pages 111–112, 118, 121, 131–133, 137–138, 142–143, 155–156, 159–160; $\epsilon_{\lambda s}$, *See* pages 88–92, 205–207.)
ϵ_{313a}, ϵ_{366a}, ϵ_{313c}, ϵ_{366c}	Molar absorptivity of an actinometer and a chemical at 313 and 366 nm (M^{-1} cm^{-1}). (*See* pages 108–110, 147–148, 151, 153, 157, 159, 177, 182, 192–193.)
$\eta_{o\lambda}$	Refractive index of air at wavelength λ. (*See* page 18.)
$E_{o\lambda}^{av}$	Average irradiance of a light source (photons [or einsteins] cm^{-2} s^{-1}). (*See* pages 80, 82–83, 89–91, 122–123, 166.)
fw	Formula weight of a chemical (g/mol). (*See* pages 122, 230.)
F_{λ}	Fraction of light absorbed by the ferrioxalate actinometer at wavelength λ. (*See* pages 149–150.)
$F_{c\lambda}$	Fraction of light absorbed by a chemical at wavelength λ. (*See* pages 10, 25, 78–79, 84, 88, 143–144, 152.)
$F_{s\lambda}$	Fraction of light absorbed by the system at wavelength λ. (*See* pages 8, 10, 25–26, 78–79, 84, 87, 143–144, 152.)
F_{d}	Fraction of sunlight in the atmosphere that is transmitted directly. (*See* page 18.)
F_{scat}	Fraction of sunlight in the atmosphere that is scattered and diffused. (*See* page 18.)
FFA	Furfuryl alcohol. (*See* pages 202–203, 210, 224, 232, 267–269.)

γ — The geometry factor. It corresponds to the ratio of the direct photoreaction rate constant of a chemical in pure water in tubes to the direct photoreaction rate constant of the chemical in a pure water body in the environment (e.g., a lake). (*See* pages 110–111, 123–124, 211, 271.)

$\gamma(x)$ — The fraction of precursor molecules following pathway x. It represents the quantum efficiency of pathway x. (*See* pages 267, 268, 271.)

GCSOLAR — Updated computer program of Zepp and Cline to compute the solar irradiance $E^{av}_{o\lambda}$. (*See* pages 46, 74, 82, 114, 123, 168.)

h — Planck's constant (J-s). (*See* page 6.)

h — Thickness of a polluted layer in the atmosphere. (*See* pages 21–24, 33, 34.)

$h\nu$ — Energy of a quantum (or photon) of light (J or kJ). (*See* page 6.)

$^1H_0$ — The ground singlet state of humic acid sensitizers. (*See* pages 270–272.)

$^1H_1$ — The first excited singlet state of humic acid sensitizers. (*See* pages 270–272.)

$^3H_1$ — The first excited triplet state of humic acid sensitizers. (*See* pages 270–272.)

i — The constant in the equation $\overline{L}_s = ih$. (*See* pages 25, 34.)

$I_{o\lambda}$ — Irradiance (intensity) of incident light at wavelength λ (photons [or einsteins] $cm^{-2}\ s^{-1}$). (*See* pages 4–5, 8, 10, 14, 17, 20, 25, 26, 27, 39, 42, 78, 87–88, 103, 143.)

$I^*_{a\lambda}$ — Irradiance (intensity) of absorbed light at wavelength λ (photons [or einsteins] $cm^{-2}\ s^{-1}$). (*See* pages 8, 39, 42.)

$I_{t\lambda}$ — Irradiance (intensity) of transmitted light at wavelength λ (photons [or einsteins] $cm^{-2}\ s^{-1}$). (*See* pages 3–5, 17.)

$I_{d\lambda}$ — Irradiance (intensity) of direct sunlight at wavelength λ (photons $cm^{-2}\ s^{-1}$). (*See* pages 20, 26, 39–41, 50.)

$I_{s\lambda}$ — Irradiance (intensity) of skylight from the sun at wavelength λ (photons $cm^{-2}\ s^{-1}$). (*See* pages 20, 26–27, 39, 40, 41, 42, 50.)

I_λ — Irradiance (intensity) of incident light in photons (or einsteins) per unit volume per unit time at a fixed wavelength λ. (*See* pages 79–80, 84–

85, 88–89, 91–92, 144–146, 152, 155, 161, 164.)

$I_{a\lambda}$ — Irradiance (intensity) of light absorbed by a chemical undergoing photoreaction in photons (or einsteins) per unit volume per unit time at a fixed wavelength λ. (*See* pages 9–11, 26–27, 39–41.)

I_a — Irradiance (intensity) of light absorbed by a chemical undergoing photoreaction in photons (or einsteins) per unit volume and per unit time summed over all wavelengths of light absorbed. (*See* page 27.)

$I^d_{a\lambda}$ — Irradiance (intensity) of light absorbed by a chemical undergoing direct photoreaction in photons (or einsteins) per unit volume per unit time at wavelength λ. (*See* pages 50, 78–80, 84.)

$I^s_{a\lambda}$ — Irradiance (intensity) of light absorbed by a sensitizer causing a chemical to undergo indirect (or sensitized) photoreaction in photons (or einsteins) per unit volume per unit time at wavelength λ. (*See* pages 87–89, 268–271.)

I_s, $I_s(z)$ — If the sky is a uniformly bright hemisphere, I_s and $I_s(z)$ represent the total sky irradiance and the incremental sky irradiance at zenith angle z (photons [or einsteins] cm^{-2} s^{-1}), respectively. (*See* pages 22–23.)

j — The conversion factor used to make the units of light intensity compatible with the units of C and ϵ_λ. (*See* pages 11, 26–27, 40–43, 80, 84–85, 89–92, 101, 218.)

J_λ — Solar irradiance (or solar flux) in the atmosphere that corresponds to the sunlight intensity at wavelength λ (photons cm^{-2} s^{-1} or photons cm^{-2} day^{-1}). (*See* pages 28–30.)

k_d — Direct photoreaction rate constant in the units of $time^{-1}$. It is used as the direct sunlight photoreaction rate constant. It represents the integrated rate constant over all wavelengths λ. (*See* pages 81–82, 85, 93, 166, 204, 211, 214, 218, 228.)

k_{da}, k_{dc} — Direct photoreaction rate constant for an actinometer and a chemical, respectively, at wavelength λ in the units of $time^{-1}$. (*See* pages 108–

111, 121–122, 137, 155–164, 176, 184, 195, 229–230, 238–239, 253–254.)

k_{dE} — Direct sunlight photoreaction rate constant in a natural water body in the environment in the units of time^{-1}. (*See* pages 51–52, 101, 122, 165–166.)

$k_{s\lambda}$ — Rate constant for the indirect photoreaction of a chemical at wavelength λ in the units time^{-1}. (*See* pages 90, 92, 206, 217, 219–223.)

k_s, k_{sc} — Indirect (or sensitized) photoreaction rate constant of a chemical in the units time^{-1}. It is used as the sunlight indirect photoreaction rate constant integrated over all wavelengths λ. (*See* pages 90–92, 204–205, 211, 214, 228, 234, 237.)

k_{sE} — Indirect (or sensitized) sunlight photoreaction rate constant in a natural water body in the environment in the units of time^{-1}. (*See* pages 94, 224, 228, 252, 254.)

k_p, k_{pa} — Net photoreaction rate constant of a chemical in the units of time^{-1}. It is used as the net sunlight photoreaction rate constant. It represents the sum of k_d and k_s. (*See* pages 93–94, 204, 228.)

k_i — The rate constant for the isomerization of *cis*- and *trans*-pentadiene in the units time^{-1}. (*See* pages 212, 277.)

k_{pE} — Net sunlight photoreaction rate constant in a natural water body in the environment in the units of time^{-1}. It represents the sum of k_{dE} and k_{sE}. (*See* pages 94, 228, 235–236, 240, 251, 252.)

$(k_{dE})_{\max}$ — Maximum direct sunlight photoreaction rate constant in a natural water body in the environment in the units of time^{-1}. (*See* pages 102, 113, 132–133, 143, 166, 193.)

$k_{a\lambda}$ — Specific light absorption rate constant of a chemical at wavelength λ. (*See* pages 40–41, 50, 52, 55, 66, 82, 101, 142.)

k_a — Specific light absorption rate constant integrated over all wavelengths absorbed by the chemical. (*See* pages 27, 51, 52–53, 55, 63–64, 82, 101, 108, 110, 121, 136–137, 256.)

$(k_{pc})_{\mathrm{SHW}}$ — Net photoreaction rate constant of a chemical in synthetic humic acid water in the units of time^{-1}. (*See* pages 235–236, 251–252.)

$(k_{dc})_W$ — Direct photoreaction rate constant of a chemical in buffered distilled water in the units of time^{-1}. (*See* pages 236, 239, 251.)

$^1k_{ic}$ — The rate constant for internal conversion (nonradiative decay) from the first excited singlet state to the ground singlet state. (*See* pages 266, 270.)

$^1k_{isc}$ — The rate constant for the intersystem crossing from the first excited singlet state to the first triplet state. (*See* pages 266, 270.)

$^3k_{ic}$ — The rate constant for the internal conversion (nonradiative decay) from the first triplet state to the ground singlet state. (*See* pages 266, 270.)

k_q — The rate constant for the quenching of an excited state. (*See* pages 266–270.)

k_c, k_{ct} — The first-order rate constant for the conversion of *cis*-1,3-pentadiene to *trans*-1,3-pentadiene. (*See* pages 271–273, 275.)

k_t, k_{tc} — The first-order rate constant for the conversion of *trans*-1,3-pentadiene to *cis*-1,3-pentadiene. (*See* pages 271–273, 275.)

k_{ox} — The rate constant for the reaction of the triplet excited state of humic acids with triplet oxygen to form singlet oxygen. (*See* pages 266–270.)

k — First-order bleaching rate constant of synthetic humic acid water. (*See* pages 234, 238–239, 253.)

k_{so}, k_{soc} — Initial sensitized (or indirect) first-order photoreaction rate constant of a chemical at time $t = 0$. (*See* page 234, 236–237, 253–254.)

k_h — Specific absorption coefficient (L[mg of organic carbon]$^{-1}$ m^{-1}).

$k_{control}$ — First-order rate loss in control samples in photoreaction experiments. It corresponds to the first-order rate loss from processes other than photoreaction (e.g., volatilization, biodegradation, and hydrolysis). (*See* page 120.)

$(k_{dc})_{corr}$ — First-order direct photoreaction rate constant of a chemical corrected for a small loss by volatilization, biodegradation, hydrolysis, etc. (*See* page 120.)

$(k_{dc})_{\text{obs}}$	First-order direct photoreaction rate constant of a chemical that is observed (measured). (*See* pages 120.)
λ	Wavelength of light (nm).
λ(center)	Average wavelength (nm) of light based on the wavelength at two intervals. (*See* pages 103–104, 118, 132, 182, 192.)
$\Delta\lambda$	Wavelength interval (nm).
ln	Natural (Napierian) logarithm (base *e*).
$\ln([C_o]/[C_t])_{\text{control}}$	First-order loss term corresponding to a small loss of chemical other than by photoreaction. (*See* page 120.)
$\ln([C_o]/[C_t])_{\text{corr}}$	First-order term corrected for a small loss of a chemical via volatilization, biodegradation, hydrolysis, etc. (*See* page 120.)
$\ln([C_o]/[C_t])_{\text{obs}}$	First-order term that is observed (measured).
$\ln(C_o/C_t)_{\text{SHW}}$, $\ln(C_o/C_t)_{\text{W}}$	First-order term for a chemical in synthetic humic acid water and buffered distilled water, respectively. (*See* pages 235, 237–238, 251.)
$\log_{10}$	Decadic logarithm (base 10).
l	Path length of light; the distance traveled by a beam of light passing through a medium. (*See* pages 3–5, 8, 10, 16, 25–26, 39, 41, 79–80, 84–89, 91–92, 103, 132, 144–145, 152–155, 192, 206.)
l_d	Path length of direct solar radiation in water. (*See* pages 34–35.)
$\bar{l}_s$	Average path length of sky solar radiation in water. (*See* pages 36–37.)
L_d	Path length of direct solar radiation in the atmosphere. (*See* pages 21, 34.)
$\bar{L}_s$	Average path length of solar sky radiation in the atmosphere. (*See* pages 24–27.)
L_λ	Solar irradiance in water at shallow depths which corresponds to day-averaged, midseason, values of sunlight intensity at wavelength λ (millieinsteins cm^{-2} day^{-1}). (*See* pages 46–47, 83, 101–104, 110–112, 121–125, 133, 137–138, 142–143, 165–166, 193, 197, 255–264.)
m	Air mass in the atmosphere (dimensionless). (*See* pages 16–20, 32.)
n	Refractive index of water (dimensionless). (*See* pages 33–37.)
N	Avogadro's number, the number of molecules per mole. (*See* page 6.)

NHW — Natural humic acid water. (*See* pages 201–202, 208, 211–212, 220–224, 231, 241.)

ν — Frequency of light in s^{-1}.

$(90 - z)°$ — Solar altitude, where z is the solar zenith angle.

o-NB — *ortho*-Nitrobenzaldehyde. (*See* pages 150–151, 163, 181.)

$^3O_2$, $^1O_2$ — Ground triplet state and first excited singlet state of oxygen, respectively. (*See* pages 204, 209–211, 219–224, 266–272.)

ϕ — Reaction quantum yield of a chemical that is independent of wavelength λ (dimensionless).

ϕ_λ — Reaction quantum yield at wavelength λ (dimensionless). (*See* page 9.)

$\phi_{d\lambda}$ — Reaction quantum yield for direct photoreaction of a chemical at wavelength λ (dimensionless).

ϕ_d, ϕ_{dc} — Direct photoreaction quantum yield of a chemical that is independent of λ (dimensionless). (*See* pages 167, 196.)

$\phi_{s\lambda}$ — Reaction quantum yield of a chemical undergoing indirect (or sensitized) photoreaction at wavelength λ (dimensionless). (*See* pages 87, 89, 265, 269, 272.)

ϕ_{ct} — Reaction quantum yield for the isomerization of *cis*- to *trans*-pentadiene (dimensionless).

ϕ_{da} — Direct photoreaction quantum yield of an actinometer (dimensionless). (*See* pages 105–106, 147–148, 151, 196.)

ϕ_{isc}, Φ_{isc} — Quantum yield for intersystem crossing of a molecule from the first singlet excited state to the first triplet state (dimensionless). (*See* pages 271–272.)

$\Phi(^1O_2)$ — Quantum yield for production of the singlet excited state of oxygen (dimensionless). (*See* page 269.)

$\Phi(AO_2)$ — Quantum yield for the reaction of singlet oxygen with an acceptor molecule A (dimensionless). (*See* page 269.)

cP, [cP] — *cis*-1,3-Pentadiene and its corresponding molar concentration. (*See* pages 203, 208, 212–213, 269–273, 275.)

tP, [tP] — *trans*-1,3-Pentadiene and its corresponding molar concentration. (*See* pages 203, 208, 212–213, 269–273, 275.)

PAHs — Polynuclear aromatic hydrocarbons. (*See* page 72.)

POB	Photochemical optical bench. (*See* pages 160–164, 173.)
PMGRR	Photochemical "merry-go-round" reactor. (*See* pages 147–148, 153–154, 156–164, 172, 174–180, 191, 195.)
PNA, [PNA]	*p*-Nitroanisole and its corresponding molar concentration. (*See* pages 106–108, 122, 125–126, 148, 175–181, 229, 245.)
PNA–PYR	*p*-Nitroanisole–pyridine actinometer. (*See* pages 106–108, 122, 125–126, 148, 175–181, 229, 245.)
PNAP, [PNAP]	*p*-Nitroacetophenone and its corresponding molar concentration. (*See* pages 108–110, 122, 125–126, 147, 175–181, 194–195, 229, 245, 252, 253.)
PNAP–PYR	*p*-Nitroacetophenone–pyridine actinometer. (*See* pages 108–110, 122, 125–126, 147, 175–181, 194–195, 229, 245, 252, 253.)
PYR, [PYR]	Pyridine and its corresponding molar concentration.
$\prod_x f(x)$	Product of a function $f(x)$ for all values of x.
Q, Q*	A ground-state quenching molecule and its excited state, respectively. (*See* pages 270.)
Q_λ	The proportionality constant in the expression $\phi_{s\lambda} = Q_\lambda[C]$. (*See* pages 89–92, 206–207, 265, 269, 272.)
R	The ratio of the net photoreaction rate constant of a chemical in humic acid water to the direct photoreaction rate constant of the chemical in pure water. (*See* pages 236, 249.)
σ_λ	Attenuation coefficient of the medium. (*See* pages 3–4, 17.)
$\sigma_{m\lambda}$, $\sigma_{p\lambda}$, $\sigma_{a\lambda}$	Attenuation coefficient of air at wavelength λ in the Napierian Bouger–Lambert equation. The symbols m, p, and a correspond to molecular scattering, particulate scattering, and absorption, respectively. (*See* page 17.)
$S_{m\lambda}$, $S_{p\lambda}$, $S_{a\lambda}$	Attenuation coefficient of air at wavelength λ in the decadic Bouger–Lambert equation. The symbols m, p, and a correspond to molecular scattering, particulate scattering, and absorption, respectively. (*See* pages 17–19.)
SOLAR	Computer program of Zepp and Cline to compute solar irradiance $E_{o\lambda}^{av}$. (*See* pages 43, 46, 54, 61, 72, 73, 82, 114.)

$(S_{m\lambda})_0$	Theoretical attenuation coefficient for molecular scattering derived by Raleigh. (*See* page 18.)
S, $[S]$	A sensitizer, its concentration, or its molar concentration. (*See* pages 88–90, 92, 205–207, 219.)
S_0, S_1	Singlet ground and first singlet excited states of a molecule. (*See* page 266.)
SHW	Synthetic humic acid water. (*See* pages 231–241, 243–248, 250–253.)
S_λ	Light-screening factor at wavelength λ. It corresponds to the ratio of the direct photoreaction rate constant in humic acid water to the direct photoreaction rate constant in pure water at λ. (*See* pages 85–86, 89.)
$S1$	The ratio of the first-order rate constants; it is equivalent to k_{soc}/k. (*See* pages 238, 253–254.)
$S2$	The ratio of the first-order rate constants; it is equivalent to k/k_{da}. (*See* pages 238, 253–254.)
$S3$	The ratio of the first-order rate constants; it is equivalent to k_{dc}/k_{da}. (*See* pages 239, 253–254.)
$\sum_x f(x)$	The sum of a function $f(x)$ for all values of x.
θ	The angle of refraction in water. (*See* pages 33, 34–35, 41, 50.)
θ, θ(z)	If the sky is a hemisphere, θ and θ(z) represent the fraction of sky area and the fraction of sky area at zenith angle z, respectively. (*See* pages 22–23.)
$t_{(½)c}$, $t_{(½)a}$	Half-life of a chemical and an actinometer, respectively, undergoing photoreaction in units of time. (*See* pages 157, 177–178, 179, 194.)
$t_{(½)E}$	Half-life of a chemical undergoing sunlight photoreaction in a natural water body in the environment in units of time. (*See* pages 51, 53, 63–64, 72, 83, 91, 94, 101, 138, 197, 228, 251, 254.)
$[t_{(½)E}]_{min}$	Minimum half-life of a chemical undergoing sunlight photoreaction in a natural water body in the environment in units of time. (*See* pages 102, 135, 143, 193.)
t_0, t_t, t_∞	The fraction of *trans*-1,3-pentadiene at time zero, t, and photostationary state, respectively for the cis–trans photoisomerization of 1,3-pentadiene. (*See* pages 212, 278.)

T_λ — Transmissivity (fraction of light transmitted), at wavelength λ (dimensionless). (*See* pages 4, 17, 103.)

$T_{m\lambda}, T_{p\lambda}, T_{a\lambda}$ — Transmissivity for molecular scattering, particulate scattering, and absorption, respectively (dimensionless). (*See* pages 17–20, 27.)

$T_{s\lambda}$ — Transmissivity corresponding to molecular and particulate scattering (dimensionless). It represents the product of $T_{s\lambda}$ and $T_{m\lambda}$. (*See* pages 17–18, 27.)

T_1 — First excited triplet state of a molecule. (*See* pages 266, 270–271.)

τ — The lifetime (or decay time) of an excited state in units of time^{-1}. (*See* page 272.)

TMAB — *p*-Trimethylaminobenzophenone chloride, an efficient singlet oxygen sensitizer. (*See* pages 203, 213.)

UV-A — Ultraviolet radiation in the spectral region 320–400 nm. (*See* pages 32, 44.)

UV-B — Ultraviolet radiation in the spectral region 280–320 nm. (*See* pages 32, 43–44, 55–56, 58–59.)

W — Buffered distilled water. (*See* pages 235–237, 250–253.)

W_λ — Solar irradiance in water when essentially all the light is absorbed. (*See* pages It corresponds to midday, midseason, values of sunlight intensity in photons cm^{-2} s^{-1} at wavelength λ. (*See* pages 40, 42–43.)

$X_{s\lambda}$ — Response function at λ. It corresponds to the ratio of the indirect (sensitized) photoreaction rate constant ($k_{s\lambda}$) divided by the irradiance $E^{\mathrm{av}}_{o\lambda}$. (*See* pages 207, 222–224.)

X — $(\mathrm{rate})_C/(\mathrm{rate})_{\mathrm{max}}$. (*See* page 153.)

z — Solar zenith angle, the angle formed from a point on the earth's surface between the sun and the normal. (*See* pages 15, 19, 21–22, 27, 35–37.)

Z_λ — Solar irradiance in water at shallow depths which corresponds to midday, midseason, values of sunlight intensity at wavelength λ (photons cm^{-2} s^{-1}). (*See* pages 41, 43, 50, 82, 101, 123.)

Index

A

B

C

D

E

F

G

H

I

J

K

L

M

N

O

P

Q

R

S

T

U

V

W

X

Z

Calvert and Pitts
Photochemistry 1966
pp 783-786

Copy editing and indexing by Linda Romaine Ross
Managing Editor Janet S. Dodd
Typeset by Techna Type, Inc., York, PA
Printed and bound by Maple Press Company, York, PA